Lecture Notes in Physics

Founding Editors: W. Beiglböck, J. Ehlers, K. Hepp, H. Weidenmüller

T0155766

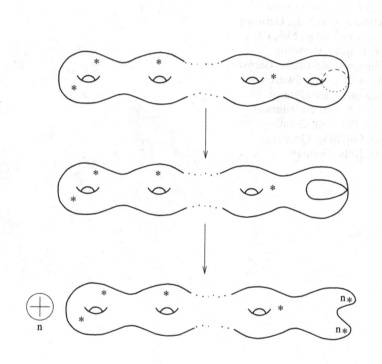

M. Schottenloher

A Mathematical Introduction to Conformal Field Theory

Second Edition

 Springer

Martin Schottenloher
Mathematisches Institut
Ludwig-Maximilians-Universität München
Theresienstr. 39
80333 München
Germany
schotten@mathematik.uni-muenchen.de

Schottenloher, M. *A Mathematical Introduction to Conformal Field Theory*, Lect. Notes
Phys. 759 (Springer, Berlin Heidelberg 2008), DOI 10.1007/978-3-540-68628-6

ISBN: 978-3-642-08815-5   e-ISBN: 978-3-540-68628-6

DOI 10.1007/978-3-540-68628-6

Lecture Notes in Physics ISSN: 0075-8450

Cover design: eStudio Calamar S.L., F. Steinen-Broo, Pau/Girona, Spain

Printed on acid-free paper

9 8 7 6 5 4 3 2 1

springer.com

To Barbara

Preface to the Second Edition

The second edition of these notes has been completely rewritten and substantially expanded with the intention not only to improve the use of the book as an introductory text to conformal field theory, but also to get in contact with some recent developments. In this way we take a number of remarks and contributions by readers of the first edition into consideration who appreciated the rather detailed and self-contained exposition in the first part of the notes but asked for more details for the second part. The enlarged edition also reflects experiences made in seminars on the subject.

The interest in conformal field theory has grown during the last 10 years and several texts and monographs reflecting different aspects of the field have been published as, e.g., the detailed physics-oriented introduction of Di Francesco, Mathieu, and Sénéchal [DMS96*],[1] the treatment of conformal field theories as vertex algebras by Kac [Kac98*], the development of conformal field theory in the context of algebraic geometry as in Frenkel and Ben-Zvi [BF01*] and more general by Beilinson and Drinfeld [BD04*]. There is also the comprehensive collection of articles by Deligne, Freed, Witten, and others in [Del99*] aiming to give an introduction to strings and quantum field theory for mathematicians where conformal field theory is one of the main parts of the text. The present expanded notes complement these publications by giving an elementary and comparatively short mathematics-oriented introduction focusing on some main principles.

The notes consist of 11 chapters organized as before in two parts. The main changes are two new chapters, Chap. 8 on Wightman's axioms for quantum field theory and Chap. 10 on vertex algebras, as well as the incorporation of several new statements, examples, and remarks throughout the text. The volume of the text of the new edition has doubled. Half of this expansion is due to the two new chapters.

We have included an exposition of Wightman's axioms into the notes because the axioms demonstrate in a convincing manner how a consistent quantum field theory in principle should be formulated even regarding the fact that no four-dimensional model with properly interacting fields satisfying the axioms is known to date. We investigate in Chap. 8 the axioms in their different appearances as postulates on operator-valued distributions in the relativistic case as well as postulates on the

---

[1] The "*" indicates that the respective reference has been added to the References in the second edition of these notes.

corresponding correlation functions on Minkowski and on Euclidean spaces. The presentation of the axioms serves as a preparation and motivation for Chap. 9 as well as for Chap. 10.

Chapter 9 deals with an axiomatic approach to two-dimensional conformal field theory. In comparison to the first edition we have added the conformal Ward identities, the state field correspondence, and some changes with respect to the presentation of the operator product expansion. The concepts and methods in this chapter were quite isolated in the first edition, and they can now be understood in the context of Wightman's axioms in its various forms and they also can be linked to the theory of vertex algebras.

Vertex algebras have turned out to be extremely useful in many areas of mathematics and physics, and they have become the main language of two-dimensional conformal field theory in the meantime. Therefore, the new Chap. 10 in these notes provides a presentation of basic concepts and methods of vertex algebras together with some examples. In this way, a number of manipulations in Chap. 9 are explained again, and the whole presentation of vertex algebras in these notes can be understood as a kind of formal and algebraic continuation of the axiomatic treatment of conformal field theory.

Furthermore, many new examples have been included which appear at several places in these notes and may serve as a link between the different viewpoints (for instance, the Heisenberg algebra H as an example of a central extension of Lie algebras in Chap. 4, as a symmetry algebra in the context of quantization of strings in Chap. 7, and as a first main example of a vertex algebra in Chap. 10). Similarly, Kac–Moody algebras are introduced, as well as the free bosonic field and the restricted unitary group in the context of quantum electrodynamics. Several of the elementary but important statements of the first edition have been explained in greater detail, for instance, the fact that the conformal groups of the Euclidean spaces are finite dimensional, even in the two-dimensional case, the fact that there does not exist a complex Virasoro group and that the unitary group $U(\mathbb{H})$ of an infinite-dimensional Hilbert space $\mathbb{H}$ is a topological group in the strong topology.

Moreover, several new statements have been included, for instance, about a detailed description of some classical groups, about the quantization of the harmonic oscillator and about general principles used throughout the notes as, for instance, the construction of representations of Lie algebras as induced representations or the use of semidirect products.

The general concept of presenting a rather brief and at the same time rigorous introduction to conformal field theory is maintained in this second edition as well as the division of the notes in two parts of a different nature: The first is quite elementary and detailed, whereas the second part requires more mathematical prerequisites, in particular, from functional analysis, complex analysis, and complex algebraic geometry.

Due to the complexity of the treatment of Wightman's axioms in the second part of the notes not all results are proven, but there are many more proofs in the second part than in the original edition. In particular, the chapter on vertex algebras is self-contained.

The final chapter on the Verlinde formula in the context of algebraic geometry, which is now Chap. 11, has nearly not been changed except for a comment on fusion rings and on the connection of the Verlinde algebra with twisted K-theory recently discovered by Freed, Hopkins, and Teleman [FHT03*].

In a brief appendix we mention further developments with respect to boundary conformal field theory, to stochastic Loewner evolution, and to modularity together with some references.

München, March 2008   *Martin Schottenloher*

References

[BD04*]   A. Beilinson and V. Drinfeld *Chiral Algebras*. AMS Colloquium Publications **51** AMS, Providence, RI, 2004.

[BF01*]   D. Ben-Zvi and E. Frenkel *Vertex Algebras and Algebraic Curves*. AMS, Providence, RI, 2001.

[Del99*] P. Deligne et al. *Quantum Fields and Strings: A Course for Mathematicians I, II*. AMS, Providence, RI, 1999.

[DMS96*] P. Di Francesco, P. Mathieu and D. Sénéchal. *Conformal Field Theory*. Springer-Verlag, 1996.

[FHT03*] D. Freed, M. Hopkins, and C. Teleman. Loop groups and twisted K-theory III. arXiv:math/0312155v3 (2003).

[Kac98*] V. Kac. *Vertex Algebras for Beginners*. University Lecture Series **10**, AMS, Providencs, RI, 2nd ed., 1998.

Preface to the First Edition

The present notes consist of two parts of approximately equal length. The first part gives an elementary, detailed, and self-contained mathematical exposition of classical conformal symmetry in n dimensions and its quantization in two-dimensions. Central extensions of Lie groups and Lie algebras are studied in order to explain the appearance of the Virasoro algebra in the quantization of two-dimensional conformal symmetry. The second part surveys some topics related to conformal field theory: the representation theory of the Virasoro algebra, some aspects of conformal symmetry in string theory, a set of axioms for a two-dimensional conformally invariant quantum field theory, and a mathematical interpretation of the Verlinde formula in the context of semi-stable holomorphic vector bundles on a Riemann surface. In contrast to the first part only few proofs are provided in this less elementary second part of the notes.

These notes constitute – except for corrections and supplements – a translation of the prepublication "Eine mathematische Einführung in die konforme Feldtheorie" in the preprint series *Hamburger Beiträge zur Mathematik*, Volume 38 (1995). The notes are based on a series of lectures I gave during November/December of 1994 while holding a *Gastdozentur* at the *Mathematisches Seminar der Universität Hamburg* and on similar lectures I gave at the *Université de Nice* during March/April 1995.

It is a pleasure to thank H. Brunke, R. Dick, A. Jochens, and P. Slodowy for various helpful comments and suggestions for corrections. Moreover, I want to thank A. Jochens for writing a first version of these notes and for carefully preparing the LaTeX file of an expanded English version. Finally, I would like to thank the Springer production team for their support.

Munich, September 1996   *Martin Schottenloher*

Contents

Introduction

Conformal field theory in two dimensions has its roots in statistical physics (cf. [BPZ84] as a fundamental work and [Gin89] for an introduction) and it has close connections to string theory and other two-dimensional field theories in physics (cf., e.g., [LPSA94]). In particular, all massless fields are conformally invariant.

The special feature of conformal field theory in two dimensions is the existence of an infinite number of independent symmetries of the system, leading to corresponding invariants of motion which are also called conserved quantities. This is the content of Noether's theorem which states that a symmetry of a physical system given by a local one-parameter group or by an infinitesimal version thereof induces an invariant of motion of the system. Any collection of invariants of motion simplifies the system in question up to the possibility of obtaining a complete solution. For instance, in a typical system of classical mechanics an invariant of motion reduces the number of degrees of freedom. If the original phase space has dimension $2n$ the application of an invariant of motion leads to a system with a phase space of dimension $2(n-1)$. In this way, an independent set of n invariants of motion can lead to a zero-dimensional phase space that means, in general, to a complete solution.

Similarly, in the case of conformal field theory the invariants of motion which are induced by the infinitesimal conformal symmetries reduce the infinite dimensional system completely. As a consequence, the structure constants which determine the system can be calculated explicitly, at least in principle, and one obtains a complete solution. This is explained in Chap. 9, in particular in Proposition 9.12.

These symmetries in a conformal field theory can be understood as infinitesimal conformal symmetries of the Euclidean plane or, more generally, of surfaces with a conformal structure, that is Riemann surfaces. Since conformal transformations on an open subset U of the Euclidean plane are angle preserving, the conformal orientation-preserving transformations on U are holomorphic functions with respect to the natural complex structure induced by the identification of the Euclidean plane with the space $\mathbb{C}$ of complex numbers. As a consequence, there is a close connection between conformal field theory and function theory. A good portion of conformal field theory is formulated in terms of holomorphic functions using many results of function theory. On the other hand, this interrelation between conformal field theory and function theory yields remarkable results on moduli spaces of vector bundles

Schottenloher, M.: *Introduction*. Lect. Notes Phys. **759**, 1–3 (2008)
DOI 10.1007/978-3-540-68628-6_1   © Springer-Verlag Berlin Heidelberg 2008

over compact Riemann surfaces and therefore provides an interesting example of how physics can be applied to mathematics.

The original purpose of the lectures on which the present text is based was to describe and to explain the role the Virasoro algebra plays in the quantization of conformal symmetries in two dimensions. In view of the usual difficulties of a mathematician reading research articles or monographs on conformal field theory, it was an essential concern of the lectures not to rely on background knowledge of standard methods in physics. Instead, the aim was to try to present all necessary concepts and methods on a purely mathematical basis. This explains the adjective "mathematical" in the title of these notes. Another motivation was to discuss the sometimes confusing use of language by physicists, who for example emphasize that the group of holomorphic maps of the complex plane is infinite dimensional – which is not true. What is meant by this statement is that a certain Lie algebra closely related to conformal symmetry, namely the Witt algebra or its central extension, the Virasoro algebra, is infinite dimensional.

Clearly, with these objectives the lectures could hardly cover an essential part of actual conformal field theory. Indeed, in the course of the present text, conformally invariant quantum field theory does not appear before Chap. 6, which treats the representation theory of the Virasoro algebra as a first topic of conformal field theory. These notes should therefore be seen as a preparation for or as an introduction to conformal field theory for mathematicians focusing on some background material in geometry and algebra. Physicists may find the detailed investigation in Part I useful, where some elementary geometric and algebraic prerequisites for conformal field theory are studied, as well as the more advanced mathematical description of fundamental structures and principles in the context of quantum field theory in Part II.

In view of the above-mentioned tasks, it makes sense to start with a detailed description of the conformal transformations in arbitrary dimensions and for arbitrary signatures (Chap. 1) and to determine the associated conformal groups (Chap. 2) with the aid of the conformal compactification of spacetime. In particular, the conformal group of the Minkowski plane turns out to be infinite dimensional, it is essentially isomorphic to $\mathrm{Diff}_+(\mathbb{S}^1) \times \mathrm{Diff}_+(\mathbb{S}^1)$, while the conformal group of the Euclidean plane is finite-dimensional, it is the group of Möbius transformations isomorphic to $\mathrm{SL}(2,\mathbb{C})/\{\pm 1\}$.

The next two chapters (Chaps. 3 and 4) are concerned with central extensions of groups and Lie algebras and their classification by cohomology. These two chapters contain several examples appearing in physics and mathematics. Central extensions are needed in physics, because the symmetry group of a quantized system usually is a central extension of (the universal covering of) the classical symmetry group, and in the same way the infinitesimal symmetry algebra of the quantum system is, in general, a central extension of the classical symmetry algebra.

Chapter 5 leads to the Virasoro algebra as the unique nontrivial central extension of the Witt algebra. The Witt algebra is the essential component of the classical infinitesimal conformal symmetry in two dimensions for the Euclidean plane as well as for the Minkowski plane. This concludes the first part of the text which is comparatively elementary except for some aspects in the examples.

The second part presents several different approaches to conformal field theory. We start this program with the representation theory of the Virasoro algebra including the Kac formula (Chap. 6) in order to describe the unitary representations.

In Chap. 7 we give an elementary introduction into the quantization of the bosonic string and explain how the conformal symmetry is present in classical and in quantized string theory. The quantization induces a natural representation of the Virasoro algebra on the Fock space of the Heisenberg algebra which is of interest in later considerations concerning examples of vertex algebras.

The next two chapters are dedicated to axiomatic quantum field theory. In Chap. 8 we provide an exposition of the relativistic case in any dimension by presenting the Wightman axioms for the field operators as well as the equivalent axioms for the correlation functions called Wightman distributions. The Wightman distributions are boundary values of holomorphic functions which can be continued analytically into a large domain in complexified spacetime and thereby provide the correlation functions of a Euclidean version of the axioms, the Osterwalder–Schrader axioms. In Chap. 9 we concentrate on the two-dimensional Euclidean case with conformal symmetry. We aim to present an axiomatic approach to conformal field theory along the suggestion of [FFK89] and the postulates of the groundbreaking paper of Belavin, Polyakov, and Zamolodchikov [BPZ84].

Many papers on conformal field theory nowadays use the language of vertex operators and vertex algebras. Chapter 10 gives a brief introduction to the basic concepts of vertex algebras and some fundamental results. Several concepts and constructions reappear in this chapter – sometimes in a slightly different form – so that one has a common view of the different approaches to conformal field theory presented in the preceding chapters.

Finally we discuss the Verlinde formula as an application of conformal field theory to mathematics (Chap. 11).

References

[BPZ84]   A. A. Belavin, A. M. Polyakov, and A. B. Zamolodchikov. Infinite conformal symmetry in two-dimensional quantum field theory. *Nucl. Phys.* **B 241** (1984), 333–380.

[FFK89]   G. Felder, J. Fröhlich, and J. Keller. On the structure of unitary conformal field theory, I. Existence of conformal blocks. *Comm. Math. Phys.* **124** (1989), 417–463.

[Gin89]   P. Ginsparg. Introduction to conformal field theory. *Fields, Strings and Critical Phenomena*, Les Houches 1988, Elsevier, Amsterdam 1989.

[LPSA94] R. Langlands, P. Pouliot, and Y. Saint-Aubin. Conformal invariance in two-dimensional percolation. *Bull. Am. Math. Soc.* **30** (1994), 1–61.

Part I
Mathematical Preliminaries

The first part of the notes begins with an elementary and detailed exposition of the notion of a conformal transformation in the case of the flat spaces $\mathbb{R}^{p,q}$ (Chap. 1) and a thorough investigation of the conformal groups, that is the groups of all conformal transformations on the corresponding compactified spaces $N^{p,q}$ (Chap. 2). As a result, the conformal groups are finite-dimensional Lie groups except for the case of the Minkowski plane. In the case of the Minkowski plane one obtains (two copies of) the infinite dimensional Witt algebra as a complexified Lie algebra of infinitesimal conformal transformations.

Chapters 3 and 4 deal with central extensions of groups and Lie algebras. Central extensions occur in a natural way if one studies projective representations and wants to compare them with true representation in the linear space to which the projective space is associated. Since quantization represents observables as linear operators in a linear (mostly Hilbert) space W and the space of quantum states is the associated projectivation $\mathbb{P}(W)$, it is unavoidable that central extensions of Lie groups and Lie algebras naturally appear as the quantization of classical symmetries.

The first part of the notes concludes with an elementary description of the Virasoro algebra as the only nontrivial central extension of the Witt algebra (Chap. 5).

As a consequence, in a two-dimensional conformally invariant quantum field theory the Virasoro algebra shall be a symmetry algebra providing the theory with an infinite collection of invariants of motion.

Chapter 1
Conformal Transformations
and Conformal Killing Fields

This chapter presents the notion of a conformal transformation on general semi-Riemannian manifolds and gives a complete description of all conformal transformations on an open connected subset $M \subset \mathbb{R}^{p,q}$ in the flat spaces $\mathbb{R}^{p,q}$. Special attention is given to the two-dimensional cases, that is to the Euclidean plane $\mathbb{R}^{2,0}$ and to the Minkowski plane $R^{1,1}$.

1.1 Semi-Riemannian Manifolds

Definition 1.1. A *semi-Riemannian manifold* is a pair (M, g) consisting of a smooth[1] manifold M of dimension n and a smooth tensor field g which assigns to each point $a \in M$ a nondegenerate and symmetric bilinear form on the tangent space $T_a M$:

$$g_a : T_a M \times T_a M \longrightarrow \mathbb{R}.$$

In local coordinates $x^1, \ldots, x^n$ of the manifold M (given by a chart $\phi : U \to V$ on an open subset U in M with values in an open subset $V \subset \mathbb{R}^n$, $\phi(a) = (x^1(a), \ldots, x^n(a))$, $a \in M$) the bilinear form g_a on $T_a M$ can be written as

$$g_a(X, Y) = g_{\mu\nu}(a) X^\mu Y^\nu.$$

Here, the tangent vectors $X = X^\mu \partial_\mu$, $Y = Y^\nu \partial_\nu \in T_a M$ are described with respect to the basis

$$\partial_\mu := \frac{\partial}{\partial x^\mu}, \quad \mu = 1, \ldots, n,$$

of the tangent space $T_a M$ which is induced by the chart ϕ.

By assumption, the matrix

$$(g_{\mu\nu}(a))$$

is nondegenerate and symmetric for all $a \in U$, that is one has

---

[1] We restrict our study to smooth (that is to $\mathscr{C}^\infty$ or infinitely differentiable) mappings and manifolds.

Schottenloher, M.: *Conformal Transformations and Conformal Killing Fields.* Lect. Notes Phys. **759**, 7–21 (2008)
DOI 10.1007/978-3-540-68628-6_1

$$\det(g_{\mu\nu}(a)) \neq 0 \quad \text{and} \quad (g_{\mu\nu}(a))^T = (g_{\mu\nu}(a)).$$

Moreover, the differentiability of g implies that the matrix $(g_{\mu\nu}(a))$ depends differentiably on a. This means that in its dependence on the local coordinates x^j the coefficients $g_{\mu\nu} = g_{\mu\nu}(x)$ are smooth functions.

In general, however, the condition $g_{\mu\nu}(a)X^\mu X^\nu > 0$ does not hold for all $X \neq 0$, that is the matrix $(g_{\mu\nu}(a))$ is not required to be positive definite. This property distinguishes Riemannian manifolds from general semi-Riemannian manifolds. The Lorentz manifolds are specified as the semi-Riemannian manifolds with $(p,q) = (n-1,1)$ or $(p,q) = (1,n-1)$.

Examples:

- $\mathbb{R}^{p,q} = (\mathbb{R}^{p+q}, g^{p,q})$ for $p,q \in \mathbb{N}$ where

$$g^{p,q}(X,Y) := \sum_{i=1}^{p} X^i Y^i - \sum_{i=p+1}^{p+q} X^i Y^i.$$

  Hence

$$(g_{\mu\nu}) = \begin{pmatrix} 1_p & 0 \\ 0 & -1_q \end{pmatrix} = \mathrm{diag}(1,\ldots,1,-1,\ldots,-1).$$

- $\mathbb{R}^{1,3}$ or $\mathbb{R}^{3,1}$: the usual *Minkowski space*.
- $\mathbb{R}^{1,1}$: the two-dimensional Minkowski space (the *Minkowski plane*).
- $\mathbb{R}^{2,0}$: the *Euclidean plane*.
- $\mathbb{S}^2 \subset \mathbb{R}^{3,0}$: compactification of $\mathbb{R}^{2,0}$; the structure of a Riemannian manifold on the 2-sphere $\mathbb{S}^2$ is induced by the inclusion in $\mathbb{R}^{3,0}$.
- $\mathbb{S} \times \mathbb{S} \subset \mathbb{R}^{2,2}$: compactification of $\mathbb{R}^{1,1}$. More precisely, $\mathbb{S} \times \mathbb{S} \subset \mathbb{R}^{2,0} \times \mathbb{R}^{0,2} \cong \mathbb{R}^{2,2}$ where the first circle $\mathbb{S} = \mathbb{S}^1$ is contained in $\mathbb{R}^{2,0}$, the second one in $\mathbb{R}^{0,2}$ and where the structure of a semi-Riemannian manifold on $\mathbb{S} \times \mathbb{S}$ is induced by the inclusion into $\mathbb{R}^{2,2}$.
- Similarly, $\mathbb{S}^p \times \mathbb{S}^q \subset \mathbb{R}^{p+1,0} \times \mathbb{R}^{0,q+1} \cong \mathbb{R}^{p+1,q+1}$, with the p-sphere $\mathbb{S}^p = \{X \in \mathbb{R}^{p+1} : g^{p+1,0}(X,X) = 1\} \subset \mathbb{R}^{p+1,0}$ and the q-sphere $\mathbb{S}^q \subset \mathbb{R}^{0,q+1}$, as a generalization of the previous example, yields a compactification of $\mathbb{R}^{p,q}$ for $p,q \geq 1$. This compact semi-Riemannian manifold will be denoted by $\mathbb{S}^{p,q}$ for all $p,q \geq 0$.

In the following, we will use the above examples of semi-Riemannian manifolds and their open subspaces only—except for the quadrics $N^{p,q}$ occurring in Sect. 2.1. (These quadrics are locally isomorphic to $\mathbb{S}^{p,q}$ from the point of view of conformal geometry.)

1.2 Conformal Transformations

Definition 1.2. Let (M,g) and (M',g') be two semi-Riemannian manifolds of the same dimension n and let $U \subset M, V \subset M'$ be open subsets of M and M', respectively. A smooth mapping $\varphi : U \to V$ of maximal rank is called a *conformal transformation*, or *conformal map*, if there is a smooth function $\Omega : U \to \mathbb{R}_+$ such that

$$\varphi^* g' = \Omega^2 g,$$

where $\varphi^* g'(X,Y) := g'(T\varphi(X), T\varphi(Y))$ and $T\varphi : TU \to TV$ denotes the tangent map (derivative) of φ. Ω is called the *conformal factor* of φ. Sometimes a conformal transformation $\varphi : U \to V$ is additionally required to be bijective and/or orientation preserving.

In local coordinates of M and M'

$$(\varphi^* g')_{\mu\nu}(a) = g'_{ij}(\varphi(a)) \partial_\mu \varphi^i \partial_\nu \varphi^j.$$

Hence, φ is conformal if and only if

$$\Omega^2 g_{\mu\nu} = (g'_{ij} \circ \varphi) \partial_\mu \varphi^i \partial_\nu \varphi^j \tag{1.1}$$

in the coordinate neighborhood of each point.

Note that for a conformal transformation φ the tangent maps $T_a\varphi : T_a M \to T_{\varphi(a)} M'$ are bijective for each point $a \in U$. Hence, by the inverse mapping theorem a conformal transformation is always locally invertible as a smooth map.

Examples:

- Local isometries, that is smooth mappings φ with $\varphi^* g' = g$, are conformal transformations with conformal factor $\Omega = 1$.
- In order to study conformal transformations on the Euclidean plane $\mathbb{R}^{2,0}$ we identify $\mathbb{R}^{2,0} \cong \mathbb{C}$ and write $z = x + iy$ for $z \in \mathbb{C}$ with "real coordinates" $(x,y) \in \mathbb{R}$. Then a smooth map $\varphi : M \to \mathbb{C}$ on a connected open subset $M \subset \mathbb{C}$ is conformal according to (1.1) with conformal factor $\Omega : M \to \mathbb{R}_+$ if and only if for $u = \operatorname{Re} \varphi$ and $v = \operatorname{Im} \varphi$

$$u_x^2 + v_x^2 = \Omega^2 = u_y^2 + v_y^2 \neq 0, \; u_x u_y + v_x v_y = 0. \tag{1.2}$$

These equations are, of course, satisfied by the holomorphic (resp. antiholomorphic) functions from M to $\mathbb{C}$ because of the Cauchy–Riemann equations $u_x = v_y, u_y = -v_x$ (resp. $u_x = -v_y, u_y = v_x$) if $u_x^2 + v_x^2 \neq 0$. For holomorphic or antiholomorphic functions, $u_x^2 + v_x^2 \neq 0$ is equivalent to $\det D\varphi \neq 0$ where $D\varphi$ denotes the Jacobi matrix representing the tangent map $T\varphi$ of φ.

Conversely, for a general conformal transformation $\varphi = (u,v)$ the equations (1.2) imply that (u_x, v_x) and (u_y, v_y) are perpendicular vectors in $\mathbb{R}^{2,0}$ of equal length $\Omega \neq 0$. Hence, $(u_x, v_x) = (-v_y, u_y)$ or $(u_x, v_x) = (v_y, -u_y)$, that is φ is holomorphic or antiholomorphic with nonvanishing $\det D\varphi$.

As a **first important result**, we have shown that the conformal transformations $\varphi : M \to \mathbb{C}$ with respect to the Euclidean structure on $M \subset \mathbb{C}$ are the locally invertible holomorphic or antiholomorphic functions. The conformal factor of φ is $|\det D\varphi|$.

- With the same identification $\mathbb{R}^{2,0} \cong \mathbb{C}$ a linear map $\varphi : \mathbb{R}^{2,0} \to \mathbb{R}^{2,0}$ with representing matrix

$$A = A_\varphi = \begin{pmatrix} a & b \\ c & d \end{pmatrix}$$

is conformal if and only if $a^2 + c^2 \neq 0$ and $a = d$, $b = -c$ or $a = -d$, $b = c$. As a consequence, for $\zeta = a + ic \neq 0$, φ is of the form $z \mapsto \zeta z$ or $z \mapsto \zeta \bar{z}$.

These conformal linear transformations are *angle preserving* in the following sense: for points $z, w \in \mathbb{C} \setminus \{0\}$ the number

$$\omega(z, w) := \frac{z\bar{w}}{|zw|}$$

determines the (Euclidean) angle between z and w up to orientation. In the case of $\varphi(z) = \zeta z$ it follows that

$$\omega(\varphi(z), \varphi(w)) = \frac{\zeta z \overline{\zeta w}}{|\zeta z \zeta w|} = \omega(z, w),$$

and the same holds for $\varphi(z) = \zeta \bar{z}$.

Conversely, the linear maps φ with $\omega(\varphi(z), \varphi(w)) = \omega(z, w)$ for all $z, w \in \mathbb{C} \setminus \{0\}$ or $\omega(\varphi(z), \varphi(w)) = -\omega(z, w)$ for all $z, w \in \mathbb{C} \setminus \{0\}$ are conformal transformations. We conclude that an $\mathbb{R}$-linear map $\varphi : \mathbb{R}^{2,0} \to \mathbb{R}^{2,0}$ is a conformal transformation for the Euclidean plane if and only if it is angle preserving.

- We have shown that an orientation-preserving $\mathbb{R}$-linear map $\varphi : \mathbb{R}^{2,0} \to \mathbb{R}^{2,0}$ is a conformal transformation for the Euclidean plane if and only if it is the multiplication with a complex number $\zeta \neq 0$: $z \mapsto \zeta z$. In the case of $\zeta = r \exp i\alpha$ with $r \in \mathbb{R}_+$ and with $\alpha \in {]0, 2\pi]}$, we obtain the following interpretation: α induces a rotation with angle α and $z \mapsto (\exp i\alpha)z$ is an isometry, while r induces a dilatation $z \mapsto rz$.

Consequently, the group of orientation-preserving $\mathbb{R}$-linear and conformal maps $\mathbb{R}^{2,0} \to \mathbb{R}^{2,0}$ is isomorphic to $\mathbb{R}_+ \times \mathbb{S} \cong \mathbb{C} \setminus \{0\}$. The group of orientation-preserving $\mathbb{R}$-linear isometries is isomorphic to $\mathbb{S}$ while the group of dilatations is isomorphic to $\mathbb{R}_+$ (with the multiplicative structure) and therefore isomorphic to the additive group $\mathbb{R}$ via $t \to r := \exp t$, $t \in \mathbb{R}$.

- The above considerations also show that the conformal transformations $\varphi : M \to \mathbb{C}$, where M is an open subset of $\mathbb{R}^{2,0}$, can also be characterized as those mappings which preserve the angles infinitesimally: let $z(t), w(t)$ be smooth curves in M with $z(0) = w(0) = a$ and $\dot{z}(0) \neq 0 \neq \dot{w}(0)$, where $\dot{z}(0) = \frac{d}{dt} z(t)|_{t=0}$ is the derivative of $z(t)$ at $t = 0$. Then $\omega(\dot{z}(0), \dot{w}(0))$ determines the angle between the curves $z(t)$ and $w(t)$ at the common point a. Let $z_\varphi = \varphi \circ z$ and $w_\varphi = \varphi \circ w$ be the image curves. By definition, φ is called to preserve angles infinitesimally if and only if $\omega(\dot{z}(0), \dot{w}(0)) = \omega(\dot{z}_\varphi(0), \dot{w}_\varphi(0))$ for all points $a \in M$ and all curves $z(t), w(t)$ in M through $a = z(0) = w(0)$ with $\dot{z}(0) \neq 0 \neq \dot{w}(0)$. Note that $\dot{z}_\varphi(0) = D\varphi(a)(\dot{z}(0))$ by the chain rule. Hence, by the above characterization of the linear conformal transformations, φ preserves angles infinitesimally if and only if $D\varphi(a)$ is a linear conformal transformation for all $a \in M$ which by (1.2) is equivalent to φ being a conformal transformation.

- Again in the case of $\mathbb{R}^{2,0} \cong \mathbb{C}$ one can deduce from the above results that the conformal, orientation-preserving, and bijective transformations $\mathbb{R}^{2,0} \to \mathbb{R}^{2,0}$ are the entire holomorphic functions $\varphi : \mathbb{C} \to \mathbb{C}$ with holomorphic inverse functions $\varphi^{-1} : \mathbb{C} \to \mathbb{C}$, that is the biholomorphic functions $\varphi : \mathbb{C} \to \mathbb{C}$. These functions are simply the complex-linear affine maps of the form

$$\varphi(z) = \zeta z + \tau, \, z \in \mathbb{C},$$

with $\zeta, \tau \in \mathbb{C}, \, \zeta \neq 0$.

The group of all conformal, orientation-preserving invertible transformations $\mathbb{R}^{2,0} \to \mathbb{R}^{2,0}$ of the Euclidean plane can thus be identified with $(\mathbb{C} \setminus \{0\}) \times \mathbb{C}$, where the group law is given by

$$(\zeta, \tau)(\zeta', \tau') = (\zeta\zeta', \zeta\tau' + \tau).$$

In particular, this group is a four-dimensional real manifold.

This is an example of a semidirect product of groups. See Sect. 3.1 for the definition.

- The orientation-preserving and $\mathbb{R}$-linear conformal transformations $\psi : \mathbb{R}^{1,1} \to \mathbb{R}^{1,1}$ can be identified by elementary matrix multiplication. They are represented by matrices of the form

$$A = A_\psi = A(s,t) = \exp t \begin{pmatrix} \cosh s & \sinh s \\ \sinh s & \cosh s \end{pmatrix}$$

with $(s,t) \in \mathbb{R}^2$ (see Corollary 1.14 for details).

- Consider $\mathbb{R}^2$ endowed with the metric on $\mathbb{R}^2$ given by the bilinear form

$$\langle (x,y), (x',y') \rangle := \frac{1}{2}(xy' + yx').$$

This is a Minkowski metric g on $\mathbb{R}^2$, for which the coordinate axes coincide with the light cone

$$L = \{(x,y) : \langle (x,y), (x,y) \rangle = 0\}$$

in $0 \in \mathbb{R}^2$. With this metric, $(\mathbb{R}^2, g)$ is isometrically isomorphic to $\mathbb{R}^{1,1}$ with respect to the isomorphism $\psi : \mathbb{R}^{1,1} \to \mathbb{R}^2$,

$$(x,y) \mapsto (x+y, x-y).$$

- The *stereographic projection*

$$\pi : \mathbb{S}^2 \setminus \{(0,0,1)\} \to \mathbb{R}^{2,0},$$

$$(x,y,z) \mapsto \frac{1}{1-z}(x,y)$$

is conformal with $\Omega = \frac{1}{1-z}$. In order to prove this it suffices to show that the inverse map $\varphi := \pi^{-1} : \mathbb{R}^{2,0} \to \mathbb{S}^2 \subset \mathbb{R}^{3,0}$ is a conformal transformation. We have

$$\varphi(\xi,\eta) = \frac{1}{1+r^2}(2\xi, 2\eta, r^2 - 1),$$

for $(\xi,\eta) \in \mathbb{R}^2$ and $r = \sqrt{\xi^2 + \eta^2}$. For the tangent vectors $X_1 = \frac{\partial}{\partial \xi}, X_2 = \frac{\partial}{\partial \eta}$ we get

$$T\varphi(X_1) = \frac{d}{dt}\varphi(\xi + t, \eta)|_{t=0}$$

$$= 2\left(\frac{1}{1+r^2}\right)^2 (r^2 + 1 - 2\xi^2, -2\xi\eta, 2\xi),$$

$$T\varphi(X_2) = 2\left(\frac{1}{1+r^2}\right)^2 (-2\xi\eta, r^2 + 1 - 2\eta^2, 2\eta).$$

Hence

$$g'(T\varphi(X_i), T\varphi(X_j)) = \left(\frac{2}{1+r^2}\right)^2 (\delta_{ij}),$$

that is $\Lambda = \frac{2}{1+r^2}$ is the conformal factor of φ. Thus, $\pi = \varphi^{-1}$ has the conformal factor $\Omega = \Lambda^{-1} = \frac{1}{2}(1+r^2) = \frac{1}{1-z}$.

Similarly, the stereographic projection of the n-sphere,

$$\pi : \mathbb{S}^n \setminus \{(0, \ldots, 0, 1)\} \to \mathbb{R}^{n,0},$$

$$(x^0, \ldots, x^n) \mapsto \frac{1}{1-x^n}(x^0, \ldots, x^{n-1}),$$

is a conformal map.
- In Proposition 2.5 we present another natural conformal map in detail, the conformal embedding

$$\tau : \mathbb{R}^{p,q} \to \mathbb{S}^p \times \mathbb{S}^q \subset \mathbb{R}^{p+1,q+1}$$

into the non-Riemannian version of $\mathbb{S}^p \times \mathbb{S}^q$. $\mathbb{S}^p \times \mathbb{S}^q$ has been described in the preceding section.
- The composition of two conformal maps is conformal.
- If $\varphi : M \to M'$ is a bijective conformal transformation with conformal factor Ω then φ is a diffeomorphism (that is φ^{-1} is smooth) and, moreover, $\varphi^{-1} : M' \to M$ is conformal with conformal factor $\frac{1}{\Omega}$. This property has been used in the investigation of the above example on the stereographic projection.

1.3 Conformal Killing Fields

In the following, we want to study the conformal maps $\varphi : M \to M'$ between open subsets $M, M' \subset \mathbb{R}^{p,q}$, $p + q = n > 1$. To begin with, we will classify them by an infinitesimal argument:

Let $X : M \subset \mathbb{R}^{p,q} \to \mathbb{R}^n$ be a smooth vector field. Then

$$\dot{\gamma} = X(\gamma)$$

for smooth curves $\gamma = \gamma(t)$ in M is an autonomous differential equation. The *local one-parameter group* $(\varphi_t^X)_{t \in \mathbb{R}}$ corresponding to X satisfies

$$\frac{d}{dt}(\varphi^X(t,a)) = X(\varphi^X(t,a))$$

with initial condition $\varphi^X(0,a) = a$. Moreover, for every $a \in U$, $\varphi^X(\cdot, a)$ is the unique maximal solution of $\dot{\gamma} = X(\gamma)$ defined on the maximal interval $]t_a^-, t_a^+[$. Let $M_t :=$ $\{a \in M : t_a^- < t < t_a^+\}$ and $\varphi_t^X(a) := \varphi^X(t,a)$ for $a \in M_t$. Then $M_t \subset M$ is an open subset of M and $\varphi_t^X : M_t \to M_{-t}$ is a diffeomorphism. Furthermore, we have $\varphi_t^X \circ$ $\varphi_s^X(a) = \varphi_{s+t}^X(a)$ if $a \in M_{t+s} \cap M_s$ and $\varphi_s^X(a) \in M_t$, and, of course, $\varphi_0^X = \mathrm{id}_M, M_0 =$ M. In particular, the local one-parameter group $(\varphi_t^X)_{t \in \mathbb{R}}$ satisfies the *flow equation*

$$\frac{d}{dt}(\varphi_t^X)|_{t=0} = X.$$

Definition 1.3. A vector field X on $M \subset \mathbb{R}^{p,q}$ is called a *conformal Killing field* if φ_t^X is conformal for all t in a neighborhood of 0.

Theorem 1.4. *Let $M \subset \mathbb{R}^{p,q}$ be open, $g = g^{p,q}$ and X a conformal Killing field with coordinates*

$$X = (X^1, \ldots, X^n) = X^\nu \partial_\nu$$

with respect to the canonical cartesian coordinates on $\mathbb{R}^n$. Then there is a smooth function $\kappa : M \to \mathbb{R}$, so that

$$X_{\mu,\nu} + X_{\nu,\mu} = \kappa g_{\mu\nu}.$$

Here we use the notation: $f_{,\nu} := \partial_\nu f$, $X_\mu := g_{\mu\nu} X^\nu$.

Proof. Let X be a conformal Killing field, (φ_t) the associated local one-parameter group, and $\Omega_t : M_t \to \mathbb{R}^+$, such that

$$(\varphi_t^* g)_{\mu\nu}(a) = g_{ij}(\varphi_t(a)) \partial_\mu \varphi_t^i \partial_\nu \varphi_t^j = (\Omega_t(a))^2 g_{\mu\nu}(a).$$

By differentiation with respect to t at $t = 0$ we get (g_{ij} is constant!)

$$\frac{d}{dt}(\Omega_t^2(a)\,g_{\mu\nu}(a))|_{t=0} = \frac{d}{dt}\left(g_{ij}(\varphi_t(a))\partial_\mu\varphi_t^i\,\partial_\nu\varphi_t^j\right)\Big|_{t=0}$$

$$= g_{ij}\partial_\mu\dot{\varphi}_0^i\,\partial_\nu\varphi_0^j + g_{ij}\partial_\mu\varphi_0^i\,\partial_\nu\dot{\varphi}_0^j$$

$$= g_{ij}\,\partial_\mu X^i(a)\,\delta_\nu^j + g_{ij}\,\delta_\mu^i\,\partial_\nu X^j(a)$$

$$= \partial_\mu X_\nu(a) + \partial_\nu X_\mu(a).$$

Hence, the statement follows with $\kappa(a) = \dfrac{d}{dt}\,\Omega_t^2(a)\big|_{t=0}$. $\qquad\qquad\square$

If $g_{\mu\nu}$ is not constant, we have

$$(L_X g)_{\mu\nu} = X_{\mu;\nu} + X_{\nu;\mu} = \kappa g_{\mu\nu}.$$

Here, L_X is the *Lie derivative* and a semicolon in the index denotes the covariant derivative corresponding to the Levi-Civita connection for g.

Definition 1.5. A smooth function $\kappa : M \subset \mathbb{R}^{p,q} \to \mathbb{R}$ is called a *conformal Killing factor* if there is a conformal Killing field X, such that

$$X_{\mu,\nu} + X_{\nu,\mu} = \kappa g_{\mu\nu}.$$

(Similarly, for general semi-Riemannian manifolds on coordinate neighborhoods:

$$X_{\mu;\nu} + X_{\nu;\mu} = \kappa g_{\mu\nu}.)$$

Theorem 1.6. $\kappa : M \to \mathbb{R}$ *is a conformal Killing factor if and only if*

$$(n-2)\kappa_{,\mu\nu} + g_{\mu\nu}\Delta_g\kappa = 0,$$

where $\Delta_g = g^{kl}\partial_k\partial_l$ *is the* Laplace–Beltrami operator *for* $g = g^{p,q}$.

Proof. "$\Rightarrow$": Let $\kappa : M \to \mathbb{R}$ and $X_{\mu,\nu} + X_{\nu,\mu} = \kappa g_{\mu\nu}$ $(M \subset \mathbb{R}^{p,q}, g = g^{p,q})$. Then from

$$\partial_k\partial_l(X_{\mu,\nu}) = \partial_\nu\partial_k(X_{\mu,l}), \quad \text{etc.},$$

it follows that

$$0 = \partial_k\partial_l(X_{\mu,\nu} + X_{\nu,\mu}) - \partial_l\partial_\mu(X_{k,\nu} + X_{\nu,k})$$
$$+ \partial_\mu\partial_\nu(X_{k,l} + X_{l,k}) - \partial_\nu\partial_k(X_{\mu,l} + X_{l,\mu}).$$

Since κ is a conformal Killing factor, one can deduce

$$\partial_k\partial_l(X_{\mu,\nu} + X_{\nu,\mu}) = \kappa_{,kl}\,g_{\mu\nu}, \quad \text{etc.}$$

Hence

$$0 = g_{\mu\nu}\,\kappa_{,kl} - g_{k\nu}\,\kappa_{,l\mu} + g_{kl}\,\kappa_{,\mu\nu} - g_{\mu l}\,\kappa_{,\nu k}.$$

By multiplication with g^{kl} (defined by $g^{\mu\lambda}g_{\lambda\nu} = \delta_\nu^\mu$) we get

$$0 = g^{kl} g_{\mu\nu} \, \kappa_{,kl} - g^{kl} g_{k\nu} \, \kappa_{,l\mu} + g^{kl} g_{kl} \, \kappa_{,\mu\nu} - g^{kl} g_{\mu l} \, \kappa_{,\nu k}$$

$$= g^{kl} (g_{\mu\nu} \, \kappa_{,kl}) - \delta_\nu^l \, \kappa_{,l\mu} + n \, \kappa_{,\mu\nu} - \delta_\mu^k \, \kappa_{,l\mu}$$

$$= g_{\mu\nu} \Delta_g \kappa + (n-2) \kappa_{,\mu\nu}.$$

The reverse implication "⇐" follows from the discussion in Sect. 1.4.   □

The theorem also holds for open subsets M in semi-Riemannian manifolds with ";" instead of ",".

Important Observation. In the case $n = 2$, κ is conformal if and only if $\Delta_g \kappa = 0$. For $n > 2$, however, there are many additional conditions. More precisely, these are

$$\kappa_{,\mu\nu} = 0 \text{ for } \mu \neq \nu,$$

$$\kappa_{,\mu\mu} = \pm(n-2)^{-1}\Delta_g \kappa.$$

1.4 Classification of Conformal Transformations

With the help of the implication "⇒" of Theorem 1.6, we will determine all conformal Killing fields and hence all conformal transformations on connected open sets $M \subset \mathbb{R}^{p,q}$.

1.4.1 Case 1: $n = p + q > 2$

From the equations $g_{\mu\mu}(n-2)\kappa_{,\mu\mu} + \Delta_g \kappa = 0$ for a conformal Killing factor κ we get $(n-2)\Delta_g \kappa + n\Delta_g \kappa = 0$ by summation, hence $\Delta_g \kappa = 0$ (as in the case $n = 2$). Using again $g_{\mu\mu}(n-2)\kappa_{,\mu\mu} + \Delta_g \kappa = 0$, it follows that $\kappa_{,\mu\mu} = 0$. Consequently, $\kappa_{,\mu\nu} = 0$ for all μ, ν. Hence, there are constants $\alpha_\mu \in \mathbb{R}$ such that

$$\kappa_{,\mu}(q^1, \ldots, q^n) = \alpha_\mu, \quad \mu = 1, \ldots, n.$$

It follows that the solutions of $(n-2)\kappa_{,\mu\nu} + g_{\mu\nu}\Delta_g \kappa = 0$ are the affine-linear maps

$$\kappa(q) = \lambda + \alpha_\nu q^\nu, \quad q = (q^\nu) \in M \subset \mathbb{R}^n,$$

with $\lambda, \alpha_\nu \in \mathbb{R}$.

To begin with a complete description of all conformal Killing fields on connected open subsets $M \subset \mathbb{R}^{p,q}$, $p + q > 2$, we first determine the conformal Killing fields X with conformal Killing factor $\kappa = 0$ (that is the proper Killing fields, which belong to local isometries). $X_{\mu,\mu} + X_{\mu,\mu} = 0$ means that X^μ does not depend on q^μ. $X_{\mu,\nu} + X_{\nu,\mu} = 0$ implies $X^\mu_{,\nu} = 0$. Thus X^μ can be written as

$$X^\mu(q) = c^\mu + \omega_\nu^\mu q^\nu$$

with $c^\mu \in \mathbb{R}$, $\omega_\nu^\mu \in \mathbb{R}$.

If all the coefficients ω_v^μ vanish, the vector field $X^\mu(q) = c^\mu$ determines the differential equation

$$\dot{q} = c,$$

with the (global) one-parameter group $\varphi^X(t,q) = q + tc$ as its flow. The associated conformal transformation ($\varphi^X(t,q)$ for $t = 1$) is the *translation*

$$\varphi_c(q) = q + c.$$

For $c = 0$ and general $\omega = (\omega_v^\mu)$ the equations

$$X_{\mu,v} + X_{v,\mu} = g_{\mu v} \kappa = 0$$

imply

$$g_{v\rho}\,\omega_\mu^\rho + g_{\mu\rho}\,\omega_v^\rho = 0,$$

that is $\omega^T g + g\,\omega = 0$. Hence, these solutions are given by the elements of the Lie algebra $\mathfrak{o}(p,q) := \{\omega : \omega^T g^{p,q} + g^{p,q}\omega = 0\}$. The associated conformal transformations ($\varphi^X(t,q) = e^{t\omega}q$ for $t = 1$) are the *orthogonal transformations*

$$\varphi_\Lambda : \mathbb{R}^{p,q} \to \mathbb{R}^{p,q}, \quad q \mapsto \Lambda q,$$

with

$$\Lambda = e^\omega \in O(p,q) := \{\Lambda \in \mathbb{R}^{n \times n} : \Lambda^T g^{p,q} \Lambda = g^{p,q}\}$$

(equivalently, $O(p,q) = \{\Lambda \in \mathbb{R}^{n \times n} : \langle \Lambda x, \Lambda x' \rangle = \langle x, x' \rangle\}$ with the symmetric bilinear form $\langle \cdot, \cdot \rangle$ given by $g^{p,q}$).

We have thus determined all local isometries on connected open subsets $M \subset \mathbb{R}^{p,q}$. They are the restrictions of maps

$$\varphi(q) = \varphi_\Lambda(q) + c, \quad \Lambda \in O(p,q), \quad c \in \mathbb{R}^n,$$

and form a finite-dimensional Lie group, the group of motions belonging to $g^{p,q}$. This group can also be described as a semidirect product (cf. Sect. 3.1) of $O(p,q)$ and $\mathbb{R}^n$.

The constant conformal Killing factors $\kappa = \lambda \in \mathbb{R} \setminus \{0\}$ correspond to the conformal Killing fields $X(q) = \lambda q$ belonging to the conformal transformations

$$\varphi(q) = e^\lambda q, \quad q \in \mathbb{R}^n,$$

which are the *dilatations*.

All the conformal transformations on $M \subset \mathbb{R}^{p,q}$ considered so far have a unique conformal continuation to $\mathbb{R}^{p,q}$. Hence, they are essentially conformal transformations on all of $\mathbb{R}^{p,q}$ associated to global one-parameter groups (φ_t). This is no longer true for the following conformal transformations.

In view of the preceding discussion, every conformal Killing factor $\kappa \neq 0$ without a constant term is linear and thus can be written as

$$\kappa(q) = 4\langle q, b \rangle, \quad q \in \mathbb{R}^n,$$

with $b \in \mathbb{R}^n \setminus \{0\}$ and $\langle q, b \rangle = g_{\mu\nu}^{p,q} q^\mu b^\nu$. A direct calculation shows that

$$X^\mu(q) := 2\langle q, b \rangle q^\mu - \langle q, q \rangle b^\mu, \quad q \in \mathbb{R}^n,$$

is a solution of $X_{\mu,\nu} + X_{\nu,\mu} = \kappa g_{\mu\nu}$. (This proves the implication "$\Leftarrow$" in Theorem 1.6 for $n > 2$.) As a consequence, for every conformal Killing field X with conformal Killing factor

$$\kappa(q) = \lambda + x_\mu q^\mu = \lambda + 4\langle q, b \rangle,$$

the vector field $Y(q) = X(q) - 2\langle q, b \rangle q^\mu - \langle q, q \rangle b^\mu - \lambda q$ is a conformal Killing field with conformal Killing factor 0. Hence, by the preceding discussion, it has the form $Y(q) - c + \omega q$. To sum up, we have proven

Theorem 1.7. *Every conformal Killing field X on a connected open subset M of $\mathbb{R}^{p,q}$ (in case of $p + q = n > 2$) is of the form*

$$X(q) = 2\langle q, b \rangle q^\mu - \langle q, q \rangle b^\mu + \lambda q + c + \omega q$$

with suitable $b, c \in \mathbb{R}^n$, $\lambda \in \mathbb{R}$ and $\omega \in \mathfrak{o}(p,q)$.

Exercise 1.8. The Lie bracket of two conformal Killing fields is a conformal Killing field. The Lie algebra of all the conformal Killing fields is isomorphic to $\mathfrak{o}(p+1, q+1)$ (cf. Exercise 1.10).

The conformal Killing field $X(q) = 2\langle q, b \rangle q - \langle q, q \rangle b$, $b \neq 0$, has no global one-parameter group of solutions for the equation $\dot{q} = X(q)$. Its solutions form the following local one-parameter group

$$\varphi_t(q) = \frac{q - \langle q, q \rangle tb}{1 - 2\langle q, tb \rangle + \langle q, q \rangle \langle tb, tb \rangle}, \quad t \in \,]t_q^-, t_q^+[\,,$$

where $]t_q^-, t_q^+[$ is the maximal interval around 0 contained in

$$\{t \in \mathbb{R} \mid 1 - 2\langle q, tb \rangle + \langle q, q \rangle \langle tb, tb \rangle \neq 0\}.$$

Hence, the associated conformal transformation $\varphi := \varphi_1$

$$\varphi(q) = \frac{q - \langle q, q \rangle b}{1 - 2\langle b, q \rangle + \langle q, q \rangle \langle b, b \rangle}$$

– which is called a *special conformal transformation* – has (as a map into $\mathbb{R}^{p,q}$) a continuation at most to M_t at $t = 1$, that is to

$$M = M_1 := \{ q \in \mathbb{R}^{p,q} \mid 1 - 2\langle b,q\rangle + \langle q,q\rangle\langle b,b\rangle \neq 0 \}. \qquad (1.3)$$

In summary, we have

Theorem 1.9. *Every conformal transformation* $\varphi : M \to \mathbb{R}^{p,q}$, $n = p+q \geq 3$, *on a connected open subset* $M \subset \mathbb{R}^{p,q}$ *is a composition of*

- *a translation* $q \mapsto q + c$, $c \in \mathbb{R}^n$,
- *an orthogonal transformation* $q \mapsto \Lambda q$, $\Lambda \in O(p,q)$,
- *a dilatation* $q \mapsto e^{\lambda} q$, $\lambda \in \mathbb{R}$, *and*
- *a special conformal transformation*

$$q \mapsto \frac{q - \langle q,q\rangle b}{1 - 2\langle q,b\rangle + \langle q,q\rangle\langle b,b\rangle}, \quad b \in \mathbb{R}^n.$$

To be precise, we have just shown that every conformal transformation $\varphi : M \to \mathbb{R}^{p,q}$ on a connected open subset $M \subset \mathbb{R}^{p,q}$, $p + q > 2$, which is an element $\varphi = \varphi_{t_0}$ of a one-parameter group (φ_t) of conformal transformations, is of the type stated in the theorem. (Then Λ is an element of $SO(p,q)$, where $SO(p,q)$ is the component containing the identity $1 = \mathrm{id}$ in $O(p,q)$.) The general case can be derived from this.

Exercise 1.10. The conformal transformations described in Theorem 1.9 form a group with respect to composition (in spite of the singularities, it is not a subgroup of the bijections $\mathbb{R}^n \to \mathbb{R}^n$), which is isomorphic to $O(p+1,q+1)/\{\pm 1\}$ (cf. Theorem 2.9).

1.4.2 Case 2: Euclidean Plane ($p = 2$, $q = 0$)

This case has already been discussed as an example (cf. 1.2).

Theorem 1.11. *Every holomorphic function*

$$\varphi = u + iv : M \to \mathbb{R}^{2,0} \cong \mathbb{C}$$

on an open subset $M \subset \mathbb{R}^{2,0}$ *with nowhere-vanishing derivative is an orientation-preserving conformal mapping with conformal Killing factor* $\Omega^2 = u_x^2 + u_y^2 = \det D\varphi = |\varphi'|^2$. *Conversely, every conformal and orientation-preserving transformation* $\varphi : M \to \mathbb{R}^{2,0} \cong \mathbb{C}$ *is such a holomorphic function.*

This follows immediately from the Cauchy–Riemann differential equations (cf. 1.2). Of course, a corresponding result holds for the antiholomorphic functions. In the case of a connected open subset M of the Euclidean plane the collection of all the holomorphic and antiholomorphic functions exhausts the conformal transformations on M.

We want to describe the conformal transformations again by analyzing conformal Killing fields and conformal Killing factors: Every conformal Killing field $X = (u,v) : M \to \mathbb{C}$ on a connected open subset M of $\mathbb{C}$ with conformal Killing factor κ satisfies $\Delta\kappa = 0$ as well as $u_y + v_x = 0$ and $u_x = \frac{1}{2}\kappa = v_y$. In particular, X fulfills the Cauchy–Riemann equations and is a holomorphic function.

In the special case of a conformal Killing field corresponding to a vanishing conformal Killing factor $\kappa = 0$, one gets

$$X(z) = c + i\theta z, \quad z \in M,$$

with $c \in \mathbb{C}$ and $\theta \in \mathbb{R}$. Here we again use the notation $z = x + iy \in \mathbb{C} \cong \mathbb{R}^{2,0}$. The respective conformal transformations are the Euclidean motions (that is the isometries of $\mathbb{R}^{2,0}$)

$$\varphi(z) = c + e^{i\theta}z.$$

For constant conformal Killing factors $\kappa \neq 0$, $\kappa = \lambda \in \mathbb{R}$, one gets the dilatations

$$X(z) = \lambda z \quad \text{with} \quad \varphi(z) = e^{\lambda}z.$$

Moreover, for $\mathbb{R}$-linear κ in the form $\kappa = 4\mathrm{Re}(z\bar{b}) = 4(xb_1 + yb_2)$ one gets the "inversions". For instance, in the case of $b = (b_1, b_2) = (1, 0)$ we obtain

$$\varphi(z) = \frac{z - |z|^2}{1 - 2x + |z|^2} = \frac{-1 + 2x - |z|^2 - x + 1 + iy}{|z-1|^2}$$

$$= -1 - \frac{z-1}{|z-1|^2} = -\frac{z}{z-1}.$$

We conclude

Proposition 1.12. *The linear conformal Killing factors κ describe precisely the Möbius transformations (cf. 2.12).*

For general conformal Killing factors $\kappa \neq 0$ on a connected open subset M of the complex plane, the equation $\Delta\kappa = 0$ implies that locally there exist holomorphic $X = (u, v)$ with $u_y + v_x = 0$, $u_x = \frac{1}{2}\kappa = v_y$, that is

$$u_x = v_y, \quad u_y = -v_x.$$

(This proves the implication "$\Leftarrow$" in Theorem 1.6 for $p = 2, q = 0$, if one localizes the definition of a conformal Killing field.) In this situation, the one-parameter groups (φ_t) for X are also holomorphic functions with nowhere-vanishing derivative.

1.4.3 Case 3: Minkowski Plane (p = q = 1)

In analogy to Theorem 1.11 we have

Theorem 1.13. *A smooth map $\varphi = (u, v) : M \to \mathbb{R}^{1,1}$ on a connected open subset $M \subset \mathbb{R}^{1,1}$ is conformal if and only if*

$$u_x^2 > v_x^2, \quad \text{and} \quad u_x = v_y, u_y = v_x \text{ or } u_x = -v_y, u_y = -v_x.$$

Proof. The condition $\varphi^* g = \Omega^2 g$ for $g = g^{1,1}$ is equivalent to the equations

$$u_x^2 - v_x^2 = \Omega^2, \quad u_x u_y - v_x v_y = 0, \quad u_y^2 - v_y^2 = -\Omega^2, \quad \Omega^2 > 0.$$

"$\Leftarrow$" : these three equations imply $u_x^2 = \Omega^2 + v_x^2 > v_x^2$ and

$$0 = \Omega^2 + 2u_x u_y - 2v_x v_y - \Omega^2 = (u_x + u_y)^2 - (v_x + v_y)^2.$$

Hence $u_x + u_y = \pm(v_x + v_y)$. In the case of the sign "$+$" it follows that

$$\begin{aligned}
0 &= u_x^2 - u_x^2 + v_x v_y - u_x u_y \\
&= u_x^2 - u_x(u_x + u_y) + v_x v_y \\
&= u_x^2 - u_x(v_x + v_y) + v_x v_y \\
&= (u_x - v_x)(u_x - v_y),
\end{aligned}$$

that is $u_x = v_x$ or $u_x = v_y$. $u_x = v_x$ is a contradiction to $u_x^2 - v_x^2 = \Omega^2 > 0$. Therefore we have $u_x = v_y$ and $u_y = v_x$.

Similarly, the sign "$-$" yields $u_x = -v_y$ and $u_y = -v_x$.

"$\Rightarrow$" : with $\Omega^2 := u_x^2 - v_x^2 > 0$ we get by substitution

$$u_y^2 - v_y^2 = v_x^2 - u_x^2 = -\Omega^2 \qquad \text{and} \qquad u_x u_y - v_x v_y = 0.$$

Hence φ is conformal. In the case of $u_x = v_y, u_y = v_x$ it follows that

$$\det D\varphi = u_x v_y - u_y v_x = u_x^2 - v_x^2 > 0,$$

that is φ is orientation preserving. In the case of $u_x = -v_y, u_y = -v_x$ the map φ reverses the orientation. $\qquad\square$

The solutions of the wave equation $\Delta \kappa = \kappa_{xx} - \kappa_{yy} = 0$ in $1 + 1$ dimensions can be written as

$$\kappa(x,y) = f(x+y) + g(x-y)$$

with smooth functions f and g of one real variable in the light cone variables $x_+ = x + y$, $x_- = x - y$. Hence, any conformal Killing factor κ has this form in the case of $p = q = 1$. Let F and G be integrals of $\frac{1}{2}f$ and $\frac{1}{2}g$, respectively. Then

$$X(x,y) = (F(x_+) + G(x_-), F(x_+) - G(x_-))$$

is a conformal Killing field with $X_{\mu,\nu} + X_{\nu,\mu} = g_{\mu\nu}\kappa$. (This eventually completes the proof of the implication "$\Leftarrow$" in Theorem 1.6.) The associated one-parameter group (φ_t) of conformal transformations consists of orientation-preserving maps with $u_x = v_y, u_y = v_x$ for $\varphi_t = (u,v)$.

Corollary 1.14. *The orientation-preserving linear and conformal maps* $\psi : \mathbb{R}^{1,1} \to \mathbb{R}^{1,1}$ *have matrix representations of the form*

$$A = A_\psi = A_+(s,t) = \exp t \begin{pmatrix} \cosh s & \sinh s \\ \sinh s & \cosh s \end{pmatrix}$$

or

$$A = A_\psi = A_-(s,t) = \exp t \begin{pmatrix} -\cosh s & \sinh s \\ \sinh s & -\cosh s \end{pmatrix}$$

with $(s,t) \in \mathbb{R}^2$.

Proof. Let A_ψ be the matrix representing $\psi = (u,v)$ with respect to the standard basis in $\mathbb{R}^2$:

$$A_\psi = \begin{pmatrix} a & b \\ c & d \end{pmatrix}.$$

Then $u = ax + by$, $v = cx + dy$, hence $u_x = a, u_y = b, v_x = c, v_y = d$. Our Theorem 1.13 implies $a^2 > c^2$ and $a = d, b = c$ (the choice of the sign comes from $\det A > 0$). There is a unique $t \in \mathbb{R}$ with $\exp 2t = a^2 - c^2$ and also a unique $s \in \mathbb{R}$ with $\sinh s = (\exp -t)c$, hence $c^2 = \exp 2t \sinh^2 s$. It follows $a^2 = \exp 2t(1 + \sinh^2 s) = (\exp t \cosh s)^2$, and we conclude $a = \exp t \cosh s = d$ or $a = -\exp t \cosh s = d$, and $b = \exp t \sinh s = d$. □

There is again an interpretation of the action of t (dilatation) and s (boost) similar to the Euclidean case.

The representation in Corollary 1.14 respects the composition: The well-known identities for sinh and cosh imply $A_+(s,t)A_+(s',t') = A_+(s+s',t+t')$.

Remark 1.15. As a consequence, the identity component of the group of linear conformal mappings $\mathbb{R}^{1,1} \to \mathbb{R}^{1,1}$ is isomorphic to the additive group $\mathbb{R}^2$. Moreover, the Lorentz group $L = L(1,1)$ (the identity component of the linear isometries) is isomorphic to $\mathbb{R}$. The corresponding Poincaré group $P = P(1,1)$ is the semidirect product $L \ltimes \mathbb{R}^2 \cong \mathbb{R} \ltimes \mathbb{R}^2$ with respect to the action $\mathbb{R} \to GL(2,\mathbb{R})$, $s \mapsto A_+(s,0)$.

Chapter 2
The Conformal Group

Definition 2.1. The *conformal group* $Conf(\mathbb{R}^{p,q})$ is the connected component containing the identity in the group of conformal diffeomorphisms of the conformal compactification of $\mathbb{R}^{p,q}$.

In this definition, the group of conformal diffeomorphisms is considered as a topological group with the topology of compact convergence, that is the topology of uniform convergence on the compact subsets. More precisely, the topology of compact convergence on the space $\mathscr{C}(X,Y)$ of continuous maps $X \to Y$ between topological spaces X, Y is generated by all the subsets

$$\{f \in \mathscr{C}(X,Y) : f(K) \subset V\},$$

where $K \subset X$ is compact and $V \subset Y$ is open.

First of all, to understand the definition we have to introduce the concept of conformal compactification. The conformal compactification as a hyperquadric in five-dimensional projective space has been used already by Dirac [Dir36*] in order to study conformally invariant field theories in four-dimensional spacetime. The concept has its origin in general geometric principles.

2.1 Conformal Compactification of $\mathbb{R}^{p,q}$

To study the collection of all conformal transformations on an open connected subset $M \subset \mathbb{R}^{p,q}$, $p+q \geq 2$, a conformal compactification $N^{p,q}$ of $\mathbb{R}^{p,q}$ is introduced, in such a way that the conformal transformations $M \to \mathbb{R}^{p,q}$ become everywhere-defined and bijective maps $N^{p,q} \to N^{p,q}$. Consequently, we search for a "minimal" compactification $N^{p,q}$ of $\mathbb{R}^{p,q}$ with a natural semi-Riemannian metric, such that every conformal transformation $\varphi : M \to \mathbb{R}^{p,q}$ has a continuation to $N^{p,q}$ as a conformal diffeomorphism $\widehat{\varphi} : N^{p,q} \to N^{p,q}$ (cf. Definition 2.7 for details).

Note that conformal compactifications in this sense do only exist for $p+q > 2$. We investigate the two-dimensional case in detail in the next two sections below. We show that the spaces $N^{p,q}$ still can be defined as compactifications of $\mathbb{R}^{p,q}, p+q = 2$, with a natural conformal structure inducing the original conformal structure on $\mathbb{R}^{p,q}$.

Schottenloher, M.: *The Conformal Group*. Lect. Notes Phys. **759**, 23–38 (2008)
DOI 10.1007/978-3-540-68628-6_3   © Springer-Verlag Berlin Heidelberg 2008

However, the spaces $N^{p,q}$ do not possess the continuation property mentioned above in full generality: there exist many conformal transformations $\varphi : M \to \mathbb{R}^{p,q}$ which do not have a conformal continuation to all of $N^{p,q}$.

Let $n = p + q \geq 2$. We use the notation $\langle x \rangle_{p,q} := g^{p,q}(x,x), x \in \mathbb{R}^{p,q}$. For short, we also write $\langle x \rangle = \langle x \rangle_{p,q}$ if p and q are evident from the context. $\mathbb{R}^{p,q}$ can be embedded into the $(n+1)$-dimensional projective space $\mathbb{P}_{n+1}(\mathbb{R})$ by the map

$$\iota : \mathbb{R}^{p,q} \to \mathbb{P}_{n+1}(\mathbb{R}),$$

$$x = (x^1, \ldots, x^n) \mapsto \left(\frac{1 - \langle x \rangle}{2} : x^1 : \ldots : x^n : \frac{1 + \langle x \rangle}{2} \right).$$

Recall that $\mathbb{P}_{n+1}(\mathbb{R})$ is the quotient

$$(\mathbb{R}^{n+2} \setminus \{0\}) / \sim$$

with respect to the equivalence relation

$$\xi \sim \xi' \quad \Longleftrightarrow \quad \xi = \lambda \xi' \text{ for a } \lambda \in \mathbb{R} \setminus \{0\}.$$

$\mathbb{P}_{n+1}(\mathbb{R})$ can also be described as the space of one-dimensional subspaces of $\mathbb{R}^{n+2}$. $\mathbb{P}_{n+1}(\mathbb{R})$ is a compact $(n+1)$-dimensional smooth manifold (cf. for example [Scho95]). If $\gamma : \mathbb{R}^{n+2} \setminus \{0\} \to \mathbb{P}_{n+1}(\mathbb{R})$ is the quotient map, a general point $\gamma(\xi) \in \mathbb{P}_{n+1}(\mathbb{R})$, $\xi = (\xi^0, \ldots, \xi^{n+1}) \in \mathbb{R}^{n+2}$, is denoted by $(\xi^0 : \ldots : \xi^{n+1}) := \gamma(\xi)$ with respect to the so-called *homogeneous coordinates*. Obviously, we have

$$(\xi^0 : \cdots : \xi^{n+1}) = (\lambda \xi^0 : \cdots : \lambda \xi^{n+1}) \quad \text{for all } \lambda \in \mathbb{R} \setminus \{0\}.$$

We are looking for a suitable compactification of $\mathbb{R}^{p,q}$. As a candidate we consider the closure $\overline{\iota(\mathbb{R}^{p,q})}$ of the image of the smooth embedding $\iota : \mathbb{R}^{p,q} \to \mathbb{P}_{n+1}(\mathbb{R})$.

Remark 2.2. $\overline{\iota(\mathbb{R}^{p,q})} = N^{p,q}$, where $N_{p,q}$ is the *quadric*

$$N^{p,q} := \{ (\xi^0 : \cdots : \xi^{n+1}) \in \mathbb{P}_{n+1}(\mathbb{R}) | \langle \xi \rangle_{p+1,q+1} = 0 \}$$

in the real projective space $\mathbb{P}_{n+1}(\mathbb{R})$.

Proof. By definition of ι we have $\langle \iota(x) \rangle_{p+1,q+1} = 0$ for $x \in \mathbb{R}^{p,q}$, that is $\overline{\iota(\mathbb{R}^{p,q})} \subset N^{p,q}$.

For the converse inclusion, let $(\xi^0 : \cdots : \xi^{n+1}) \in N^{p,q} \setminus \iota(\mathbb{R}^{p,q})$. Then $\xi^0 + \xi^{n+1} = 0$, since

$$\iota(\lambda^{-1}(\xi^1, \ldots, \xi^n)) = (\xi^0 : \cdots : \xi^{n+1}) \in \iota(\mathbb{R}^{p,q})$$

for $\lambda := \xi^0 + \xi^{n+1} \neq 0$. Given $(\xi^0 : \cdots : \xi^{n+1}) \in N^{p,q}$ there always exist sequences $\epsilon_k \to 0, \delta_k \to 0$ with $\epsilon_k \neq 0 \neq \delta_k$ and $2\xi^1 \epsilon_k + \epsilon_k^2 = 2\xi^{n+1} \delta_k + \delta_k^2$. For $p \geq 1$ we have

$$P_k := (\xi^0 : \xi^1 + \epsilon_k : \xi^2 : \cdots : \xi^n : \xi^{n+1} + \delta_k) \in N^{p,q}.$$

Moreover, $\xi^0 + \xi^{n+1} + \delta_k = \delta_k \neq 0$ implies $P_k \in \iota(\mathbb{R}^{p,q})$. Finally, since $P_k \to (\xi^0 : \ldots : \xi^{n+1})$ for $k \to \infty$ it follows that $(\xi^0 : \cdots : \xi^{n+1}) \in \overline{\iota(\mathbb{R}^{p,q})}$, that is $N^{p,q} \subset \overline{\iota(\mathbb{R}^{p,q})}$. $\qquad\square$

We therefore choose $N^{p,q}$ as the underlying manifold of the conformal compactification. $N^{p,q}$ is a regular quadric in $\mathbb{P}_{n+1}(\mathbb{R})$. Hence it is an n-dimensional compact submanifold of $\mathbb{P}_{n+1}(\mathbb{R})$. $N^{p,q}$ contains $\iota(\mathbb{R}^{p,q})$ as a dense subset.

We get another description of $N^{p,q}$ using the quotient map γ on $\mathbb{R}^{p+1,q+1}$ restricted to $\mathbb{S}^p \times \mathbb{S}^q \subset \mathbb{R}^{p+1,q+1}$.

Lemma 2.3. *The restriction of γ to the product of spheres*

$$\mathbb{S}^p \times \mathbb{S}^q := \left\{ \xi \in \mathbb{R}^{n+2} : \sum_{j=0}^{p} (\xi^j)^2 = 1 = \sum_{j=p+1}^{n+1} (\xi^j)^2 \right\} \subset \mathbb{R}^{n+2}$$

gives a smooth 2-to-1 covering

$$\pi := \gamma|_{\mathbb{S}^p \times \mathbb{S}^q} : \mathbb{S}^p \times \mathbb{S}^q \to N^{p,q}.$$

Proof. Obviously $\gamma(\mathbb{S}^p \times \mathbb{S}^q) \subset N^{p,q}$. For $\xi, \xi' \in \mathbb{S}^p \times \mathbb{S}^q$ it follows from $\gamma(\xi) = \gamma(\xi')$ that $\xi = \lambda \xi'$ with $\lambda \in \mathbb{R} \setminus \{0\}$. $\xi, \xi' \in \mathbb{S}^p \times \mathbb{S}^q$ implies $\lambda \in \{1, -1\}$. Hence, $\gamma(\xi) = \gamma(\xi')$ if and only if $\xi = \xi'$ or $\xi = -\xi'$. For $P = (\xi^0 : \ldots : \xi^{n+1}) \in N^{p,q}$ the two inverse images with respect to π can be specified as follows: $P \in N^{p,q}$ implies $\langle \xi \rangle = 0$, that is $\sum_{j=0}^{p} (\xi^j)^2 = \sum_{j=p+1}^{n+1} (\xi^j)^2$. Let

$$r := \left(\sum_{j=0}^{p} (\xi^j)^2 \right)^{\frac{1}{2}}$$

and $\eta := \frac{1}{r}(\xi^0, \ldots, \xi^{n+1}) \in \mathbb{S}^p \times \mathbb{S}^q$. Then η and $-\eta$ are the inverse images of ξ. Hence, π is surjective and the description of the inverse images shows that π is a local diffeomorphism. $\qquad\square$

With the aid of the map $\pi : \mathbb{S}^p \times \mathbb{S}^q \to N^{p,q}$, which is locally a diffeomorphism, the metric induced on $\mathbb{S}^p \times \mathbb{S}^q$ by the inclusion $\mathbb{S}^p \times \mathbb{S}^q \subset \mathbb{R}^{p+1,q+1}$, that is the semi-Riemannian metric of $\mathbb{S}^{p,q}$ described in the examples of Sect. 1.1 on page 8, can be carried over to $N^{p,q}$ in such a way that $\pi : \mathbb{S}^{p,q} \to N^{p,q}$ becomes a (local) isometry.

Definition 2.4. $N^{p,q}$ with this semi-Riemannian metric will be called the *conformal compactification* of $\mathbb{R}^{p,q}$.

In particular, it is clear what the conformal transformations $N^{p,q} \to N^{p,q}$ are. In this way, $N^{p,q}$ obtains a *conformal structure* (that is the equivalence class of semi-Riemannian metrics).

We know that $\iota : \mathbb{R}^{p,q} \to N^{p,q}$ is an embedding (injective and regular) and that $\iota(\mathbb{R}^{p,q})$ is dense in the compact manifold $N^{p,q}$. In order to see that this embedding is conformal we compare ι with the natural map $\tau : \mathbb{R}^{p,q} \to \mathbb{S}^p \times \mathbb{S}^q$ defined by

$$\tau(x) = \frac{1}{r(x)} \left(\frac{1 - \langle x \rangle}{2}, x^1, \ldots, x^n, \frac{1 + \langle x \rangle}{2} \right),$$

where

$$r(x) = \frac{1}{2} \sqrt{1 + 2 \sum_{j=1}^{n} (x^j)^2 + \langle x \rangle^2} \geq \frac{1}{2}.$$

τ is well-defined because of

$$r(x)^2 = \left(\frac{1 - \langle x \rangle}{2} \right)^2 + \sum_{j=1}^{p} (x^j)^2 = \sum_{j=p+1}^{n} (x^j)^2 + \left(\frac{1 + \langle x \rangle}{2} \right)^2,$$

and we have

Proposition 2.5. $\tau : \mathbb{R}^{p,q} \to \mathbb{S}^p \times \mathbb{S}^q$ *is a conformal embedding with* $\iota = \pi \circ \tau$.

Proof. For the proof we only have to confirm that τ is indeed a conformal map. This can be checked in a similar manner as in the case of the stereographic projection on p. 12 in Chap. 1. We denote the factor $\frac{1}{r}$ by ρ and will observe that the result is independent of the special factor in question. For an index $1 \leq i \leq n$ we denote by τ_i, ρ_i the partial derivatives with respect to the coordinate x^i of $\mathbb{R}^{p,q}$. We have for $i \leq p$

$$\tau_i = \left(\rho_i \frac{1 - \langle x \rangle}{2} - \rho x^i, \rho_i x^1, \ldots \rho_i x^i + \rho, \ldots, \rho_i x^n, \rho_i \frac{1 + \langle x \rangle}{2} + \rho x^i \right)$$

and a similar formula for $j > p$ with only two changes in signs. For $i \leq p$ we obtain in $\mathbb{R}^{p+1,q+1}$

$$\langle \tau_i, \tau_i \rangle = \left(\rho_i \frac{1 - \langle x \rangle}{2} - \rho x^i \right)^2 + (\rho_i x^1)^2 + \ldots + (\rho_i x^i + \rho)^2 +$$

$$+ \ldots - (\rho_i x^n)^2 - \left(\rho_i \frac{1 + \langle x \rangle}{2} + \rho x^i \right)^2$$

$$= -2\rho_i \left(\rho_i \frac{\langle x \rangle}{2} + \rho x^i \right) + (\rho_i x^1)^2 + \ldots + (\rho_i x^i)^2 + 2\rho_i x^1 \rho +$$

$$+ \rho^2 - (\rho_i x^{p+1})^2 \ldots - (\rho_i x^n)^2$$

$$= -\rho_i^2 \langle x \rangle + \rho_i^2 \langle x \rangle - 2\rho_i x^1 \rho + 2\rho_i x^1 \rho$$

$$= \rho^2,$$

and for $j > p$ we obtain $\langle \tau_j, \tau_j \rangle = -\rho^2$ in the same way. Similarly, one checks $\langle \tau_i, \tau_j \rangle = 0$ for $i \neq j$. Hence, $\langle \tau_i, \tau_j \rangle = \rho^2 \eta_{ij}$ where $\eta = \text{diag}(1, \ldots 1, -1, \ldots, -1)$ is the diagonal matrix of the standard Minkowski metric of $\mathbb{R}^{p,q}$. This property is equivalent to τ being a conformal map. $\square$

We now want to describe the collection of all conformal transformations $N_{p,q} \to N_{p,q}$.

Theorem 2.6. *For every matrix* $\Lambda \in O(p+1, q+1)$ *the map* $\psi = \psi_\Lambda : N^{p,q} \to N^{p,q}$
defined by

$$\psi_\Lambda(\xi^0 : \ldots : \xi^{n+1}) := \gamma(\Lambda\xi), \quad (\xi^0 : \ldots : \xi^{n+1}) \in N^{p,q}$$

is a conformal transformation and a diffeomorphism. The inverse transformation
$\psi^{-1} = \psi_{\Lambda^{-1}}$ *is also conformal. The map* $\Lambda \mapsto \psi_\Lambda$ *is not injective. However,* $\psi_\Lambda = \psi_{\Lambda'}$ *implies* $\Lambda = \Lambda'$ *or* $\Lambda = -\Lambda'$.

Proof. For $\xi \in \mathbb{R}^{n+2} \setminus \{0\}$ with $\langle x \rangle = 0$ and $\Lambda \in O(p+1, q+1)$ we have $\langle \Lambda\xi \rangle = g(\Lambda\xi, \Lambda\xi) = g(\xi, \xi) = \langle \xi \rangle = 0$, that is $\gamma(\Lambda\xi) \in N^{p,q}$. $\gamma(\Lambda\xi)$ does not depend on the representative ξ as we can easily check: $\xi \sim \xi'$, that is $\xi' = r\xi$ with $r \in \mathbb{R} \setminus \{0\}$, implies $\Lambda\xi' = r\Lambda\xi$, that is $\Lambda\xi' \sim \Lambda\xi$. Altogether, $\psi : N^{p,q} \to N^{p,q}$ is well-defined. Because of the fact that the metric on $\mathbb{R}^{p+1, q+1}$ is invariant with respect to Λ, ψ_Λ turns out to be conformal. For $P \in N^{p,q}$ one calculates the conformal factor $\Omega^2(P) = \sum_{j=0}^{n+1} (\Lambda_k^j \xi^k)^2$ if P is represented by $\xi \in \mathbb{S}^p \times \mathbb{S}^q$. (In general, $\Lambda(\mathbb{S}^p \times \mathbb{S}^q)$ is not contained in $\mathbb{S}^p \times \mathbb{S}^q$, and the (punctual) deviation from the inclusion is described precisely by the conformal factor $\Omega(P)$:

$$\frac{1}{\Omega(P)} \Lambda(\xi) \in \mathbb{S}^p \times \mathbb{S}^q \text{ for } \xi \in \mathbb{S}^p \times \mathbb{S}^q \text{ and } P = \gamma(\xi).$$

Obviously, $\psi_\Lambda = \psi_{-\Lambda}$ and $\psi_\Lambda^{-1} = \psi_{\Lambda^{-1}}$. In the case $\psi_\Lambda = \psi_{\Lambda'}$ for $\Lambda, \Lambda' \in O(p+1, q+1)$ we have $\gamma(\Lambda\xi) = \gamma(\Lambda'\xi)$ for all $\xi \in \mathbb{R}^{n+2}$ with $\langle \xi \rangle = 0$. Hence, $\Lambda = r\Lambda'$ with $r \in \mathbb{R} \setminus \{0\}$. Now $\Lambda, \Lambda' \in O(p+1, q+1)$ implies $r = 1$ or $r = -1$. $\square$

The requested continuation property for conformal transformations can now be formulated as follows:

Definition 2.7. Let $\varphi : M \to \mathbb{R}^{p,q}$ be a conformal transformation on a connected open subset $M \subset \mathbb{R}^{p,q}$. Then $\widehat{\varphi} : N^{p,q} \to N^{p,q}$ is called a *conformal continuation* of φ, if $\widehat{\varphi}$ is a conformal diffeomorphism (with conformal inverse) and if $\iota(\varphi(x)) = \widehat{\varphi}(\iota(x))$ for all $x \in M$. In other words, the following diagram is commutative:

$$
\begin{array}{ccc}
M & \xrightarrow{\varphi} & \mathbb{R}^{p,q} \\
\downarrow{\iota} & & \downarrow{\iota} \\
N^{p,q} & \xrightarrow{\widehat{\varphi}} & N^{p,q}
\end{array}
$$

Remark 2.8. In a more conceptual sense the notion of a conformal compactification should be defined and used in the following general formulation. A *conformal compactification* of a connected semi-Riemannian manifold X is a compact semi-Riemannian manifold N together with a conformal embedding $\iota : X \to N$ such that

1. $\iota(X)$ is dense in N.
2. Every conformal transformation $\varphi : M \to X$ (that φ is injective and conformal) on an open and connected subset $M \subset X, M \neq \emptyset$, has a conformal continuation $\widehat{\varphi} : N \to N$.

A conformal compactification is unique up to isomorphism if it exists.

In the case of $X = \mathbb{R}^{p,q}$ the construction of $\iota : \mathbb{R}^{p,q} \to N^{p,q}$ so far together with Theorem 2.9 asserts that $N^{p,q}$ is indeed a conformal compactification in this general sense.

2.2 The Conformal Group of $\mathbb{R}^{p,q}$ for $p+q > 2$

Theorem 2.9. *Let $n = p+q > 2$. Every conformal transformation on a connected open subset $M \subset \mathbb{R}^{p,q}$ has a unique conformal continuation to $N^{p,q}$. The group of all conformal transformations $N^{p,q} \to N^{p,q}$ is isomorphic to $O(p+1, q+1)/\{\pm 1\}$. The connected component containing the identity in this group – that is, by Definition 2.1 the conformal group $\mathrm{Conf}(\mathbb{R}^{p,q})$ – is isomorphic to $SO(p+1, q+1)$ (or $SO(p+1, q+1)/\{\pm 1\}$ if -1 is in the connected component of $O(p+1, q+1)$ containing 1, for example, if p and q are odd.)*

Here, $SO(p+1, q+1)$ is defined to be the connected component of the identity in $O(p+1, q+1)$. $SO(p+1, q+1)$ is contained in

$$\{\Lambda \in O(p+1, q+1) | \det \Lambda = 1\}.$$

However, it is, in general, different from this subgroup, e.g., for the case $(p, q) = (2, 1)$ or $(p, q) = (3, 1)$.

Proof. It suffices to find conformal continuations $\widehat{\varphi}$ to $N^{p,q}$ (according to Definition 2.7) of all the conformal transformations φ described in Theorem 1.9 and to represent these continuations by matrices $\Lambda \in O(p+1, q+1)$ according to Lemma 2.3:

1. Orthogonal transformations. The easiest case is the conformal continuation of an orthogonal transformation $\varphi(x) = \Lambda'x$ represented by a matrix $\Lambda' \in O(p, q)$ and defined on all of $\mathbb{R}^{p,q}$. For the block matrix

$$\Lambda = \begin{pmatrix} 1 & 0 & 0 \\ 0 & \Lambda' & 0 \\ 0 & 0 & 1 \end{pmatrix},$$

one obviously has $\Lambda \in O(p+1, q+1)$, because of $\Lambda^T \eta \Lambda = \eta$, where $\eta = \mathrm{diag}(1, \ldots, 1, -1, \ldots, -1)$ is the matrix representing $g^{p+1, q+1}$. Furthermore,

$$\Lambda \in SO(p+1, q+1) \iff \Lambda' \in SO(p, q).$$

We define a conformal map $\widehat{\varphi} : N^{p,q} \to N^{p,q}$ by $\widehat{\varphi} := \psi_\Lambda$, that is

$$\widehat{\varphi}(\xi^0 : \ldots : \xi^{n+1}) = (\xi^0 : \Lambda'\xi : \xi^{n+1})$$

for $(\xi^0 : \ldots : \xi^{n+1}) \in N^{p,q}$ (cf. Theorem 2.6). For $x \in \mathbb{R}^{p,q}$ we have

$$\widehat{\varphi}(\iota(x)) = \left(\frac{1-\langle x\rangle}{2} : \Lambda'x : \frac{1+\langle x\rangle}{2} \right)$$

$$= \left(\frac{1-\langle \Lambda'x\rangle}{2} : \Lambda'x : \frac{1+\langle \Lambda'x\rangle}{2} \right),$$

since $\Lambda' \in O(p,q)$ implies $\langle x\rangle = \langle \Lambda'x\rangle$. Hence, $\widehat{\varphi}(\iota(x)) = \iota(\varphi(x))$ for all $x \in \mathbb{R}^{p,q}$.

2. Translations. For a translation $\varphi(x) = x+c, c \in \mathbb{R}^n$, one has the continuation

$$\widehat{\varphi}(\xi^0 : \ldots : \xi^{n+1}) := (\xi^0 - \langle\xi',c\rangle - \xi^+\langle c\rangle : \xi' + 2\xi^+c$$
$$: \xi^{n+1} + \langle\xi',c\rangle + \xi^+\langle c\rangle)$$

for $(\xi^0 : \ldots : \xi^{n+1}) \in N^{p,q}$. Here,

$$\xi^+ = \frac{1}{2}(\xi^{n+1} + \xi^0) \quad \text{and} \quad \xi' = (\xi^1, \ldots, \xi^n).$$

We have

$$\widehat{\varphi}(\iota(x)) = \left(\frac{1-\langle x\rangle}{2} - \langle x,c\rangle - \frac{\langle c\rangle}{2} : x+c : \frac{1+\langle x\rangle}{2} + \langle x,c\rangle + \frac{\langle c\rangle}{2} \right),$$

since $\iota(x)^+ = \frac{1}{2}$, and therefore

$$\widehat{\varphi}(\iota(x)) = \left(\frac{1-\langle x+c\rangle}{2} : x+c : \frac{1+\langle x+c\rangle}{2} \right) = \iota(\varphi(x)).$$

Since $\widehat{\varphi} = \psi_\Lambda$ with $\Lambda \in SO(p+1,q+1)$ can be shown as well, $\widehat{\varphi}$ is a well-defined conformal map, that is a conformal continuation of φ. The matrix we look for can be found directly from the definition of $\widehat{\varphi}$. It can be written as a block matrix:

$$\Lambda_c = \begin{pmatrix} 1-\frac{1}{2}\langle c\rangle & -(\eta'c)^T & -\frac{1}{2}\langle c\rangle \\ c & E_n & c \\ \frac{1}{2}\langle c\rangle & (\eta'c)^T & 1+\frac{1}{2}\langle c\rangle \end{pmatrix}.$$

Here, E_n is the $(n \times n)$ unit matrix and

$$\eta' = \text{diag}(1, \ldots, 1, -1, \ldots, -1)$$

is the $(n \times n)$ diagonal matrix representing $g^{p,q}$. The proof of $\Lambda_c \in O(p+1,q+1)$ requires some elementary calculation. $\Lambda_c \in SO(p+1,q+1)$ can be shown by looking at the curve $t \mapsto \Lambda_{tc}$ connecting E_{n+2} and Λ_c.

3. Dilatations. The following matrices belong to the dilatations $\varphi(x) = rx$, $r \in \mathbb{R}_+$:

$$\Lambda_r = \begin{pmatrix} \frac{1+r^2}{2r} & 0 & \frac{1-r^2}{2r} \\ 0 & E_n & 0 \\ \frac{1-r^2}{2r} & 0 & \frac{1+r^2}{2r} \end{pmatrix}$$

($\Lambda_r \in O(p+1,q+1)$ requires a short calculation again).
$\Lambda_r \in SO(p+1,q+1)$ follows as above using the curve $t \mapsto \Lambda_{tr}$. The conformal transformation $\widehat{\varphi} = \psi_\Lambda$ actually is a conformal continuation of φ, as can be seen by substitution:

$$\widehat{\varphi}(\xi^0 : \ldots : \xi^{n+1})$$

$$= \left(\frac{1+r^2}{2r}\xi^0 + \frac{1-r^2}{2r}\xi^{n+1} : \xi' : \frac{1+r^2}{2r}\xi^{n+1} + \frac{1-r^2}{2r}\xi^0 \right)$$

$$= \left(\frac{1+r^2}{2}\xi^0 + \frac{1-r^2}{2}\xi^{n+1} : r\xi' : \frac{1+r^2}{2}\xi^{n+1} + \frac{1-r^2}{2}\xi^0 \right).$$

For $\xi = \iota(x)$, that is $\xi' = x$, $\xi^0 = \frac{1}{2}(1 - \langle x \rangle)$, $\xi^{n+1} = \frac{1}{2}(1 + \langle x \rangle)$, one has

$$\widehat{\varphi}(\iota(x)) = \left(\frac{1 - \langle x \rangle r^2}{2} : rx : \frac{1 + \langle x \rangle r^2}{2} \right)$$

$$= \left(\frac{1 - \langle rx \rangle}{2} : rx : \frac{1 + \langle rx \rangle}{2} \right) = \iota(\varphi(x)).$$

4. Special conformal transformations. Let $b \in \mathbb{R}^n$ and

$$\varphi(x) = \frac{x - \langle x \rangle b}{1 - 2\langle x, b \rangle + \langle x \rangle \langle b \rangle}, \qquad x \in M_1 \subsetneq \mathbb{R}^{p,q}.$$

With $N = N(x) = 1 - 2\langle x, b \rangle + \langle x \rangle \langle b \rangle$ the equation $\langle \varphi(x) \rangle = \frac{\langle x \rangle}{N}$ implies

$$\iota(\varphi(x)) = \left(\frac{1 - \langle \varphi(x) \rangle}{2} : \frac{x - \langle x \rangle b}{N} : \frac{1 + \langle \varphi(x) \rangle}{2} \right)$$

$$= \left(\frac{N - \langle x \rangle}{2} : x - \langle x \rangle b : \frac{N + \langle x \rangle}{2} \right).$$

This expression also makes sense for $x \in \mathbb{R}^{p,q}$ with $N(x) = 0$. It furthermore leads to the continuation

$$\widehat{\varphi}(\xi^0 : \ldots : \xi^{n+1}) = (\xi^0 - \langle \xi', b \rangle + \xi^- \langle b \rangle : \xi' - 2\xi^- b$$
$$: \xi^{n+1} - \langle \xi', b \rangle + \xi^- \langle b \rangle),$$

where $\xi^- = \frac{1}{2}(\xi^{n+1} - \xi^0)$. Because of $\iota(x)^- = \frac{1}{2}\langle x \rangle$, one finally gets

$$\widehat{\varphi}(\iota(x)) = \left(\frac{N - \langle x \rangle}{2} : x - \langle x \rangle b : \frac{N + \langle x \rangle}{2} \right) = \iota(\varphi(x))$$

for all $x \in \mathbb{R}^{p,q}$, $N(x) \neq 0$. The mapping $\widehat{\varphi}$ is conformal, since $\widehat{\varphi} = \psi_\Lambda$ with

$$\Lambda = \begin{pmatrix} 1 - \frac{1}{2}\langle b \rangle & -(\eta' b)^T & \frac{1}{2}\langle b \rangle \\ b & E_n & -b \\ -\frac{1}{2}\langle b \rangle & -(\eta' b)^T & 1 + \frac{1}{2}\langle b \rangle \end{pmatrix} \in SO(p+1, q+1).$$

In particular, $\widehat{\varphi}$ is a conformal continuation of φ.

To sum up, for all conformal transformations φ on open connected $M \subset \mathbb{R}^{p,q}$ we have constructed conformal continuations in the sense of Definition 2.7 $\widehat{\varphi} : N^{p,q} \to N^{p,q}$ of the type $\widehat{\varphi}(\xi^0 : \ldots : \xi^{n+1}) = \gamma(\Lambda \xi)$ with $\Lambda \in SO(p+1, q+1)$ having a conformal inverse $\widehat{\varphi}^{-1} = \psi_{\Lambda^{-1}}$. The map $\varphi \mapsto \widehat{\varphi}$ turns out to be injective (at least if φ is conformally continued to a maximal domain M in $\mathbb{R}^{p,q}$, that is $M = \mathbb{R}^{p,q}$ or $M = M_1$, cf. Theorem 1.9). Conversely, every conformal transformation $\psi : N^{p,q} \to N^{p,q}$ is of the type $\psi = \widehat{\varphi}$ with a conformal transformation φ on $\mathbb{R}^{p,q}$, since there exist open nonempty subsets $U, V \subset \iota(\mathbb{R}^{p,q})$ with $\psi(U) = V$ and the map

$$\varphi := \iota^{-1} \circ \psi \circ \iota : \iota^{-1}(U) \to \iota^{-1}(V)$$

is conformal, that is φ has a conformal continuation $\widehat{\varphi}$, which must be equal to ψ. Furthermore, the group of conformal transformations $N^{p,q} \to N^{p,q}$ is isomorphic to $O(p+1, q+1)/\{\pm 1\}$, since $\widehat{\varphi}$ can be described by the uniquely determined set $\{\Lambda, -\Lambda\}$ of matrices in $O(p+1, q+1)$. This is true algebraically in the first place, but it also holds for the topological structures. Finally, this implies that the connected component containing the identity in the group of all conformal transformations $N^{p,q} \to N^{p,q}$, that is the conformal group $\mathrm{Conf}(\mathbb{R}^{p,q})$, is isomorphic to $SO(p+1, q+1)$. This completes the proof of the theorem. $\qquad\square$

2.3 The Conformal Group of $\mathbb{R}^{2,0}$

By Theorem 1.11, the orientation-preserving conformal transformations $\varphi : M \to \mathbb{R}^{2,0} \cong \mathbb{C}$ on open subsets $M \subset \mathbb{R}^{2,0} \cong \mathbb{C}$ are exactly those holomorphic functions with nowhere-vanishing derivative. This immediately implies that a conformal compactification according to Remark 2.2 and Definition 2.7 cannot exist, because there are many noninjective conformal transformations, e.g.,

$$\mathbb{C} \setminus \{0\} \to \mathbb{C}, \quad z \mapsto z^k, \quad \text{for} \quad k \in \mathbb{Z} \setminus \{-1, 0, 1\}.$$

There are also many injective holomorphic functions without a suitable holomorphic continuation, like

$$z \mapsto \sqrt{z}, \quad z \in \{w \in \mathbb{C} : \mathrm{Re} w > 0\},$$

or the principal branch of the logarithm on the plane that has been slit along the negative real axis $\mathbb{C} \setminus \{-x : x \in \mathbb{R}_+\}$. However, there is a useful version of the ansatz from Sect. 2.3 for the case $p = 2, q = 0$, which leads to a result similar to Theorem 2.9.

Definition 2.10. A *global* conformal transformation on $\mathbb{R}^{2,0}$ is an injective holomorphic function, which is defined on the entire plane $\mathbb{C}$ with at most one exceptional point.

The analysis of conformal Killing factors (cf. Sect. 1.4.2) shows that the global conformal transformations and all those conformal transformations, which admit a (necessarily unique) continuation to a global conformal transformation are exactly the transformations which have a linear conformal Killing factor or can be written as a composition of a transformation having a linear conformal Killing factor with a reflection $z \mapsto \bar{z}$. Using this result, the following theorem can be proven in the same manner as Theorem 2.9.

Theorem 2.11. *Every global conformal transformation φ on $M \subset \mathbb{C}$ has a unique conformal continuation $\widehat{\varphi} : N^{2,0} \to N^{2,0}$, where $\widehat{\varphi} = \varphi_\Lambda$ with $\Lambda \in \mathrm{O}(3,1)$. The group of conformal diffeomorphisms $\psi : N^{2,0} \to N^{2,0}$ is isomorphic to $\mathrm{O}(3,1)/\{\pm 1\}$ and the connected component containing the identity is isomorphic to $\mathrm{SO}(3,1)$.*

In view of this result, it is justified to call the connected component containing the identity the conformal group $\mathrm{Conf}(\mathbb{R}^{2,0})$ of $\mathbb{R}^{2,0}$. Another reason for this comes from the impossibility of enlarging this group by additional conformal transformations discussed below.

A comparison of Theorems 2.9 and 2.11 shows the following exceptional situation of the case $p + q > 2$: every conformal transformation, which is defined on a connected open subset $M \subset \mathbb{R}^{p,q}$, is injective and has a unique continuation to a global conformal transformation. (A global conformal transformation in the case of $\mathbb{R}^{p,q}$, $p + q > 2$, is a conformal transformation $\varphi : M \to \mathbb{R}^{p,q}$, which is defined on the entire set $\mathbb{R}^{p,q}$ with the possible exception of a hyperplane. By the results of Sect. 1.4.2, the domain M of definition of a global conformal transformation is $M = \mathbb{R}^{p,q}$ or $M = M_1$, see (1.3).)

Now, $N^{2,0}$ is isometrically isomorphic to the 2-sphere $\mathbb{S}^2$ (in general, one has $N^{p,0} \cong \mathbb{S}^p$, since $\mathbb{S}^p \times \mathbb{S}^0 = \mathbb{S}^p \times \{1, -1\}$) and hence $N^{2,0}$ is conformally isomorphic to the Riemann sphere $\mathbb{P} := \mathbb{P}_1(\mathbb{C})$.

Definition 2.12. A Möbius transformation is a holomorphic function φ, for which there is a matrix

$$\begin{pmatrix} a & b \\ c & d \end{pmatrix} \in \mathrm{SL}(2, \mathbb{C}) \quad \text{such that} \quad \varphi(z) = \frac{az + b}{cz + d}, cz + d \neq 0.$$

The set Mb of these Möbius transformations is precisely the set of all orientation-preserving global conformal transformations (in the sense of Definition 2.10). Mb forms a group with respect to composition (even though it is not a subgroup of the bijections of $\mathbb{C}$). For the exact definition of the group multiplication of φ and ψ one usually needs a continuation of $\varphi \circ \psi$ (cf. Lemma 2.13). This group operation coincides with the matrix multiplication in $SL(2, \mathbb{C})$. Hence, Mb is also isomorphic to the group $PSL(2, \mathbb{C}) := SL(2, \mathbb{C})/\{\pm 1\}$. Moreover, by Theorem 2.11, Mb is isomorphic to the group of orientation-preserving and conformal diffeomorphisms of $N^{2,0} \cong \mathbb{P}$, that is Mb is isomorphic to the group $\text{Aut}(\mathbb{P})$ of all biholomorphic maps $\psi : \mathbb{P} \to \mathbb{P}$ of the Riemann sphere $\mathbb{P}$. This transition from the group Mb to $\text{Aut}(\mathbb{P})$ using the compactification $\mathbb{C} \to \mathbb{P}$ has been used as a model for the compactification $N^{p,q}$ of $\mathbb{R}^{p,q}$ and the respective Theorem 2.9. Theorem 2.11 says even more: Mb is also isomorphic to the proper Lorentz group $SO(3, 1)$. An interpretation of the isomorphism $\text{Aut}(\mathbb{P}) \cong SO(3, 1)$ from a physical viewpoint was given by Penrose, cf., e.g., [Scho95, p. 210]. In summary, we have

$$\text{Mb} \cong PSL(2, \mathbb{C}) \cong \text{Aut}(\mathbb{P}) \cong SO(3, 1) \cong \text{Conf}(\mathbb{R}^{2,0}).$$

2.4 In What Sense Is the Conformal Group Infinite Dimensional?

We have seen in the preceding section that from the point of view of mathematics the conformal group of the Euclidean plane or the Euclidean 2-sphere is the group $\text{Mb} \cong SO(3, 1)$ of Möbius transformations.

However, throughout physics texts on two-dimensional conformal field theory one finds the claim that the group $\mathcal{G}$ of conformal transformations on $\mathbb{R}^{2,0}$ is infinite dimensional, e.g.,

"The situation is somewhat better in two dimensions. The main reason is that the conformal group is infinite dimensional in this case; it consists of the conformal analytical transformations..." and later "...the conformal group of the 2-dimensional space consists of all substitutions of the form

$$z \mapsto \xi(z), \quad \bar{z} \mapsto \bar{\xi}(\bar{z}),$$

where ξ and $\bar{\xi}$ are arbitrary analytic functions." [BPZ84, p. 335]

"Two dimensions is an especially promising place to apply notions of conformal field invariance, because there the group of conformal transformations is infinite dimensional. Any analytical function mapping the complex plane to itself is conformal." [FQS84, p. 420]

"The conformal group in 2-dimensional Euclidean space is infinite dimensional and has an algebra consisting of two commuting copies of the Virasoro algebras." [GO89, p. 333]

At first sight, the statements in these citations seem to be totally wrong. For instance, the class of all holomorphic (that is analytic) and injective functions $z \mapsto \xi(z)$ does not form a group – in contradiction to the first citation – since for two general holomorphic functions $f : U \to V$, $g : W \to Z$ with open subsets $U, V, W, Z \subset \mathbb{C}$, the composition $g \circ f$ can be defined at best if $f(U) \cap W \neq \emptyset$. Moreover, the non injective holomorphic functions are not invertible. If we restrict ourselves to the set J of all injective holomorphic functions the composition cannot define a group structure on

J because of the fact that $f(U) \subset W$ will, in general, be violated; even $f(U) \cap W = \emptyset$ can occur. Of course, J contains groups, e.g., Mb and the group of biholomorphic $f : U \to U$ on an open subset $U \subset \mathbb{C}$. However, these groups Aut(U) are not infinite dimensional, they are finite-dimensional Lie groups. If one tries to avoid the difficulties of $f(U) \cap W = \emptyset$ and requires – as the second citation [FQS84] seems to suggest – the transformations to be global, one obtains the finite-dimensional Möbius group. Even if one admits more than 1-point singularities, this yields no larger group than the group of Möbius transformations, as the following lemma shows:

Lemma 2.13. *Let* $f : \mathbb{C} \setminus S \to \mathbb{C}$ *be holomorphic and injective with a discrete set of singularities* $S \subset \mathbb{C}$. *Then,* f *is a restriction of a Möbius transformation. Consequently, it can be holomorphically continued on* $\mathbb{C}$ *or* $\mathbb{C} \setminus \{p\}$, $p \in S$.

Proof. By the theorem of Casorati–Weierstraß, the injectivity of f implies that all singularities are poles. Again from the injectivity it follows by the Riemann removable singularity theorem that at most one of these poles is not removable and this pole is of first order. □

The omission of larger parts of the domain or of the range also yields no infinite-dimensional group: doubtless, Mb should be a subgroup of the conformal group $\mathscr{G}$. For a holomorphic function $f : U \to V$, such that $\mathbb{C} \setminus U$ contains the disc D and $\mathbb{C} \setminus V$ contains the disc D', there always exists a Möbius transformation h with $h(V) \subset D'$ (inversion with respect to the circle $\partial D'$). Consequently, there is a Möbius transformation g with $g(V) \subset D$. But then Mb $\cup \{f\}$ can generate no group, since f cannot be composed with $g \circ f$ because of $(g \circ f(U)) \cap U = \emptyset$. A similar statement is true for the remaining $f \in J$.

As a result, there can be no infinite dimensional conformal group $\mathscr{G}$ for the Euclidean plane.

What do physicists mean when they claim that the conformal group is infinite dimensional? The misunderstanding seems to be that physicists mostly think and calculate infinitesimally, while they write and talk globally. Many statements become clearer, if one replaces "group" with "Lie algebra" and "transformation" with "infinitesimal transformation" in the respective texts.

If, in the case of the Euclidean plane, one looks at the conformal Killing fields instead of conformal transformations (cf. Sect. 1.4.2), one immediately finds many infinite dimensional Lie algebras within the collection of conformal Killing fields. In particular, one finds the *Witt algebra*. In this context, the Witt algebra W is the complex vector space with basis $(L_n)_{n \in \mathbb{Z}}$, $L_n := -z^{n+1} \frac{d}{dz}$ or $L_n := z^{1-n} \frac{d}{dz}$ (cf. Sect. 5.2), and the Lie bracket

$$[L_n, L_m] = (n - m)L_{n+m}.$$

The Witt algebra will be studied in detail in Chap. 5 together with the Virasoro algebra.

In two-dimensional conformal field theory usually only the infinitesimal conformal invariance of the system under consideration is used. This implies the existence of an infinite number of independent constraints, which yields the exceptional feature of two-dimensional conformal field theory.

In this context the question arises whether there exists an abstract Lie group $\mathscr{G}$ such that the corresponding Lie algebra Lie $\mathscr{G}$ is essentially the algebra of infinitesimal conformal transformations. We come back to this question in Sect. 5.4 after having introduced and studied the Witt algebra and the Virasoro algebra in Chap. 5.

Another explanation for the claim that the conformal group is infinite dimensional can perhaps be given by looking at the Minkowski plane instead of the Euclidean plane. This is not the point of view in most papers on conformal field theory, but it fits in with the type of conformal invariance naturally appearing in string theory (cf. Chap. 2). Indeed, conformal symmetry was investigated in string theory, before the actual work on conformal field theory had been done. For the Minkowski plane, there is really an infinite dimensional conformal group, as we will show in the next section. The associated complexified Lie algebra is again essentially the Witt algebra (cf. Sect. 5.1).

Hence, on the infinitesimal level the cases $(p,q) = (2,0)$ and $(p,q) = (1,1)$ seem to be quite similar. However, in the interpretation and within the representation theory there are differences, which we will not discuss here in detail. We shall just mention that the Lie algebra $\mathfrak{sl}(2,\mathbb{C})$ belongs to the Witt algebra in the Euclidean case since it agrees with the Lie algebra of Mb generated by $L_{-1}, L_0, L_1 \in W$, while in the Minkowski case $\mathfrak{sl}(2,\mathbb{C})$ is generated by complexification of $\mathfrak{sl}(2,\mathbb{R})$ which is a subalgebra of the infinitesimal conformal transformations of the Minkowski plane.

2.5 The Conformal Group of $\mathbb{R}^{1,1}$

By Theorem 1.13 the conformal transformations $\varphi : M \to \mathbb{R}^{1,1}$ on domains $M \subset \mathbb{R}^{1,1}$ are precisely the maps $\varphi = (u,v)$ with

$$u_x = v_y, u_y = v_x \text{ or } u_x = -v_y, u_y = -v_x,$$

and, in addition,

$$u_x^2 > v_x^2.$$

For $M = \mathbb{R}^{1,1}$ the global orientation-preserving conformal transformations can be described by using light cone coordinates $x_\pm = x \pm y$ in the following way:

Theorem 2.14. *For $f \in C^\infty(\mathbb{R})$ let $f_\pm \in C^\infty(\mathbb{R}^2, \mathbb{R})$ be defined by $f_\pm(x,y) := f(x \pm y)$. The map*

$$\Phi : C^\infty(\mathbb{R}) \times C^\infty(\mathbb{R}) \longrightarrow C^\infty(\mathbb{R}^2, \mathbb{R}^2),$$

$$(f,g) \longmapsto \frac{1}{2}(f_+ + g_-, f_+ - g_-)$$

has the following properties:

1. $\operatorname{im} \Phi = \{(u,v) \in C^\infty(\mathbb{R}^2, \mathbb{R}^2) : u_x = v_y, u_y = v_x\}$.
2. $\Phi(f,g)$ *conformal* $\Longleftrightarrow f' > 0, g' > 0$ *or* $f' < 0, g' < 0$.

3. $\Phi(f,g)$ *bijective* $\Longleftrightarrow$ f *and* g *bijective.*
4. $\Phi(f \circ h, g \circ k) = \Phi(f,g) \circ \Phi(h,k)$ *for* $f,g,h,k \in C^\infty(\mathbb{R})$.

Hence, the group of orientation-preserving conformal diffeomorphisms

$$\varphi : \mathbb{R}^{1,1} \to \mathbb{R}^{1,1}$$

is isomorphic to the group

$$\left(\mathrm{Diff}_+(\mathbb{R}) \times \mathrm{Diff}_+(\mathbb{R})\right) \cup \left(\mathrm{Diff}_-(\mathbb{R}) \times \mathrm{Diff}_-(\mathbb{R})\right).$$

The group of all conformal diffeomorphisms $\varphi : \mathbb{R}^{1,1} \to \mathbb{R}^{1,1}$, endowed with the topology of uniform convergence of φ and all its derivatives on compact subsets of $\mathbb{R}^2$, consists of four components. Each component is homeomorphic to $\mathrm{Diff}_+(\mathbb{R}) \times \mathrm{Diff}_+(\mathbb{R})$. Here, $\mathrm{Diff}_+(\mathbb{R})$ denotes the group of orientation-preserving diffeomorphisms $f : \mathbb{R} \to \mathbb{R}$ with the topology of uniform convergence of f and all its derivatives on compact subsets $K \subset \mathbb{R}$.

Proof.

1. Let $(u,v) = \Phi(f,g)$. From

$$u_x = \frac{1}{2}(f'_+ + g'_-), u_y = \frac{1}{2}(f'_+ - g'_-),$$

$$v_x = \frac{1}{2}(f'_+ - g'_-), v_y = \frac{1}{2}(f'_+ + g'_-),$$

it follows immediately that $u_x = v_y, u_y = v_x$. Conversely, let

$$(u,v) \in C^\infty(\mathbb{R}^2, \mathbb{R}^2)$$

with $u_x = v_y, u_y = v_x$. Then $u_{xx} = v_{yx} = u_{yy}$. Now, a solution of the one-dimensional wave equation u has the form $u(x,y) = \frac{1}{2}(f_+(x,y) + g_-(x,y))$ with suitable $f,g \in C^\infty(\mathbb{R})$. Because of $v_x = u_y = \frac{1}{2}(f'_+ - g'_-)$ and $v_y = u_x = \frac{1}{2}(f'_+ + g'_-)$, we have $v = \frac{1}{2}(f_+ - g_-)$ where f and g possibly have to be changed by a constant.
2. For $(u,v) = \Phi(f,g)$ one has $u_x^2 - v_x^2 = f'_+ g'_-$. Hence

$$u_x^2 - v_x^2 > 0 \Longleftrightarrow f'_+ g'_- > 0 \Longleftrightarrow f'g' > 0.$$

3. Let f and g be injective. For $\varphi = \Phi(f,g)$ we have as follows:
   $\varphi(x,y) = \varphi(x',y')$ implies

$$f(x+y) + g(x-y) = f(x'+y') + g(x'-y'),$$
$$f(x+y) - g(x-y) = f(x'+y') - g(x'-y').$$

Hence, $f(x+y) = f(x'+y')$ and $g(x-y) = g(x'-y')$, that is $x+y = x'+y'$ and $x-y = x'-y'$. This implies $x = x', y = y'$. So φ is injective if f and g are injective. From the preceding discussion one can see that if φ is injective then f and g are injective too. Let now f and g be surjective and $\varphi = \Phi(f,g)$. For $(\xi,\eta) \in \mathbb{R}^2$

there exist $s, t \in \mathbb{R}$ with $f(s) = \xi + \eta$, $g(t) = \xi - \eta$. Then $\varphi(x, y) = (\xi, \eta)$ with $x := \frac{1}{2}(s \mid t)$, $y := \frac{1}{2}(s - t)$. Conversely, the surjectivity of f and g follows from the surjectivity of φ.

4. With $\varphi = \Phi(f, g)$, $\psi = \Phi(h, k)$ we have $\varphi \circ \psi = \frac{1}{2}(f_+ \circ \psi + g_- \circ \psi, f_+ \circ \psi - g_- \circ \psi)$ and $f_+ \circ \psi = f(\frac{1}{2}(h_+ + k_-) + \frac{1}{2}(h_+ - k_-)) = f \circ h_+ = (f \circ h)_+$, etc. Hence

$$\varphi \circ \psi = \frac{1}{2}((f \circ h)_+ + (g \circ k)_-, (f \circ h)_+ - (g \circ k)_-) = \Phi(f \circ h, g \circ k). \qquad \square$$

As in the case $p = 2, q = 0$, there is no theorem similar to Theorem 2.9. For $p = q = 1$, the global conformal transformations need no continuation at all, hence a conformal compactification is not necessary. In this context it would make sense to define the conformal group of $\mathbb{R}^{1,1}$ simply as the connected component containing the identity of the group of conformal transformations $\mathbb{R}^{1,1} \to \mathbb{R}^{1,1}$. This group is very large; it is by Theorem 2.14 isomorphic to $\mathrm{Diff}_+(\mathbb{R}) \times \mathrm{Diff}_+(\mathbb{R})$.

However, for various reasons one wants to work with a group of transformations on a compact manifold with a conformal structure. Therefore, one replaces $\mathbb{R}^{1,1}$ with $\mathbb{S}^{1,1}$ in the sense of the conformal compactification of the Minkowski plane which we described at the beginning (cf. page 8):

$$\mathbb{R}^{1,1} \to \mathbb{S}^{1,1} = \mathbb{S} \times \mathbb{S} \subset \mathbb{R}^{2,0} \times \mathbb{R}^{0,2}.$$

In this manner, one defines the conformal group $\mathrm{Conf}(\mathbb{R}^{1,1})$ as the connected component containing the identity in the group of all conformal diffeomorphisms $\mathbb{S}^{1,1} \to \mathbb{S}^{1,1}$. Of course, this group is denoted by $\mathrm{Conf}(\mathbb{S}^{1,1})$ as well.

In analogy to Theorem 2.14 one can describe the group of orientation-preserving conformal diffeomorphisms $\mathbb{S}^{1,1} \to \mathbb{S}^{1,1}$ using $\mathrm{Diff}_+(\mathbb{S})$ and $\mathrm{Diff}_-(\mathbb{S})$ (one simply has to repeat the discussion with the aid of 2π-periodic functions). As a consequence, the group of orientation-preserving conformal diffeomorphisms $\mathbb{S}^{1,1} \to \mathbb{S}^{1,1}$ is isomorphic to the group

$$(\mathrm{Diff}_+(\mathbb{S}) \times \mathrm{Diff}_+(\mathbb{S})) \cup (\mathrm{Diff}_-(\mathbb{S}) \times \mathrm{Diff}_-(\mathbb{S})).$$

Corollary 2.15. $\mathrm{Conf}(\mathbb{R}^{1,1}) \cong \mathrm{Diff}_+(\mathbb{S}) \times \mathrm{Diff}_+(\mathbb{S})$.

In the course of the investigation of classical field theories with conformal symmetry $\mathrm{Conf}(\mathbb{R}^{1,1})$ and its quantization one is therefore interested in the properties of the group $\mathrm{Diff}_+(\mathbb{S})$ and even more (cf. the discussion of the preceding section) in its associated Lie algebra of infinitesimal transformations.

Now, $\mathrm{Diff}_+(\mathbb{S})$ turns out to be a Lie group with models in the Fréchet space of smooth $\mathbb{R}$-valued functions $f : \mathbb{S} \to \mathbb{R}$ endowed with the uniform convergence on $\mathbb{S}$ of f and all its derivatives. The corresponding Lie algebra $\mathrm{Lie}(\mathrm{Diff}_+(\mathbb{S}))$ is the Lie algebra of smooth vector fields $\mathrm{Vect}(\mathbb{S})$. The complexification of this Lie algebra contains the Witt algebra W (mentioned at the end of the preceding section 2.4) as a dense subspace.

For the quantization of such classical field theories the symmetry groups or algebras $\mathrm{Diff}_+(\mathbb{S})$, $\mathrm{Lie}(\mathrm{Diff}_+(\mathbb{S}))$, and W have to be replaced with suitable central extensions. We will explain this procedure in general for arbitrary symmetry algebras and groups in the following two chapters and introduce after that the Virasoro algebra Vir as a nontrivial central extension of the Witt algebra W in Chap. 5.

Remark 2.16. Recall that in the case of $n = p + q, p, q \geq 1$, but $(p, q) \neq (1, 1)$, the conformal group has been identified with the group $SO(p + 1, q + 1)$ or $SO(p + 1, q + 1)/\{\pm 1\}$ using the natural compactifications of $\mathbb{R}^{p,q}$ described above. To have a finite dimensional counterpart to these conformal groups also in the case of $(p, q) = (1, 1)$ one could call the group $SO(2, 2)/\{\pm 1\} \subset \text{Conf}(\mathbb{S}^{1,1})$ the *restricted conformal group* of the (compactified) Minkowski plane and use it instead of the full infinite dimensional conformal group $\text{Conf}(\mathbb{S}^{1,1})$.

The restricted conformal group is generated by the translations and the Lorentz transformations, which form a three-dimensional subgroup, and moreover by the dilatations and the special transformations.

Introducing again light cone coordinates replacing $(x, y) \in \mathbb{R}^2$ by

$$x_+ = x + y, \quad x_- = x - y,$$

the restricted conformal group $SO(2, 2)/\{\pm 1\}$ acts in the form of two copies of $PSL(2, \mathbb{R}) = SL(2, \mathbb{R})/\{\pm 1\}$. For $SL(2, \mathbb{R})$-matrices

$$A_+ = \begin{pmatrix} a_+ & b_+ \\ c_+ & d_+ \end{pmatrix}, \quad A_- = \begin{pmatrix} a_- & b_- \\ c_- & d_- \end{pmatrix}$$

the action decouples in the following way:

$$(A_+, A_-)(x_+, x_-) = \left(\frac{a_+ x_+ + b_+}{c_+ x_+ + d_+}, \frac{a_- x_- + b_-}{c_- x_- + d_-} \right).$$

Proposition 2.17. *The action of the restricted conformal group decouples with respect to the light cone coordinates into two separate actions of* $PSL(2, \mathbb{R}) = SL(2, \mathbb{R})/\{\pm 1\}$:

$$SO(2, 2)/\{\pm 1\} \cong PSL(2, \mathbb{R}) \times PSL(2, \mathbb{R}).$$

References

[BPZ84] A.A. Belavin, A.M. Polyakov, and A.B. Zamolodchikov. In- finite conformal symmetry in two-dimensional quantum field theory. *Nucl. Phys.* **B 241** (1984), 333–380.

[Dir36*] P.A.M. Dirac. Wave equations in conformal space. *Ann. Math.* **37** (1936), 429–442.

[FQS84] D. Friedan, Z. Qiu, and S. Shenker. Conformal invariance, unitarity and two-dimensional critical exponents. In: *Vertex Operators in Mathematics and Physics*. Lepowsky et al. (Eds.), 419–449. Springer-Verlag, Berlin, 1984.

[GO89] P. Goddard and D. Olive. Kac-Moody and Virasoro algebras in relation to quantum mechanics. *Int. J. Mod. Phys.* **A1** (1989), 303–414.

[Scho95] M. Schottenloher. *Geometrie und Symmetrie in der Physik*. Vieweg, Braunschweig, 1995.

Chapter 3
Central Extensions of Groups

The notion of a central extension of a group or of a Lie algebra is of particular importance in the quantization of symmetries. We give a detailed introduction to the subject with many examples, first for groups in this chapter and then for Lie algebras in the next chapter.

3.1 Central Extensions

In this section let A be an abelian group and let G be an arbitrary group. The trivial group consisting only of the neutral element is denoted by 1.

Definition 3.1. An *extension* of G *by the group* A is given by an *exact sequence* of group homomorphisms

$$1 \longrightarrow A \xrightarrow{\iota} E \xrightarrow{\pi} G \longrightarrow 1.$$

Exactness of the sequence means that the kernel of every map in the sequence equals the image of the previous map. Hence the sequence is exact if and only if ι is injective, π is surjective, the image im ι is a normal subgroup, and

$$\ker \pi = \operatorname{im} \iota (\cong A).$$

The extension is called *central* if A is abelian and its image im ι is in the center of E, that is

$$a \in A, b \in E \Rightarrow \iota(a)b = b\iota(a).$$

Note that A is written multiplicatively and 1 is the neutral element although A is supposed to be abelian.

Examples:

- A *trivial* extension has the form

$$1 \longrightarrow A \xrightarrow{i} A \times G \xrightarrow{pr_2} G \longrightarrow 1,$$

Schottenloher, M.: *Central Extensions of Groups*. Lect. Notes Phys. **759**, 39–62 (2008)
DOI 10.1007/978-3-540-68628-6_4   © Springer-Verlag Berlin Heidelberg 2008

where $A \times G$ denotes the product group and where $i : A \to G$ is given by $a \mapsto$ $(a, 1)$. This extension is central.

- An example for a nontrivial central extension is the exact sequence

$$1 \longrightarrow \mathbb{Z}/k\mathbb{Z} \longrightarrow E = \mathrm{U}(1) \overset{\pi}{\longrightarrow} \mathrm{U}(1) \longrightarrow 1$$

with $\pi(z) := z^k$ for $k \in \mathbb{N}$, $k \geq 2$. This extension cannot be trivial, since $E = \mathrm{U}(1)$ and $\mathbb{Z}/k\mathbb{Z} \times \mathrm{U}(1)$ are not isomorphic. Another argument for this uses the fact – known for example from function theory – that a homomorphism $\tau : \mathrm{U}(1) \to E$ with $\pi \circ \tau = id_{\mathrm{U}(1)}$ does not exist, since there is no global kth root.

- A special class of group extensions is given by semidirect products. For a group G acting on another group H by a homomorphism $\tau : G \to \mathrm{Aut}(H)$ the *semidirect product* group $G \ltimes H$ is the set $H \times G$ with the multiplication given by the formula

$$(x, g).(x', g') := (x\tau(g)(x'), gg')$$

for $(g, x), (g', x') \in G \times H$. With $\pi(g, x) = x$ and $\iota(x) = (a, x)$, one obtains the group extension

$$1 \longrightarrow H \overset{\iota}{\longrightarrow} G \ltimes H \overset{\pi}{\longrightarrow} G \longrightarrow 1.$$

For example, for a vector space V the general linear group $\mathrm{GL}(V)$ of invertible linear mappings acts naturally on the additive group V, $\tau(g)(x) = g(x)$, and the resulting semidirect group $\mathrm{GL}(V) \ltimes V$ is (isomorphic to) the group of affine transformations.

With the same action $\tau : \mathrm{GL}(V) \to \mathrm{Aut}(V)$ the group of motions of $\mathbb{R}^{p,q}, n = p + q > 2$, as a semi-Riemannian space can be described as a semidirect product $\mathrm{O}(p, q) \ltimes \mathbb{R}^n$ (see the example in Sect. 1.4). As a particular case, we obtain the Poincaré group as the semidirect group $\mathrm{SO}(1, 3) \ltimes \mathbb{R}^4$ (cf. Sect. 8.1).

Observe that these examples of group extensions are not central, although the additive group V (resp. $\mathbb{R}^n$) of translations is abelian.

- The universal covering group of the Lorentz group $\mathrm{SO}(1, 3)$ (that is the identity component of the group $\mathrm{O}(1, 3)$ of all metric-preserving linear maps $\mathbb{R}^{1,3} \to \mathbb{R}^{1,3}$) is (isomorphic to) a central extension of $\mathrm{SO}(1, 3)$ by the group $\{+1, -1\}$. In fact, there is the exact sequence of Lie groups

$$1 \longrightarrow \{+1, -1\} \longrightarrow \mathrm{SL}(2, \mathbb{C}) \overset{\pi}{\longrightarrow} \mathrm{SO}(1, 3) \longrightarrow 1,$$

where π is the 2-to-1 covering.

This is a special case of the general fact that for a given connected Lie group G the universal covering group E of G is an extension of G by the group of deck transformations which in turn is isomorphic to the fundamental group $\pi(G)$ of G.

- Let V be a vector space over a field K. Then

$$1 \longrightarrow K^\times \overset{i}{\longrightarrow} \mathrm{GL}(V) \overset{\pi}{\longrightarrow} \mathrm{PGL}(V) \longrightarrow 1$$

with $i : K^\times \to \mathrm{GL}(V), \lambda \mapsto \lambda \mathrm{id}_V$, is a central extension by the (commutative) multiplicative group $K^\times = K \setminus \{0\}$ of units in K. Here, the *projective linear group* $\mathrm{PGL}(V)$ is simply the factor group $\mathrm{PGL}(V) = \mathrm{GL}(V)/K^\times$.

- The main example in the context of quantization of symmetries is the following: Let $\mathbb{H}$ be a Hilbert space and let $\mathbb{P} = \mathbb{P}(\mathbb{H})$ be the projective space of one-dimensional linear subspaces of $\mathbb{H}$, that is

$$\mathbb{P}(\mathbb{H}) := (\mathbb{H} \setminus \{0\})/\sim,$$

with the equivalence relation

$$f \sim g :\Leftrightarrow \exists \lambda \in \mathbb{C}^\times : f = \lambda g \quad \text{for} \quad f, g \in \mathbb{H}.$$

$\mathbb{P}$ is the space of states in quantum physics, that is the quantum mechanical phase space. In Lemma 3.4 it is shown that the group $\mathrm{U}(\mathbb{H})$ of unitary operators on $\mathbb{H}$ is in a natural way a nontrivial central extension of the group $\mathrm{U}(\mathbb{P})$ of (unitary) projective transformations on $\mathbb{P}$ by $\mathrm{U}(1)$

$$1 \longrightarrow \mathrm{U}(1) \overset{\iota}{\longrightarrow} \mathrm{U}(\mathbb{H}) \overset{\widehat{\gamma}}{\longrightarrow} \mathrm{U}(\mathbb{P}) \longrightarrow 1.$$

To explain this last example and for later purposes we recall some basic notions concerning Hilbert spaces. A *pre-Hilbert* space $\mathbb{H}$ is a complex vector space with a positive definite hermitian form, called an inner product or scalar product. A *hermitian form* is an $\mathbb{R}$-bilinear map

$$\langle \,,\, \rangle : \mathbb{H} \times \mathbb{H} \to \mathbb{C},$$

which is complex antilinear in the first variable (another convention is to have the form complex linear in the first variable) and satisfies

$$\langle f, g \rangle = \overline{\langle g, f \rangle}$$

for all $f, g \in \mathbb{H}$. A hermitian form is an *inner product* if, in addition,

$$\langle f, f \rangle > 0 \text{ for all } f \in \mathbb{H} \setminus \{0\}.$$

The inner product induces a norm on $\mathbb{H}$ by $\|f\| := \sqrt{\langle f, f \rangle}$ and hence a topology. $\mathbb{H}$ with the inner product is called a *Hilbert space* if $\mathbb{H}$ is complete as a normed space with respect to this norm.

Typical finite-dimensional examples of Hilbert spaces are the $\mathbb{C}^m$ with the standard inner product

$$\langle z, w \rangle := \sum_{j=1}^{m} \overline{z}_j w_j.$$

In quantum theory important Hilbert spaces are the spaces $\mathrm{L}^2(X, \lambda)$ of square-integrable complex functions $f : X \to \mathbb{C}$ on various measure spaces X with a measure λ on X, where the inner product is

$$\langle f, g \rangle := \int_X \overline{f}(x) g(x) d\lambda(x).$$

In the case of $X = \mathbb{R}^n$ with the Lebesgue measure, this space is separable, that is there exists a countable dense subset in $\mathbb{H}$. A separable Hilbert space has a countable (Schauder) basis, that is a sequence (e_n), $e_n \in \mathbb{H}$, which is mutually orthonormal, $\langle e_n, e_m \rangle = \delta_{n,m}$, and such that every $f \in \mathbb{H}$ has a unique representation as a convergent series

$$f = \sum_n \alpha_n e_n$$

with coefficients $\alpha_n \in \mathbb{C}$. These coefficients are $\alpha_n = \langle e_n, f \rangle$.

In quantum theory the Hilbert spaces describing the states of the quantum system are required to be separable. Therefore, in the sequel the Hilbert spaces are assumed to be separable.

A *unitary operator* U on $\mathbb{H}$ is a $\mathbb{C}$-linear bijective map $U : \mathbb{H} \to \mathbb{H}$ leaving the inner product invariant:

$$f, g \in \mathbb{H} \quad \Longrightarrow \quad \langle Uf, Ug \rangle = \langle f, g \rangle.$$

It is easy to see that the inverse $U^{-1} : \mathbb{H} \to \mathbb{H}$ of a unitary operator $U : \mathbb{H} \to \mathbb{H}$ is unitary as well and that the composition $U \circ V$ of two unitary operators U, V is always unitary. Hence, the composition of operators defines the structure of a group on the set of all unitary operators on $\mathbb{H}$. This group is denoted by $U(\mathbb{H})$ and called the *unitary group* of $\mathbb{H}$.

In the finite-dimensional situation ($m = \dim \mathbb{H}$) the unitary group $U(\mathbb{H})$ is isomorphic to the matrix group $U(m)$ of all complex $m \times m$-matrices B with $B^{-1} = B^*$. For example, $U(1)$ is isomorphic to $\mathbb{S}^1$. The special unitary groups are the

$$SU(m) = \{ B \in U(m) : \det B = 1 \}.$$

$SU(2)$ is isomorphic to the group of unit quaternions and can be identified with the unit sphere $\mathbb{S}^3$ and thus provides a 2-to-1 covering of the rotation group $SO(3)$ (which in turn is the three-dimensional real projective space $\mathbb{P}(\mathbb{R}^4)$).

Let $\gamma : \mathbb{H} \setminus \{0\} \to \mathbb{P}$ be the canonical map into the quotient space $\mathbb{P}(\mathbb{H}) = (\mathbb{H} \setminus \{0\})/ \sim$ with respect to the equivalence relation which identifies all points on a complex line through 0 (see above). Let $\varphi = \gamma(f)$ and $\psi = \gamma(g)$ be points in the projective space $\mathbb{P}$ with $f, g \in \mathbb{H}$. We then define the *"transition probability"* as

$$\delta(\varphi, \psi) := \frac{|\langle f, g \rangle|^2}{\|f\|^2 \|g\|^2}.$$

δ is not quite the same as a metric but it defines in the same way as a metric a topology on $\mathbb{P}$ which is the natural topology on $\mathbb{P}$. This topology is generated by the open subsets $\{ \varphi \in \mathbb{P} : \delta(\varphi, \psi) < r \}$, $r \in \mathbb{R}$, $\psi \in \mathbb{P}$. It is also the quotient topology on $\mathbb{P}$ with respect to the quotient map γ, that is a subset $W \subset \mathbb{P}$ is open if and only if $\gamma^{-1}(W) \subset \mathbb{H}$ is open in the Hilbert space topology.

Definition 3.2. A bijective map $T : \mathbb{P} \to \mathbb{P}$ with the property

$$\delta(T\varphi, T\psi) = \delta(\varphi, \psi) \quad \text{for} \quad \varphi, \psi \in \mathbb{P},$$

is called a *projective transformation* or *projective automorphism*.

Furthermore, we define the group $\mathrm{Aut}(\mathbb{P})$ of projective transformations to be the set of all projective transformations where the group structure is again given by composition. Hence, $\mathrm{Aut}(\mathbb{P})$ is the group of bijections of $\mathbb{P}$, the quantum mechanical phase space, preserving the transition probability. This means that $\mathrm{Aut}(\mathbb{P})$ is the full symmetry group of the quantum mechanical state space.

For every $U \in \mathrm{U}(\mathbb{H})$ we define a map $\widehat{\gamma}(U) : \mathbb{P} \to \mathbb{P}$ by

$$\widehat{\gamma}(U)(\varphi) := \gamma(U(f))$$

for all $\varphi = \gamma(f) \in \mathbb{P}$ with $f \in \mathbb{H}$. It is easy to show that $\widehat{\gamma}(U) : \mathbb{P} \to \mathbb{P}$ is well defined and belongs to $\mathrm{Aut}(\mathbb{P})$. This is true not only for unitary operators, but also for the so-called anti-unitary operators V, that is for the $\mathbb{R}$-linear bijective maps $V : \mathbb{H} \to \mathbb{H}$ with

$$\langle Vf, Vg \rangle = \overline{\langle f, g \rangle}, V(if) = -iV(f)$$

for all $f, g \in \mathbb{H}$.

Note that $\widehat{\gamma} : \mathrm{U}(\mathbb{H}) \to \mathrm{Aut}(\mathbb{P})$ is a homomorphism of groups.

The following theorem is a complete characterization of the projective automorphisms:

Theorem 3.3. (Wigner [Wig31], Chap. 20, Appendix) *For every projective transformation $T \in \mathrm{Aut}(\mathbb{P})$ there exists a unitary or an anti-unitary operator U with $T = \widehat{\gamma}(U)$.*

The elementary proof of Wigner has been simplified by Bargmann [Bar64].

Let

$$\mathrm{U}(\mathbb{P}) := \widehat{\gamma}(\mathrm{U}(\mathbb{H})) \subset \mathrm{Aut}(\mathbb{P}).$$

Then $\mathrm{U}(\mathbb{P})$ is a subgroup of $\mathrm{Aut}(\mathbb{P})$, called the group of unitary projective transformations. The following result is easy to show:

Lemma 3.4. *The sequence*

$$1 \longrightarrow \mathrm{U}(1) \overset{\iota}{\longrightarrow} \mathrm{U}(\mathbb{H}) \overset{\widehat{\gamma}}{\longrightarrow} \mathrm{U}(\mathbb{P}) \longrightarrow 1$$

with $\iota(\lambda) := \lambda \, \mathrm{id}_{\mathbb{H}}, \lambda \in \mathrm{U}(1)$, is an exact sequence of homomorphism and hence defines a central extension of $\mathrm{U}(\mathbb{P})$ by $\mathrm{U}(1)$.

Proof. In order to prove this statement one only has to check that $\ker \widehat{\gamma} = \mathrm{U}(1)\mathrm{id}_{\mathbb{H}}$.

Let $U \in \ker \widehat{\gamma}$, that is $\widehat{\gamma}(U) = \mathrm{id}_{\mathbb{P}}$. Then for all $f \in \mathbb{H}, \varphi := \gamma(f)$,

$$\widehat{\gamma}(U)(\varphi) = \varphi = \gamma(f) \quad \text{and} \quad \widehat{\gamma}(U)(\varphi) = \gamma(Uf),$$

hence $\gamma(Uf) = \gamma(f)$. Consequently, there exists $\lambda \in \mathbb{C}$ with $\lambda f = Uf$. Since U is unitary, it follows that $\lambda \in U(1)$. By linearity of U, λ is independent of f, that is U has the form $U = \lambda \mathrm{id}_\mathbb{H}$. Therefore, $U \in U(1)\mathrm{id}_\mathbb{H}$.

Conversely, let $\lambda \in U(1)$. Then for all $f \in \mathbb{H}$, $\varphi := \gamma(f)$, we have

$$\widehat{\gamma}(\lambda \mathrm{id}_\mathbb{H})(\varphi) = \gamma(\lambda f) = \gamma(f) = \varphi,$$

that is $\widehat{\gamma}(\lambda \mathrm{id}_\mathbb{H}) = \mathrm{id}_\mathbb{P}$ and hence, $\lambda \mathrm{id}_\mathbb{H} \in \ker \widehat{\gamma}$. $\square$

Note that this basic central extension is nontrivial, cf. Example 3.21.

The significance of Wigner's Theorem in quantum theory is the following: The states of a quantum system are represented by points in $\mathbb{P} = \mathbb{P}(\mathbb{H})$ for a suitable separable Hilbert space. A *symmetry* of such a quantum system or an *invariance principle* is a bijective transformation leaving invariant the transition probability δ, hence it is an element of the automorphism group $\mathrm{Aut}(\mathbb{P})$, that is a projective transformation. Now Wigner's Theorem 3.3 asserts that such a symmetry is always induced by either a unitary or an anti-unitary operator on the Hilbert space $\mathbb{H}$. In physical terms, "Every symmetry transformation between coherent states is implementable by a one-to-one complex-linear or antilinear isometry of $\mathbb{H}$."

In the next section we consider the same question not for a single symmetry given by only one transformation but for a group of symmetries. Note that this means that the notion of symmetry is extended from a single invariance principle to a group of symmetry operations.

3.2 Quantization of Symmetries

Examples for classical systems with a symmetry group G are

- $G = SO(3)$ for systems with rotational symmetry;
- $G = $ Galilei group, for free particles in classical nonrelativistic mechanics;
- $G = $ Poincaré group $SO(1,3) \ltimes \mathbb{R}^4$, for free particles in the special theory of relativity;
- $G = \mathrm{Diff}_+(\mathbb{S}) \times \mathrm{Diff}_+(\mathbb{S})$ in string theory and in conformal field theory on $\mathbb{R}^{1,1}$;
- $G = $ gauge group $= \mathrm{Aut}(P)$, where P is a principal fiber bundle, for gauge theories;
- $G = $ unitary group $U(\mathbb{H})$ as a symmetry of the Hilbert space $\mathbb{H}$ (resp. $U(\mathbb{P})$ as a symmetry of $P = P(\mathbb{H})$) when $\mathbb{H}$ (resp. $\mathbb{P}$) is considered as a classical phase space, for instance in the context of quantum electrodynamics (see below p. 51).

In these examples and in other classical situations the symmetry in question is manifested by a group homomorphism

$$\tau : G \to \mathrm{Aut}(Y)$$

with respect to the classical phase space Y (often represented by a manifold Y equipped with a symplectic form) and a suitable group $\mathrm{Aut}(Y)$ of transformations leaving invariant the physics of the classical system. (In case of a manifold with a symplectic form at least the symplectic form is left invariant so that the automorphisms have to be canonical transformations.) In addition, in most cases τ is supposed to be continuous for natural topologies on G and $\mathrm{Aut}(Y)$. The symmetry can also be described by the corresponding (continuous) action of the symmetry group G on Y:

$$G \times Y \to Y, (g,y) \mapsto \tau(g)(y).$$

Example: Rotationally invariant classical system with phase space $Y = \mathbb{R}^3 \times \mathbb{R}^3$ and action $\mathrm{SO}(3) \times Y \to Y, (g,(q,p)) \mapsto (g^{-1}q, g^{-1}p)$.

In general, such a group homomorphism is called a *representation* of G in Y. In case of a vector space Y and $\mathrm{Aut}(Y) = \mathrm{GL}(Y)$, the group of invertible linear maps $Y \to Y$ the representation space Y sometimes is called a *G-module*. Whether or not the representation is assumed to be continuous or more (e.g., differentiable) depends on the context.

Note, however, that the symmetry groups in the above six examples are topological groups in a natural way.

Definition 3.5. A *topological group* is a group G equipped with a topology, such that the group operation $G \times G \to G$, $(g,h) \mapsto gh$, and the inversion map $G \to G$, $g \mapsto g^{-1}$, are continuous.

The above examples of symmetry groups are even Lie groups, that is they are manifolds and the composition and inversion are differentiable maps. The first three examples are finite-dimensional Lie groups, while the last three examples are, in general, infinite dimensional Lie groups (modeled on Fréchet spaces). (The topology of $\mathrm{Diff}_+(\mathbb{S})$ will be discussed briefly at the beginning of Chap. 5, and the unitary group $\mathrm{U}(\mathbb{H})$ has a Lie group structure given by the operator norm (cf. p. 46), but it also carries another important topology, the strong topology which will be investigated below after Definition 3.6.)

Now, the quantization of a classical system Y means to find a Hilbert space $\mathbb{H}$ on which the classical observables (that is functions on Y) in which one is interested now act as (mostly self-adjoint) operators on $\mathbb{H}$ in such a way that the commutators of these operators correspond to the Poisson bracket of the classical variables, see Sect. 7.2 for further details on canonical quantization.

After quantization of a classical system with the classical symmetry $\tau: G \to \mathrm{Aut}(Y)$ a homomorphism

$$T : G \to \mathrm{U}(\mathbb{P})$$

will be induced, which in most cases is continuous for the strong topology on $\mathrm{U}(\mathbb{P})$ (see below for the definition of the strong topology).

This property cannot be proven – it is, in fact, an **assumption** concerning the quantization procedure. The reasons for making this assumption are the following. It seems to be evident from the physical point of view that each classical symmetry $g \in G$ acting on the classical phase space should induce after quantization a transformation of the quantum phase space $\mathbb{P}$. This requirement implies the existence of a map

$$T(g) : \mathbb{P} \to \mathbb{P}$$

for each $g \in G$. Again by physical arguments, $T(g)$ should preserve the transition probability, since δ is – at least in the case of classical mechanics – the quantum analogue of the symplectic form which is preserved by g. Hence, by these considerations, one obtains a map

$$T : G \to \mathrm{Aut}(\mathbb{P}).$$

In addition to these requirements it is simply reasonable and convenient to assume that T has to respect the natural additional structures on G and $\mathrm{Aut}(\mathbb{P})$, that is that T has to be a homomorphism since τ is a homomorphism, and that it is a continuous homomorphism when τ is continuous.

This (continuous) homomorphism $T : G \to \mathrm{U}(\mathbb{P})$ is sometimes called the *quantization of the symmetry* τ. See, however, Theorem 3.10 and Corollary 3.12 which yield a (continuous) homomorphism $S : E \to \mathrm{U}(\mathbb{H})$ of a central extension of G which is also called the quantization of the classical symmetry τ.

Definition 3.6. *Strong (operator) topology on* $\mathrm{U}(\mathbb{H})$: Typical open neighborhoods of $U_0 \in \mathrm{U}(\mathbb{H})$ are the sets

$$\mathscr{V}_f(U_0, r) := \{ U \in \mathrm{U}(\mathbb{H}) : \|U_0(f) - U(f)\| < r \}$$

with $f \in \mathbb{H}$ and $r > 0$. These neighborhoods form a subbasis of the strong topology: A subset $\mathscr{W} \subset \mathrm{U}(\mathbb{H})$ is by definition open if for each $U_0 \in \mathscr{W}$ there exist finitely many such $\mathscr{V}_{f_j}(U_0, r_j), j = 1, \ldots, k$, so that the intersection is contained in $\mathscr{W}$, that is

$$U_0 \subset \bigcap_{j=1}^{k} \mathscr{V}_{f_j}(U_0, r_j) \subset \mathscr{W}.$$

On $\mathrm{U}(\mathbb{P}) = \widehat{\gamma}(\mathrm{U}(\mathbb{H}))$ a topology (the quotient topology) is defined using the map $\widehat{\gamma} : \mathrm{U}(\mathbb{H}) \to \mathrm{U}(\mathbb{P})$:

$$\mathscr{V} \subset \mathrm{U}(\mathbb{P}) \text{open} \quad :\Longleftrightarrow \quad \widehat{\gamma}^{-1}(\mathscr{V}) \subset \mathrm{U}(\mathbb{H}) \text{open}.$$

We see that the strong topology is the topology of pointwise convergence in both cases. The strong topology can be defined on any subset

$$M \subset \mathscr{B}_{\mathbb{R}}(\mathbb{H}) := \{ A : \mathbb{H} \to \mathbb{H} \,|\, A \text{ is } \mathbb{R}\text{-linear and bounded} \}$$

of the space of $\mathbb{R}$-linear continuous endomorphisms, hence in particular on

$$M_u = \{ U : \mathbb{H} \to \mathbb{H} \,|\, U \text{unitary or anti-unitary} \}.$$

Note that a linear map $A : \mathbb{H} \to \mathbb{H}$ is continuous if and only if it is *bounded*, that is if its *operator norm*

$$\|A\| := \sup \{ \|Af\| : f \in \mathscr{B}_{\mathbb{R}}, \|f\| \leq 1 \}$$

is finite. And with the operator norm the space $\mathscr{B}_{\mathbb{R}}(\mathbb{H})$ is a Banach space, that is a complete normed space. Evidently, a unitary or anti-unitary operator is bounded with operator norm equal to 1.

In the same way as above the strong topology on $\mathrm{Aut}(\mathbb{P})$ is defined using δ replacing the norm.

Observe that the strong topology on $\mathrm{U}(\mathbb{H})$ and $\mathrm{U}(\mathbb{P})$ as well as on M_u and $\mathrm{Aut}(\mathbb{P})$ is the topology of pointwise convergence. So, in contrast to its name, the strong topology is rather a weak topology.

Since all these sets of mappings are uniformly bounded they are equicontinuous by the theorem of Banach–Steinhaus and hence the strong topology also agrees with the compact open topology, that is the topology of uniform convergence on the compact subsets of $\mathbb{H}$ (resp. of $\mathbb{P}$). We also conclude that in the case of a separable Hilbert space (which we always assume), the strong topology on $\mathrm{U}(\mathbb{H})$ as well as on $\mathrm{U}(\mathbb{P})$ is metrizable.

On subsets M of $\mathscr{B}_{\mathbb{R}}(\mathbb{H})$ we also have the natural norm topology induced by the operator norm. This topology is much stronger than the strong topology in the infinite dimensional case, since it is the topology of uniform convergence on the unit ball of $\mathbb{H}$.

Definition 3.7. For a topological group G a *unitary representation* R of G in the Hilbert space $\mathbb{H}$ is a continuous homomorphism

$$R : G \to \mathrm{U}(\mathbb{H})$$

with respect to the strong topology on $\mathrm{U}(\mathbb{H})$. A *projective representation* R of G is, in general, a continuous homomorphism

$$R : G \to \mathrm{U}(\mathbb{P})$$

with respect to the strong topology on $\mathrm{U}(\mathbb{P})$ ($\mathbb{P} = \mathbb{P}(\mathbb{H})$).

Note that $\mathrm{U}(\mathbb{H})$ and $\mathrm{U}(\mathbb{P})$ are topological groups with respect to the strong topology (cf. 3.11). Moreover, both these groups are connected and metrizable (see below).

The reason that in the context of representation theory one prefers the strong topology over the norm topology is that only few homomorphisms $G \to \mathrm{U}(\mathbb{H})$ turn out to be continuous with respect to the norm topology. In particular, for a compact Lie group G and its Hilbert space $\mathbb{H} = L^2(G)$ of square-integrable measurable functions with respect to Haar measure the regular representation

$$R : G \to \mathrm{U}(L^2(G)), g \mapsto (R_g : f(x) \mapsto f(xg)),$$

is not continuous in the norm topology, in general. But R is continuous in the strong topology, since all the maps $g \mapsto R_g(f)$ are continuous for fixed $f \in L^2(G)$. This last property is equivalent to the action

$$G \times L^2(G) \to L^2(G), (g, f) \mapsto R_g(f),$$

of G on $L^2(G)$ being continuous.

Another reason to use the strong topology is the fact that various related actions, e.g., the natural action of $U(\mathbb{H})$ on the space of Fredholm operators on $\mathbb{H}$ or on the Hilbert space of Hilbert–Schmidt operators, are continuous in the strong topology. Hence, the strong topology is weak enough to allow many important representations to be continuous and strong enough to ensure that natural actions of $U(\mathbb{H})$ are continuous.

Lifting Projective Representations. When quantizing a classical symmetry group G the following question arises naturally: Given a projective representation T, that is a continuous homomorphism $T : G \to U(\mathbb{P})$ with $\mathbb{P} = \mathbb{P}(\mathbb{H})$, does there exist a unitary representation $S : G \to U(\mathbb{H})$, such that the following diagram commutes?

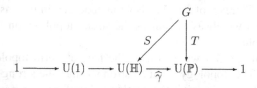

In other words, can a projective representation T always be induced by a proper unitary representation S on $\mathbb{H}$ so that $T = \widehat{\gamma} \circ S$?

The answer is no; such a lifting does not exist in general. Therefore, it is, in general, not possible to take G as the quantum symmetry group in the sense of a unitary representation $S : G \to U(\mathbb{H})$ in the Hilbert space $\mathbb{H}$. However, a lifting exists with respect to the central extension of the universal covering group of the classical symmetry group. (Here and in the following, the *universal covering group* of a connected Lie group G is the (up to isomorphism) uniquely determined connected and simply connected universal covering $\widetilde{G}$ of G with its Lie group structure.) This is well known for the rotation group $SO(3)$ where the transition from $SO(3)$ to the simply connected 2-to-1 covering group $SU(2)$ can be described in the following way:

Example 3.8. To every projective representation $T' : SO(3) \to U(\mathbb{P})$ there corresponds a unitary representation $S : SU(2) \to U(\mathbb{H})$ such that $\widehat{\gamma} \circ S = T' \circ P =: T$. The following diagram is commutative:

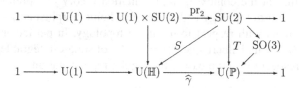

S is unique up to a scalar multiple of norm 1.

$SU(2)$ is the universal covering group of $SO(3)$ with covering map (and group homomorphism) $P : SU(2) \to SO(3)$. From a general point of view the lifting $S : SU(2) \to U(\mathbb{H})$ of $T := T' \circ P$ (that is $T = \widehat{\gamma} \circ S$) in the diagram is obtained via the lifting of T to a central extension of $SU(2)$ which always exists according to the subsequent Theorem 3.10. Since each central extension of $SU(2)$ is trivial (cf. Remark 4.10), this lifting factorizes and yields the lifting T (cf. Bargmann's Theorem 4.8).

Remark 3.9. In a similar matter one can lift every projective representation T' : $SO(1,3) \to U(\mathbb{P})$ of the Lorentz group $SO(1,3)$ to a proper unitary representation $S : SL(2,\mathbb{C}) \to U(\mathbb{H})$ in $\mathbb{H}$ of the group $SL(2,\mathbb{C})$: $T' \circ P = \widehat{\gamma} \circ S$.

Here, $P : SL(2,\mathbb{C}) \to SO(1,3)$ is the 2-to-1 covering map and homomorphism.

Because of these facts – the lifting up to the covering maps – the group $SL(2,\mathbb{C})$ is sometimes called the quantum Lorentz group and, correspondingly, $SU(2)$ is called the quantum mechanical rotation group.

Theorem 3.10. *Let G be a group and $T : G \to U(\mathbb{P})$ be a homomorphism. Then there is a central extension E of G by $U(1)$ and a homomorphism $S : E \to U(\mathbb{H})$, so that the following diagram commutes:*

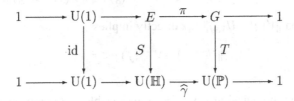

Proof. We define

$$E := \{(U,g) \in U(\mathbb{H}) \times G \mid \widehat{\gamma}(U) = Tg\}.$$

E is a subgroup of the product group $U(\mathbb{H}) \times G$, because $\widehat{\gamma}$ and T are homomorphisms. Obviously, the inclusion

$$\iota : U(1) \to E, \lambda \mapsto (\lambda \, \mathrm{id}_\mathbb{H}, 1_G)$$

and the projection $\pi := \mathrm{pr}_2 : E \to G$ are homomorphisms such that the upper row is a central extension. Moreover, the projection $S := \mathrm{pr}_1 : E \to U(\mathbb{H})$ onto the first component is a homomorphism satisfying $T \circ \pi = \widehat{\gamma} \circ S$. $\qquad\square$

Proposition 3.11. $U(\mathbb{H})$ *is a topological group with respect to the strong topology.*

This property simplifies the proof of Bargmann's Theorem (4.8) significantly. The proposition is in sharp contrast to claims in the corresponding literature on quantization of symmetries (e.g., [Sim68]) and in other publications. Since even in the latest publications it is repeated that $U(\mathbb{H})$ is not a topological group, we provide the simple proof (cf. [Scho95, p. 174]):

Proof. In order to show the continuity of the group operation $(U,U') \mapsto UU' = U \circ U'$ it suffices to show that to any pair $(U,U') \in U(\mathbb{H}) \times U(\mathbb{H})$ and to arbitrary $f \in \mathbb{H}, r > 0$, there exist open subsets $\mathcal{V}, \mathcal{V}'$ of $U(\mathbb{H})$ satisfying

$$\{VV' | V \in \mathscr{V}, V' \in \mathscr{V}'\} \subset \mathscr{V}_f(UU', r).$$

Because of

$$\|UU'(f) - VV'(f)\|$$
$$= \|UU'(f) - VU'(f) + VU'(f) - VV'(f)\|$$
$$\leq \|UU'(f) - VU'(f)\| + \|VU'(f) - VV'(f)\|$$
$$= \|UU'(f) - VU'(f)\| + \|U'(f) - V'(f)\|$$
$$= \|U(g) - V(g)\| + \|U'(f) - V'(f)\|,$$

where $g = U'(f)$, the condition is satisfied for $\mathscr{V} = \mathscr{V}_g(U, \frac{1}{2}r)$ and $\mathscr{V}' = \mathscr{V}_f(U', \frac{1}{2}r)$. To show the continuity of $U \mapsto U^{-1}$ let $g = U^{-1}(f)$ hence $f = U(g)$. Then

$$\|U^{-1}(f) - V^{-1}(f)\| = \|g - V^{-1}U(g)\| = \|V(g) - U(g)\|,$$

and the condition $\|V(g) - U(g)\| < r$ directly implies

$$\|U^{-1}(f) - V^{-1}(f)\| < r.$$

$\square$

Note that the topological group $U(\mathbb{H})$ is metrizable and complete in the strong topology and the same is true for $U(\mathbb{P})$.

Because of Proposition 3.11, it makes sense to carry out the respective investigations in the topological setting from the beginning, that is for topological groups and continuous homomorphisms. Among others we have the following properties:

1. $U(\mathbb{H})$ is connected, since $U(\mathbb{H})$ is pathwise connected with respect to the norm topology. Every unitary operator is in the orbit of a suitable one-parameter group $\exp(iAt)$.
2. $U(\mathbb{P})$ and $\text{Aut}(\mathbb{P})$ are also topological groups with respect to the strong topology.
3. $\widehat{\gamma} : U(\mathbb{H}) \to U(\mathbb{P})$ is a continuous homomorphism (with local continuous sections, cf. Lemma 4.9).
4. $U(\mathbb{P})$ is a connected metrizable group. $U(\mathbb{P})$ is the connected component containing the identity in $\text{Aut}(\mathbb{P})$.
5. Every continuous homomorphism $T : G \to \text{Aut}(\mathbb{P})$ on a connected topological group G has its image in $U(\mathbb{P})$, that is it is already a continuous homomorphism $T : G \to U(\mathbb{P})$. This is the reason why – in the context of quantization of symmetries for connected groups G – it is in most cases enough to study continuous homomorphism $T : G \to U(\mathbb{P})$ into $U(\mathbb{P})$ instead of $T : G \to \text{Aut}(\mathbb{P})$

Corollary 3.12. *If, in the situation of Theorem 3.10, G is a topological group and $T : G \to U(\mathbb{P})$ is a projective representation of G, that is T is a continuous homomorphism, then the central extension E of G by $U(1)$ has a natural structure of a topological group such that the inclusion $\iota : U(1) \to E$, the projection $\pi : E \to G$ and the lift $S : E \to U(\mathbb{H})$ are continuous. In particular, S is a unitary representation in $\mathbb{H}$.*

To show this statement one only has to observe that the product group $G \times U(\mathbb{H})$ is a topological group with respect to the product topology and thus E is a topological group with respect to the induced topology.

Remark 3.13. In view of these results a quantization of a classical symmetry group G can in general be regarded as a central extension E of the universal covering group of G by the group $U(1)$ of phases.

Quantum Electrodynamics. We conclude this section with an interesting example of a central extension of groups which occurs naturally in the context of second quantization in quantum electrodynamics. The first quantization leads to a separable Hilbert space $\mathbb{H}$ of infinite dimension, sometimes called the one-particle space, which decomposes according to the positive and negative energy states: We have two closed subspaces $\mathbb{H}_+, \mathbb{H}_- \subset \mathbb{H}$ such that $\mathbb{H}$ is the orthogonal sum of $\mathbb{H}_\pm$, that is $\mathbb{H} = \mathbb{H}_+ \oplus \mathbb{H}_-$. For example, $\mathbb{H}_\pm$ is given by the positive resp. negative or zero eigenspaces of the Dirac hamiltonian on $\mathbb{H} = L^2(\mathbb{R}^3, \mathbb{C}^4)$.

An orthogonal decomposition $\mathbb{H} = \mathbb{H}_+ \oplus \mathbb{H}_-$ with infinite dimensional components $\mathbb{H}_\pm$ is called a *polarization*.

Now, the Hilbert space $\mathbb{H}$ (or its projective space $\mathbb{P} = \mathbb{P}(\mathbb{H})$) can be viewed as a classical phase space with the imaginary part of the scalar product as the symplectic form and with the unitary group $U(\mathbb{H})$ (or $U(\mathbb{P})$) as symmetry group. In this context the observables one is interested in are the elements of the CAR algebra $\mathscr{A}(\mathbb{H})$ of $\mathbb{H}$. Second quantization is the quantization of these observables.

The CAR (Canonical Anticommutation Relation) algebra $\mathscr{A}(\mathbb{H}) = \mathscr{A}$ of a Hilbert space $\mathbb{H}$ is the universal unital C^*-algebra generated by the *annihilation operators* $a(f)$ and the *creation operators* $a^*(f), f \in \mathbb{H}$, with the following commutation relations:

$$a(f)a^*(g) + a^*(g)a(f) = \langle f, g \rangle \mathbf{1},$$

$$a^*(f)a^*(g) + a^*(g)a^*(f) = 0 = a(f)a(g) + a(g)a(f).$$

Here, $a^* : \mathbb{H} \to \mathscr{A}$ is a complex-linear map and $a : \mathbb{H} \to \mathscr{A}$ is complex antilinear (other conventions are often used in the literature). The CAR algebra $\mathscr{A}(\mathbb{H})$ can be described as a Clifford algebra using the tensor algebra of $\mathbb{H}$.

Recall that a *Banach algebra* is an associative algebra B over $\mathbb{C}$ which is a complex Banach space such that the multiplication satisfies $\|ab\| \leq \|a\| \|b\|$ for all $a, b \in B$. A *unital* Banach algebra B is a Banach algebra with a unit of norm 1. Finally, a C^*-algebra is a Banach algebra B with an antilinear involution $* : B \to B, b \mapsto b^*$ satisfying $(ab)^* = b^* a^*$ and $\|aa^*\| = \|a\|^2$ for all $a, b \in B$.

Let us now assume to have a polarization $\mathbb{H} = \mathbb{H}_+ \oplus \mathbb{H}_-$ induced by a (first) quantization (for example the quantization of the Dirac hamiltonian). For a general complex Hilbert space $\mathbb{W}$ the complex conjugate $\overline{\mathbb{W}}$ is $\mathbb{W}$ as an abelian group endowed with the "conjugate" scalar multiplication $(\lambda, w) \mapsto \overline{\lambda} w$ and with the conjugate scalar product.

The second quantization is obtained by representing the CAR algebra $\mathscr{A}$ in the *fermionic Fock space* (which also could be called *spinor space*) $S(\mathbb{H}_+) = S$ depending on the polarization. S is the Hilbert space completion of

$$\bigwedge \mathbb{H}_+ \otimes \bigwedge \overline{\mathbb{H}_-},$$

with the induced scalar product on $\bigwedge \mathbb{H}_+ \otimes \bigwedge \overline{\mathbb{H}_-}$, where

$$\bigwedge \mathbb{W} = \bigoplus \overset{p}{\bigwedge} \mathbb{W}$$

is the exterior algebra of the Hilbert space $\mathbb{W}$ equipped with the induced scalar product on $\bigwedge \mathbb{W}$.

In order to define the representation π of $\mathscr{A}$ in S, the actions of $a^*(f), a(f)$ on S are given in the following using $a^*(f) = a^*(f_+) + a^*(f_-), a(f) = a(f_+) + a(f_-)$ with respect to the decomposition $f = f_+ + f_- \in \mathbb{H}_+ \oplus \mathbb{H}_-$.

For $f_1, f_2, \ldots f_n \in \mathbb{H}_+, g_1, g_2, \ldots, g_m \in \mathbb{H}_-$, and $\xi \in \bigwedge^k \mathbb{H}_+, \eta \in \bigwedge \overline{\mathbb{H}_-}$, one defines

$$\pi(a^*)(f_+)\xi \otimes \eta := (f_+ \wedge \xi) \otimes \eta,$$

$$\pi(a^*)(f_-)(\xi \otimes g_1 \wedge \ldots g_m) := \sum_{j=1}^{j=n} (-1)^{k+j+1} \xi \otimes \langle g_j, f_- \rangle g_1 \wedge \ldots \widehat{g_j} \wedge \ldots g_m,$$

$$\pi(a)(f_+)(f_1 \wedge \ldots \wedge f_n \otimes \eta) := \sum_{j=1}^{j=n} (-1)^{j+1} \langle f_+, f_j \rangle f_1 \wedge \ldots \widehat{f_j} \wedge \ldots f_n \otimes \eta,$$

$$\pi(a)(f_-)(\xi \otimes \eta) := (-1)^k \xi \otimes f_- \wedge \eta.$$

Lemma 3.14. *This definition yields a representation*

$$\pi : \mathscr{A} \to \mathscr{B}(S)$$

of C^*-algebras satisfying the anticommutation relations.

Here, $\mathscr{B}(\mathbb{H}) \subset \operatorname{End} \mathbb{H}$ is the C^*-algebra of bounded $\mathbb{C}$-linear endomorphisms of $\mathbb{H}$.

The representation induces the field operators $\Phi : \mathbb{H} \to \mathscr{B}(S)$ by $\Phi(f) = \pi(a(f))$ and its adjoint $\Phi^*, \Phi^* = \pi \circ a^*$.

One is interested to know which unitary operators $U \in U(\mathbb{H})$ can be carried over to unitary operators in S in order to have the dynamics of the first quantization implemented in the Fock space (or spinor space) S, that is in the second quantized theory. To "carry over" means for a unitary $U \in U(\mathbb{H})$ to find a unitary operator $U^\sim \in U(S)$ in the Fock space S such that

$$U^\sim \circ \Phi(f) = \Phi(Uf) \circ U^\sim, f \in \mathbb{H},$$

with the same condition for Φ^*. In this situation $U^\sim$ is called an *implementation* of U.

A result of Shale and Stinespring [ST65*] yields the condition under which U is implementable.

Theorem 3.15. *Each unitary operator $U \in \mathrm{U}(\mathbb{H})$ has an implementation $U^\sim \in \mathrm{U}(\mathsf{S})$ if and only if in the block matrix representation of U*

$$U = \begin{pmatrix} U_{++} & U_{-+} \\ U_{+-} & U_{--} \end{pmatrix} : \mathbb{H}_+ \oplus \mathbb{H}_- \longrightarrow \mathbb{H}_+ \oplus \mathbb{H}_-$$

the off-diagonal components

$$U_{+-} : \mathbb{H}_+ \to \mathbb{H}_-, U_{-+} : \mathbb{H}_- \to \mathbb{H}_+$$

are Hilbert–Schmidt operators. Moreover, any two implementations $U^\sim$, $'U^\sim$ of such an operator U are the same up to a phase factor $\lambda \in \mathrm{U}(1)$: $'U^\sim = \lambda U^\sim$.

Recall that a bounded operator $T : \mathbb{H} \to \mathbb{W}$ between separable Hilbert spaces is *Hilbert–Schmidt* if with respect to a Schauder basis (e_n) of $\mathbb{H}$ the condition $\sum \|Te_n\|^2 < \infty$ holds.

$$\|T\|_{\mathrm{HS}} = \sqrt{\sum \|Te_n\|^2}$$

is the Hilbert–Schmidt norm.

Definition 3.16. The group $\mathrm{U}_{\mathrm{res}} = \mathrm{U}_{\mathrm{res}}(\mathbb{H}_+)$ of all implementable unitary operators on $\mathbb{H}$ is called the *restricted unitary group*.

The set of implemented operators

$$\mathrm{U}_{\mathrm{res}}^\sim = \mathrm{U}_{\mathrm{res}}^\sim(\mathbb{H}_+) = \{V \in \mathrm{U}(\mathsf{S}) | \exists U : U^\sim = V\}$$

is a subgroup of the unitary group $\mathrm{U}(\mathsf{S})$, and the natural "restriction" map

$$\pi : \mathrm{U}_{\mathrm{res}}^\sim \to \mathrm{U}_{\mathrm{res}}$$

is a homomorphism with kernel $\{\lambda \, \mathrm{id}_\mathsf{S} : \lambda \in \mathrm{U}(1)\} \cong \mathrm{U}(1)$.

As a result, with $\iota(\lambda) := \lambda \, \mathrm{id}_\mathsf{S}, \lambda \in \mathrm{U}(1)$, we obtain an exact sequence of groups

$$1 \longrightarrow \mathrm{U}(1) \overset{\iota}{\longrightarrow} \mathrm{U}_{\mathrm{res}}^\sim \overset{\pi}{\longrightarrow} \mathrm{U}_{\mathrm{res}} \longrightarrow 1, \tag{3.1}$$

and therefore another example of a central extension of groups appearing naturally in the context of quantization. This is the example we intended to present, and we want briefly to report about some properties of this remarkable central extension in the following.

We cannot expect to represent $\mathrm{U}_{\mathrm{res}}$ in the Fock space S, that is to have a homomorphism $\rho : \mathrm{U}_{\mathrm{res}} \to \mathrm{U}(\mathsf{S})$ with $\pi \circ \rho = \mathrm{id}_{\mathrm{U}_{\mathrm{res}}}$, because this would imply that the extension is trivial: such a ρ is a splitting, and the existence of a splitting implies triviality (see below in the next section). One knows, however, that the extension is not trivial (cf. [PS86*] or [Wur01*], for example).

As a compensation we obtain a homomorphism $\rho : U_{res} \to U(\mathbb{P}(S))$. The existence of ρ follows directly from the properties of the central extension (3.1).

In what sense can we expect $\rho : U_{res} \to U(\mathbb{P}(S))$ to be continuous? In other words, for which topology on U_{res} is ρ a representation? The strong topology on U_{res} is not enough. But on U_{res} there is the natural topology induced by the norm

$$\|U_{++}\| + \|U_{--}\| + \|U_{+-}\|_{HS} + \|U_{-+}\|_{HS},$$

where $\| \ \|_{HS}$ is the Hilbert–Schmidt norm. With respect to this topology the group U_{res} becomes a real Banach Lie group and ρ is continuous.

Moreover, on $U_{res}^{\sim}$ one obtains a topology such that this group is a Banach Lie group as well, and the natural projection is a Lie group homomorphism (cf. [PS86*], [Wur01*]). Altogether, the exact sequence (3.1) turns out to be an exact sequence of Lie group homomorphisms and hence a central extension of infinite dimensional Banach Lie groups.

According to Theorem 3.15 the phase of an implemented operator $U^{\sim}$ for $U \in U_{res}$ is not determined, and the possible variations are described by our exact sequence (3.1). In the search of a physically relevant phase of the second quantized theory, it is therefore natural to ask whether or not there exists a continuous map

$$s : U_{res} \to U_{res}^{\sim} \quad \text{with} \quad \pi \circ s = id_{U_{res}}.$$

We know already that there is no such homomorphism since the central extension is not trivial. And it turns out that there also does not exist such a continuous section s.

The arguments which prove this result are rather involved and do not have their place in these notes. Nevertheless, we give some indications.

First of all, we observe that the restriction map

$$\pi : U_{res}^{\sim} \to U_{res}$$

in the exact sequence (3.1) is a principal fiber bundle with structure group $U(1)$ (cf. [Diec91*] or [HJJS08*] for general properties of principal fiber bundles). This observation is in close connection with the investigation leading to Bargmann's Theorem, cf. Lemma 4.9. Note that a principal fiber bundle $\pi : P \to X$ is (isomorphic to) the trivial bundle if and only if there exists a global continuous section $s : X \to P$ satisfying $\pi \circ s = id_X$.

The existence of a continuous section $s : U_{res} \to U_{res}^{\sim}$ in our situation, that is $\pi \circ s = id_{U_{res}}$, would imply that the principal bundle is a trivial bundle and thus homeomorphic to $U_{res} \times U(1)$. Although we know already that $U_{res}^{\sim}$ cannot be isomorphic to the product group $U_{res} \times U(1)$ as a group, it is in principle not excluded that these spaces are homeomorphic, that is isomorphic as topological spaces.

But the principal bundle π cannot be trivial in the topological sense. To see this, one can use some interesting universal properties of another principal fiber bundle

$$\tau : \mathscr{E} \longrightarrow GL_{res}^0(\mathbb{H}_+),$$

which is in close connection to $\pi : U_{res}^{\sim} \to U_{res}$.

Here $GL_{res}(\mathbb{H}_+)$ is the group of all bounded invertible operators $\mathbb{H} \to \mathbb{H}$ whose off-diagonal components are Hilbert–Schmidt operators, so that $U_{res} = U(\mathbb{H}) \cap GL_{res}(\mathbb{H}_+)$. $GL_{res}(\mathbb{H}_+)$ will be equipped with the topology analogous to the topology on U_{res} respecting the Hilbert–Schmidt norms, and $GL^0_{res}(\mathbb{H}_+)$ is the connected component of $GL_{res}(\mathbb{H}_+)$ containing the identity. The group $\mathscr{E}$ is in a similar relation to $\tilde{U}_{res}$ as $GL_{res}(\mathbb{H}_+)$ to U_{res}. In concrete terms

$$\mathscr{E} := \{(T,P) \in GL^0_{res}(\mathbb{H}_+) \times GL(\mathbb{H}_+) : T - P \in \mathscr{I}_1\},$$

where $\mathscr{I}_1$ is the class of operators having a trace, that is being a trace class operator. (We refer to [RS80*] for concepts and results about operators on a Hilbert space.) $\mathscr{E}$ obtains its topology from the embedding into $GL^0_{res}(\mathbb{H}_+) \times \mathscr{I}_1(\mathbb{H}_+)$. The structure group of the principal bundle $\tau : \mathscr{E} \longrightarrow GL^0_{res}(\mathbb{H}_+)$ is the Banach Lie group $\mathscr{D}$ of invertible bounded operators having a determinant, that is of operators of the form $1 + T$ with T having a trace.

τ is simply the projection into the first component and we obtain another exact sequence of infinite dimensional Banach Lie groups as well as a principal fiber bundle

$$1 \longrightarrow \mathscr{D} \overset{\iota}{\longrightarrow} \mathscr{E} \overset{\tau}{\longrightarrow} GL^0_{res}(\mathbb{H}_+) \longrightarrow 1. \tag{3.2}$$

$\mathscr{E}$ is studied in the book of Pressley and Segal [PS86*] where, in particular, it is shown that $\mathscr{E}$ is contractible. This crucial property is investigated by Wurzbacher [Wur06*] in greater detail. The main ingredient of the proof is Kuiper's result on the homotopy type of the unitary group $U(\mathbb{H})$ of a separable and infinite dimensional Hilbert space $\mathbb{H}$: $U(\mathbb{H})$ with the norm topology is contractible and this also holds for the general linear group $GL(\mathbb{H})$ with the norm topology (cf. [Kui65*]).

By general properties of classifying spaces the contractibility of the group $\mathscr{E}$ implies that τ is a universal fiber bundle for $\mathscr{D}$ (see [Diec91*], for example). This means that every principal fiber bundle $P \to X$ with structure group $\mathscr{D}$ can be obtained as the pullback of τ with respect to a suitable continuous map $X \to GL^0_{res}(\mathbb{H})$. Since there exist nontrivial principal fiber bundles with structure group $\mathscr{D}$ the bundle $\tau : \mathscr{E} \to GL^0_{res}$ cannot be trivial, and thus there cannot exist a continuous section $GL^0_{res}(\mathbb{H}_+) \to \mathscr{E}$.

One can construct directly a nontrivial principal fiber bundle with structure group $\mathscr{D}$. Or one uses another interesting result, namely that the group $\mathscr{D}$ is homotopy equivalent to $U(\infty)$ according to a result of Palais [Pal65*]. $U(\infty)$ is the limit of the unitary groups $U(n) \subset U(n+1)$ and the above exact sequence (3.2) realizes the universal sequence

$$1 \longrightarrow U(\infty) \longrightarrow EU(\infty) \longrightarrow BU(\infty) \longrightarrow 1.$$

Since there exist nontrivial fiber bundles with structure group $U(n)$ it follows that there exist nontrivial principal fiber bundles with structure group $U(\infty)$ as well, and hence the same holds for $\mathscr{D}$ as structure group.

The closed subgroup $\mathscr{D}_1 := \{P \in \mathscr{D} : \det P = 1\}$ of $\mathscr{D}$ induces the exact sequence

$$1 \longrightarrow \mathscr{D}_1 \xrightarrow{\iota} \mathscr{D} \xrightarrow{\det} \mathbb{C}^\times \longrightarrow 1.$$

With the quotient $\mathrm{GL}^{0\sim}_{\mathrm{res}}(\mathbb{H}_+) := \mathscr{E}/\mathscr{D}_1$ one obtains another universal bundle

$$\mathrm{GL}^{0\sim}_{\mathrm{res}}(\mathbb{H}_+) \to \mathrm{GL}^0_{\mathrm{res}}(\mathbb{H}_+),$$

now with the multiplicative group $\mathbb{C}^\times$ as structure group. We have the exact sequence

$$1 \longrightarrow \mathbb{C}^\times \xrightarrow{\iota} \mathrm{GL}^{0\sim}_{\mathrm{res}}(\mathbb{H}_+) \xrightarrow{\pi} \mathrm{GL}_{\mathrm{res}}(\mathbb{H}_+) \longrightarrow 1,$$

which is another example of a central extension. Using the universality of this sequence one concludes that $\mathrm{GL}^{0\sim}_{\mathrm{res}}(\mathbb{H}_+) \to \mathrm{GL}^0_{\mathrm{res}}(\mathbb{H}_+)$ again has no continuous section. It follows in the same way that eventually our original bundle $\pi : \mathrm{U}^\sim_{\mathrm{res}} \to \mathrm{U}_{\mathrm{res}}$ (3.1) cannot have a continuous section. In summary we have

Proposition 3.17. *The exact sequence of Banach Lie groups*

$$1 \longrightarrow \mathrm{U}(1) \xrightarrow{\iota} \mathrm{U}^\sim_{\mathrm{res}} \xrightarrow{\pi} \mathrm{U}_{\mathrm{res}} \longrightarrow 1$$

is a central extension of the restricted unitary group $\mathrm{U}_{\mathrm{res}}$ *and a principal fiber bundle which does not admit a continuous section.*

In the same manner the basic central extension

$$1 \longrightarrow \mathrm{U}(1) \xrightarrow{\iota} \mathrm{U}(\mathbb{H}) \xrightarrow{\hat{\gamma}} \mathrm{U}(\mathbb{P}) \longrightarrow 1$$

introduced in Lemma 3.4 has no continuous section when endowed with the norm topology. Since $\mathrm{U}(\mathbb{H})$ is contractible [Kui65*] the bundle is universal. But we know that there exist nontrivial $\mathrm{U}(1)$-bundles, for instance the central extensions

$$1 \longrightarrow \mathrm{U}(1) \xrightarrow{\iota} \mathrm{U}(n) \xrightarrow{\hat{\gamma}} \mathrm{U}(\mathbb{P}(\mathbb{C}^n)) \longrightarrow 1$$

are nontrivial fiber bundles for $n > 1$ (cf. Example 3.21 below).

As will be seen in the next section the basic central extension also has no sections which are group homomorphisms (that is there exists no splitting map, cf. Example 3.21).

3.3 Equivalence of Central Extensions

We now come to general properties of central extensions beginning the discussion without taking topological questions into account.

Definition 3.18. Two central extensions

$$1 \longrightarrow A \xrightarrow{\iota} E \xrightarrow{\pi} G \longrightarrow 1 \,,\, 1 \longrightarrow A \xrightarrow{\iota} E' \xrightarrow{\pi} G \longrightarrow 1$$

of a group G by A are equivalent, if there exists an isomorphism $\psi : E \to E'$ of groups such that the diagram

$$
\begin{array}{ccccccccc}
1 & \longrightarrow & A & \longrightarrow & A \times G & \longrightarrow & G & \longrightarrow & 1 \\
& & \Big\downarrow \text{id} & & \Big\downarrow \psi & & \Big\downarrow \text{id} & & \\
1 & \longrightarrow & A & \longrightarrow & E & \longrightarrow & G & \longrightarrow & 1
\end{array}
$$

commutes.

Definition 3.19. An exact sequence of group homomorphisms

$$
1 \longrightarrow A \overset{\iota}{\longrightarrow} E \overset{\pi}{\longrightarrow} G \longrightarrow 1
$$

splits if there is a homomorphism $\sigma : G \to E$ such that $\pi \circ \sigma = \text{id}_G$.

Of course, by the surjectivity of π one can always find a map $\tau : G \to E$ with $\pi \circ \tau = \text{id}_G$. But this map will not be a group homomorphism, in general.

If the sequence splits with splitting map $\sigma : G \to E$, then

$$
\psi : A \times G \to E, \quad (a,g) \mapsto \iota(a)\sigma(g),
$$

is a group isomorphism leading to the trivial extension

$$
1 \longrightarrow A \longrightarrow A \times G \longrightarrow G \longrightarrow 1,
$$

which is equivalent to the original sequence: the diagram

$$
\begin{array}{ccccccccc}
1 & \longrightarrow & A & \longrightarrow & A \times G & \longrightarrow & G & \longrightarrow & 1 \\
& & \Big\downarrow \text{id} & & \Big\downarrow \psi & & \Big\downarrow \text{id} & & \\
1 & \longrightarrow & A & \longrightarrow & E & \longrightarrow & G & \longrightarrow & 1
\end{array}
$$

commutes. Conversely, if such a commutative diagram with a group isomorphism ψ exists, the sequence

$$
1 \longrightarrow A \longrightarrow E \longrightarrow G \longrightarrow 1
$$

splits with splitting map $\sigma(g) := \psi(1_A, g)$. We have shown that

Lemma 3.20. *A central extension splits if and only if it is equivalent to a trivial central extension.*

Example 3.21. There exist many nontrivial central extensions by $U(1)$. A general example of special importance in the context of quantization is given by the exact sequence (Lemma 3.4)

$$1 \longrightarrow U(1) \xrightarrow{\iota} U(n) \xrightarrow{\widehat{\gamma}} U(\mathbb{P}(\mathbb{C}^n)) \longrightarrow 1$$

for each $n \in \mathbb{N}, n > 1$, and

$$1 \longrightarrow U(1) \xrightarrow{\iota} U(\mathbb{H}) \xrightarrow{\widehat{\gamma}} U(\mathbb{P}) \longrightarrow 1$$

for infinite dimensional Hilbert spaces $\mathbb{H}$. These extensions are not equivalent to the trivial extension. They are also nontrivial as fiber bundles (with respect to both topologies on $U(\mathbb{H})$, the norm topology or the strong topology).

Proof. All these extensions are nontrivial if this holds for $n = 2$ since this extension is contained in the others induced by the natural embeddings $\mathbb{C}^2 \hookrightarrow \mathbb{C}^n$ resp. $\mathbb{C}^2 \hookrightarrow \mathbb{H}$. The nonequivalence to a trivial extension in the case $n = 2$ follows from well-known facts.

In particular, we have the following natural isomorphisms:

$$U(2) \cong U(1) \times SU(2) \text{ and } PU(2) = U(\mathbb{P}(\mathbb{C}^2)) \cong SO(3)$$

as groups (and as topological spaces). If the central extension

$$1 \longrightarrow U(1) \xrightarrow{\iota} U(2) \xrightarrow{\widehat{\gamma}} PU(2) \longrightarrow 1$$

would be equivalent to the trivial extension then there would exist a splitting homomorphism

$$\sigma : SO(3) \cong PU(2) \to U(2) \cong U(1) \times SU(2).$$

The two components of σ are homomorphisms as well, so that the second component $\sigma_2 : SO(3) \to SU(2)$ would be a splitting map of the natural central extension

$$1 \longrightarrow \{+1, -1\} \xrightarrow{\iota} SU(2) \xrightarrow{\pi} SO(3) \longrightarrow 1,$$

which also is the universal covering. This is a contradiction. For instance, the standard representation $\rho : SU(2) \hookrightarrow GL(\mathbb{C}^2)$ cannot be obtained as a lift of a representation of $SO(3)$ because of $\pi(\pm 1) = 1$.

In the same way one concludes that there is no continuous section. $\square$

Note that the nonexistence of a continuous section has the elementary proof just presented above without reference to the universal properties which have been considered at the end of the preceding section. One can give an elementary proof for Proposition 3.17 as well, with a similar ansatz using the fact that the projection $U^{\sim}_{res} \to U_{res}$ corresponds to the natural projection $\widehat{\gamma} : U(S) \to U(\mathbb{P}(S))$.

On the other hand, the basic exact sequence

$$1 \longrightarrow U(1) \xrightarrow{\iota} U(\mathbb{H}) \xrightarrow{\widehat{\gamma}} U(\mathbb{P}) \longrightarrow 1$$

is universal also for the strong topology (not only for the norm topology as mentioned in the preceding section), since the unitary group $U(\mathbb{H})$ is contractible in the strong topology as well whenever $\mathbb{H}$ is an infinite dimensional Hilbert space.

In the following remark we present a tool which helps to check which central extensions are equivalent to the trivial extension.

Remark 3.22. Let

$$1 \longrightarrow A \xrightarrow{\iota} E \xrightarrow{\pi} G \longrightarrow 1$$

be a central extension and let $\tau : G \to E$ be a map (not necessarily a homomorphism) with $\pi \circ \tau = \mathrm{id}_G$ and $\tau(1) = 1$. We set $\tau_x := \tau(x)$ for $x \in G$ and define a map

$$\omega : G \times G \longrightarrow A \cong \iota(A) \subset E,$$
$$(x,y) \longmapsto \tau_x \tau_y \tau_{xy}^{-1}.$$

(Here, $\tau_{xy}^{-1} = (\tau_{xy})^{-1} = (\tau(xy))^{-1}$ denotes the inverse element of τ_{xy} in the group E.) This map ω is well-defined since $\tau_x \tau_y \tau_{xy}^{-1} \in \ker \pi$, and it satisfies

$$\omega(1,1) = 1 \quad \text{and} \quad \omega(x,y)\omega(xy,z) = \omega(x,yz)\omega(y,z) \tag{3.3}$$

for $x, y, z \in G$.

Proof. By definition of ω we have

$$
\begin{aligned}
\omega(x,y)\omega(xy,z) &= \tau_x \tau_y \tau_{xy}^{-1} \tau_{xy} \tau_z \tau_{xyz}^{-1} \\
&= \tau_x \tau_y \tau_z \tau_{xyz}^{-1} \\
&= \tau_x \tau_y \tau_z \tau_{yz}^{-1} \tau_{yz} \tau_{xyz}^{-1} \\
&= \tau_x \omega(y,z) \tau_{yz} \tau_{xyz}^{-1} \\
&= \tau_x \tau_{yz} \tau_{xyz}^{-1} \omega(y,z) \quad (A \text{ is central}) \\
&= \omega(x,yz)\omega(y,z).
\end{aligned}
$$
$\qquad\square$

Definition 3.23. Any map $\omega : G \times G \longrightarrow A$ having the property (3.3) is called a *2-cocycle*, or simply a *cocycle* (*on G with values in A*).

The central extension of G by A *associated with a cocycle* ω is given by the exact sequence

$$1 \longrightarrow A \xrightarrow{\iota} A \times_\omega G \xrightarrow{pr_2} G \longrightarrow 1,$$
$$a \longmapsto (a,1).$$

Here, $A \times_\omega G$ denotes the product $A \times G$ endowed with the multiplication defined by

$$(a,x)(b,y) := (\omega(x,y)ab, xy)$$

for $(a,x), (b,y) \in A \times G$.

It has to be shown that this multiplication defines a group structure on $A \times_\omega G$ for which ι and pr_2 are homomorphisms. The crucial property is the associativity of the multiplication, which is guaranteed by the condition (3.3):

$$
\begin{aligned}
((a,x)(b,y))(c,z) &= (\omega(x,y)ab,xy)(c,z) \\
&= (\omega(xy,z)\omega(x,y)abc,xyz) \\
&= (\omega(x,yz)\omega(y,z)abc,xyz) \\
&= (a,x)(\omega(y,z)bc,yz) \\
&= (a,x)((b,y)(c,z)).
\end{aligned}
$$

The other properties are easy to check.

Remark 3.24. This yields a correspondence between the set of cocycles on G with values in A and the set of central extensions of G by A.

The extension E in Theorem 3.10

$$
1 \longrightarrow U(1) \longrightarrow E \overset{\pi}{\longrightarrow} G \longrightarrow 1
$$

is of the type $U(1) \times_\omega G$. How do we get a suitable map $\omega : G \times G \to U(1)$ in this situation? For every $g \in G$ by Wigner's Theorem 3.3 there is an element $U_g \in U(\mathbb{H})$ with $\widehat{\gamma}(U_g) = Tg$. Thus we have a map $\tau_g := (U_g, g)$, $g \in G$, which defines a map $\omega : G \times G \to U(1)$ satisfying (3.3) given by

$$
\omega(g,h) := \tau_g \tau_h \tau_{gh}^{-1} = (U_g U_h U_{gh}^{-1}, 1_G).
$$

Note that $g \mapsto U_g$ is not, in general, a homomorphism and also not continuous (if G is a topological group and T is continuous); however, in particular cases which turn out to be quite important ones, the U_gs can be chosen to yield a continuous homomorphism (cf. Bargmann's Theorem (4.8)).

If G and A are topological groups then for a cocycle $\omega : G \times G \to A$ which is continuous the extension $A \times_\omega G$ is a topological group and the inclusion and projection in the exact sequence are continuous homomorphisms. The reverse implication does not hold, since continuous maps $\tau : G \to E$ with $\pi \circ \tau = \mathrm{id}_G$ need not exist, in general. The central extension $p : z \longmapsto z^2$

$$
1 \longrightarrow \{+1, -1\} \longrightarrow U(1) \overset{p}{\longrightarrow} U(1) \longrightarrow 1
$$

provides a simple counterexample. A more involved counterexample is (cf. Proposition 3.17)

$$
1 \longrightarrow U(1) \overset{\iota}{\longrightarrow} U^{\sim}_{\mathrm{res}} \overset{\pi}{\longrightarrow} U_{\mathrm{res}} \longrightarrow 1.
$$

Lemma 3.25. *Let $\omega : G \times G \longrightarrow A$ be a cocycle. Then the central extension $A \times_\omega G$ associated with ω splits if and only if there is a map $\lambda : G \to A$ with*

$$
\lambda(xy) = \omega(x,y)\lambda(x)\lambda(y).
$$

Proof. The central extension splits if and only if there is a map $\sigma : G \to A \times_\omega G$ with $pr_2 \circ \sigma = \mathrm{id}_G$ which is a homomorphism. Such a map σ is of the form $\sigma(x) := (\lambda(x), x)$ for $x \in G$ with a map $\lambda : G \to A$. Now, σ is a homomorphism if and only if for all $x, y \in G$:

$$\sigma(xy) = \sigma(x)\sigma(y)$$
$$\iff (\lambda(xy), xy) = (\lambda(x), x)(\lambda(y), y)$$
$$\iff (\lambda(xy), xy) = ((\omega(x,y)\lambda(x)\lambda(y)), xy)$$
$$\iff \lambda(xy) \quad = \omega(x,y)\lambda(x)\lambda(y). \qquad \square$$

Definition 3.26.

$$H^2(G,A) := \{\omega : G \times G \to A \,|\, \omega \text{ is a cocycle}\}/\sim,$$

where the equivalence relation $\omega \sim \omega'$ holds by definition if and only if there is a $\lambda : G \to A$ with

$$\lambda(xy) = \omega(x,y)\omega'(x,y)^{-1}\lambda(x)\lambda(y).$$

$H^2(G,A)$ is called the *second cohomology group* of the group G with coefficients in A.

$H^2(G,A)$ is an abelian group with the multiplication induced by the pointwise multiplication of the maps $\omega : G \times G \to A$.

Remark 3.27. The above discussion shows that the second cohomology group $H^2(G,A)$ is in one-to-one correspondence with the equivalence classes of central extensions of G by A.

This is the reason why in the context of quantization of classical field theories with conformal symmetry $\mathrm{Diff}_+(\mathbb{S}) \times \mathrm{Diff}_+(\mathbb{S})$ one is interested in the cohomology group $H^2(\mathrm{Diff}_+(\mathbb{S}), U(1))$.

References

[Bar64]   V. Bargmann. Note on Wigner's theorem on symmetry operations. *J. Math. Phys.* **5** (1964), 862–868.
[Diec91*]   T. tom Dieck. *Topologie*. de Gruyter, Berlin, 1991.
[HJJS08*]   Husemöller, D., Joachim, M., Jurco, B., Schottenloher, M.: *Basic Bundle Theory and K-Cohomological Invariants*. Lect. Notes Phys. **726**. Springer, Heidelberg (2008)
[Kui65*]   N. Kuiper. The homotopy type of the unitary group of Hilbert space. *Topology* **3**, (1965), 19–30.
[Pal65*]   R.S. Palais. On the homotopy type of certain groups of operators. *Topology* **3** (1965), 271–279.
[PS86*]   A. Pressley and G. Segal. *Loop Groups*. Oxford University Press, Oxford, 1986.
[RS80*]   M. Reed and B. Simon. *Methods of modern Mathematical Physics, Vol. 1: Functional Analysis*. Academic Press, New York, 1980.

[Scho95]   M. Schottenloher. *Geometrie und Symmetrie in der Physik.* Vieweg, Braunschweig, 1995.

[ST65*]   D. Shale and W.F. Stinespring. Spinor representations of infinite orthogonal groups. *J. Math. Mech.* **14** (1965), 315–322.

[Sim68]   D. Simms. *Lie Groups and Quantum Mechanics.* Lecture Notes in Mathematics **52**, Springer Verlag, Berlin, 1968.

[Wig31]   E. Wigner. *Gruppentheorie.* Vieweg, Braunschweig, 1931.

[Wur01*]   T. Wurzbacher. Fermionic second quantization and the geometry of the restricted Grassmannian. *Infinite dimensional Kähler manifolds* (Oberwolfach 1995), DMV Sem. 31, 351–399. Birkhäuser, Basel, 2001.

[Wur06*]   T. Wurzbacher. An elementary proof of the homotopy equivalence between the restricted general linear group and the space of Fredholm operators. In: *Analysis, Geometry and Topology of Elliptic Operators*, 411–426, World Scientific Publishing, Hackensack, NJ, 2006.

Chapter 4
Central Extensions of Lie Algebras and Bargmann's Theorem

In this chapter some basic results on Lie groups and Lie algebras are assumed to be known, as presented, for instance, in [HN91] or [BR77]. For example, every finite-dimensional Lie group G has a corresponding Lie algebra Lie G determined up to isomorphism, and every differentiable homomorphism $R : G \to H$ of Lie groups induces a Lie algebra homomorphism Lie $R = \dot{R} :$ Lie $G \to$ Lie H. Conversely, if G is connected and simply connected, every such Lie algebra homomorphism $\rho :$ Lie $G \to$ Lie H determines a unique smooth Lie group homomorphism $R : G \to H$ with $\dot{R} = \rho$.

In addition, for the proof of Bargmann's Theorem we need a more involved result due to Montgomory and Zippin, namely the solution of one of Hilbert's problems: every topological group G, which is a finite-dimensional topological manifold (that is every $x \in G$ has an open neighborhood U with a topological map $\varphi : U \to \mathbb{R}^n$), is already a Lie group (cf. [MZ55]): G has a smooth structure (that is, it is a smooth manifold), such that the composition $(g, h) \to gh$ and the inversion $g \to g^{-1}$ are smooth mappings.

4.1 Central Extensions and Equivalence

A Lie algebra $\mathfrak{a}$ is called *abelian* if the Lie bracket of $\mathfrak{a}$ is trivial, that is $[X, Y] = 0$ for all $X, Y \in \mathfrak{a}$.

Definition 4.1. Let $\mathfrak{a}$ be an abelian Lie algebra over $\mathbb{K}$ and $\mathfrak{g}$ a Lie algebra over $\mathbb{K}$ (the case of $\dim \mathfrak{g} = \infty$ is not excluded). An exact sequence of Lie algebra homomorphisms

$$0 \longrightarrow \mathfrak{a} \longrightarrow \mathfrak{h} \overset{\pi}{\longrightarrow} \mathfrak{g} \longrightarrow 0$$

is called a *central extension* of $\mathfrak{g}$ by $\mathfrak{a}$, if $[\mathfrak{a}, \mathfrak{h}] = 0$, that is $[X, Y] = 0$ for all $X \in \mathfrak{a}$ and $Y \in \mathfrak{h}$. Here we identify $\mathfrak{a}$ with the corresponding subalgebra of $\mathfrak{h}$.

For such a central extension the abelian Lie algebra $\mathfrak{a}$ is realized as an ideal in $\mathfrak{h}$ and the homomorphism $\pi : \mathfrak{h} \to \mathfrak{g}$ serves to identify $\mathfrak{g}$ with $\mathfrak{h}/\mathfrak{a}$.

Schottenloher, M.: *Central Extensions of Lie Algebras and Bargmann's Theorem*. Lect. Notes Phys. **759**, 63–73 (2008)
DOI 10.1007/978-3-540-68628-6_5

Examples:

- Let

$$1 \longrightarrow A \xrightarrow{I} E \xrightarrow{R} G \longrightarrow 1$$

be a central extension of finite-dimensional Lie groups A, E, and G with differentiable homomorphisms I and R. Then, for $\dot{I} = \text{Lie } I$ and $\dot{R} = \text{Lie } R$ the sequence

$$0 \longrightarrow \text{Lie } A \xrightarrow{\dot{I}} \text{Lie } E \xrightarrow{\dot{R}} \text{Lie } G \longrightarrow 0$$

is a central extension of Lie algebras.

- In particular, every central extension E of the Lie group G by $\mathrm{U}(1)$

$$1 \longrightarrow \mathrm{U}(1) \longrightarrow E \xrightarrow{R} G \longrightarrow 1$$

with a differentiable homomorphism R induces a central extension

$$0 \longrightarrow \mathbb{R} \longrightarrow \text{Lie } E \xrightarrow{\dot{R}} \text{Lie } G \longrightarrow 0$$

of the Lie algebra Lie G by the abelian Lie algebra $\mathbb{R} \cong i\,\mathbb{R} \cong \text{Lie } \mathrm{U}(1)$.

- This holds for infinite dimensional Banach Lie groups and their Banach Lie algebras as well. For example, when we equip the unitary group $\mathrm{U}(\mathbb{H})$ with the norm topology it becomes a Banach Lie group as a real subgroup of the complex Banach Lie group $\mathrm{GL}(\mathbb{H})$ of all bounded and complex-linear and invertible transformations $\mathbb{H} \to \mathbb{H}$. Therefore, the central extension

$$1 \longrightarrow \mathrm{U}(1) \longrightarrow \mathrm{U}(\mathbb{H}) \xrightarrow{\hat{\gamma}} \mathrm{U}(\mathbb{P}) \longrightarrow 1$$

in Lemma 3.4 induces a central extension of Banach Lie algebras

$$0 \longrightarrow \mathbb{R} \longrightarrow \mathfrak{u}(\mathbb{H}) \longrightarrow \mathfrak{u}(\mathbb{P}) \longrightarrow 0,$$

where $\mathfrak{u}(\mathbb{H})$ is the real Lie algebra if bounded self-adjoint operators on $\mathbb{H}$, and $\mathfrak{u}(\mathbb{P})$ is the Lie algebra of $\mathrm{U}(\mathbb{P})$

In the same manner we obtain a central extension

$$0 \longrightarrow \mathbb{R} \longrightarrow \widetilde{\mathfrak{u}_{\mathrm{res}}}(\mathbb{H}) \longrightarrow \mathfrak{u}_{\mathrm{res}}(\mathbb{H}) \longrightarrow 0$$

by differentiating the corresponding exact sequence of Banach Lie groups (cf. Proposition 3.17).

- A basic example in the context of quantization is the *Heisenberg algebra* H which can be defined as the vector space

$$\mathsf{H} := \mathbb{C}[T, T^{-1}] \oplus \mathbb{C}Z$$

with *central element* Z and with the algebra of *Laurent polynomials* $\mathbb{C}[T, T^{-1}]$. (This algebra can be replaced with the algebra of convergent Laurent series $\mathbb{C}(T)$ or with the algebra of formal series $\mathbb{C}\left[\left[T, T^{-1}\right]\right]$ to obtain the same results as for $\mathbb{C}[T, T^{-1}]$.) H will be equipped with the Lie bracket

$$[f \oplus \lambda Z, g \oplus \mu Z] := \sum k f_k g_{-k} \, Z,$$

$f, g \in \mathbb{C}[T, T^{-1}], \lambda, \mu \in \mathbb{C}$, where $f = \sum f_n T^n, g = \sum g_n T^n$ for the Laurent polynomials $f, g \in \mathbb{C}[T, T^{-1}]$ with $f_n, g_n \in \mathbb{C}$. (All the sums are finite and therefore well-defined, since for $f = \sum f_n T^n \in \mathbb{C}[T, T^{-1}]$ only finitely many of the coefficients $f_n \in \mathbb{C}$ are different from zero.)

One can easily check that the maps

$$i : \mathbb{C} \to \mathsf{H}, \ \lambda \mapsto \lambda Z,$$

and

$$\mathrm{pr}_1 : \mathsf{H} \to \mathbb{C}[T, T^{-1}], \ f \oplus \lambda Z \mapsto f,$$

are Lie algebra homomorphisms with respect to the abelian Lie algebra structures on $\mathbb{C}$ and on $\mathbb{C}[T, T^{-1}]$. We thus have defined an exact sequence of Lie algebra homomorphisms

$$0 \longrightarrow \mathbb{C} \overset{i}{\longrightarrow} \mathsf{H} \overset{\mathrm{pr}_1}{\longrightarrow} \mathbb{C}[T, T^{-1}] \longrightarrow 0 \qquad (4.1)$$

with $[\lambda Z, g] = 0$. As a consequence, the Heisenberg algebra H is a central extension of the abelian Lie algebra of Laurent polynomials $\mathbb{C}[T, T^{-1}]$ by $\mathbb{C}$.

Note that the Heisenberg algebra is not abelian although it is a central extension of an abelian Lie algebra.

The map

$$\Theta : \mathbb{C}[T, T^{-1}] \times \mathbb{C}[T, T^{-1}] \to \mathbb{C}, (f, g) \mapsto \sum k f_k g_{-k},$$

is bilinear and alternating. Θ is called a cocycle in this context (cf. Definition 4.4), and the significance of the cocycle lies in the fact that the Lie algebra structure on the central extension H is determined by Θ since $[f + \lambda Z, g + \mu Z] = \Theta(f, g)Z$. The cocycle Θ can also be described by the residue of fg' at $0 \in \mathbb{C}$:

$$\Theta(f, g) = - \mathrm{Res}_{z=0} f(z) g'(z).$$

This can be easily seen by using the expansion of the product fg':

$$fg'(T) = \sum_{n \in \mathbb{Z}} \left(\sum_{k \in \mathbb{Z}} (n - k + 1) f_k g_{n-k+1} \right) T^n.$$

To describe H in a slightly different way observe that the monomials $a_n := T^n, n \in \mathbb{Z}$, form a basis of $\mathbb{C}[T, T^{-1}]$. Hence, the Lie algebra structure on the Heisenberg algebra H is completely determined by

$$[a_m, a_n] = m \delta_{m+n} Z, \ [Z, a_m] = 0.$$

Here, δ_k is used as an abbreviation of Kronecker's δ_k^0.

- Another example which will be of interest in Chap. 10 in order to obtain relevant examples of vertex algebras is the *affine Kac–Moody algebra* or *current algebra* as a non-abelian generalization of the construction of the Heisenberg algebra. We begin with a Lie algebra $\mathfrak{g}$ over $\mathbb{C}$. For any associative algebra R the Lie algebra structure on $R \otimes \mathfrak{g}$ is given by

$$[r \otimes a, s \otimes b] = rs \otimes [a,b] \text{ or } [ra, sb] = rs[a,b].$$

Two special cases are $R = \mathbb{C}[T, T^{-1}]$, the algebra of complex Laurent polynomials, and $R = \mathbb{C}(T)$, the algebra of convergent Laurent series. The following construction and its main properties are valid for both these algebras and in the same way also for the algebra of formal Laurent series of $\mathbb{C}(T)$, which is used in Chap. 10 on vertex algebras. Here, we treat the case $R = \mathbb{C}[T, T^{-1}]$ with the Lie algebra $\mathfrak{g}[T, T^{-1}] = \mathbb{C}[T, T^{-1}] \otimes \mathfrak{g}$ which is sometimes called the *loop algebra* of $\mathfrak{g}$.

We fix an invariant symmetric bilinear form on $\mathfrak{g}$, that is a symmetric bilinear

$$(,) : \mathfrak{g} \times \mathfrak{g} \to \mathbb{C}, \ a, b \mapsto (a, b),$$

on $\mathfrak{g}$ satisfying

$$([a, b], c) = (a, [b, c]).$$

The *affinization* of $\mathfrak{g}$ is the vector space

$$\hat{\mathfrak{g}} := \mathfrak{g}[T, T^{-1}] \oplus \mathbb{C}Z$$

endowed with the following Lie bracket

$$[T^m \otimes a, T^n \otimes b] := T^{m+n} \otimes [a, b] + m(a, b)\delta_{m+n}Z,$$
$$[T^m \otimes a, Z] := 0,$$

for $a, b \in \mathfrak{g}$ and $m, n \in \mathbb{Z}$. Using the abbreviations

$$a_m := T^m a, \ b_n := T^n b,$$

this definition takes the form

$$[a_m, b_n] = [a, b]_{m+n} + m(a, b)\delta_{m+n}Z.$$

It is easy to check that this defines a Lie algebra structure on $\hat{\mathfrak{g}}$ and that the two natural maps

$$i : \mathbb{C} \to \hat{\mathfrak{g}}, \ \lambda \mapsto \lambda Z,$$
$$\mathrm{pr}_1 : \hat{\mathfrak{g}} \to \mathfrak{g}[T, T^{-1}], \ f \otimes a + \mu Z \mapsto f \otimes a,$$

are Lie algebra homomorphisms. We have defined an exact sequence of Lie algebras

$$0 \longrightarrow \mathbb{C} \overset{i}{\longrightarrow} \hat{\mathfrak{g}} \overset{\mathrm{pr}_1}{\longrightarrow} \mathfrak{g}[T, T^{-1}] \longrightarrow 0. \tag{4.2}$$

This exact sequence provides another example of a central extension, namely the affinization $\hat{\mathfrak{g}}$ of $\mathfrak{g}$ as a central extension of the loop algebra $\mathfrak{g}[T, T^{-1}]$.

In the case of the abelian Lie algebra $\mathfrak{g} = \mathbb{C}$ we are back in the preceding example of the Heisenberg algebra. As in that example there is a characterizing cocycle on the loop algebra

$$\Theta : \mathfrak{g}[T, T^{-1}] \times \mathfrak{g}[T, T^{-1}] \to \mathbb{C},$$
$$(T^m a, T^n b) \mapsto m(a, b) \delta_{n+m} Z,$$

determining the Lie algebra structure on $\hat{\mathfrak{g}}$.

In the particular case of a simple Lie algebra $\mathfrak{g}$ there exists only one nonvanishing invariant symmetric bilinear form on $\mathfrak{g}$ (up to scalar multiplication), the Killing form. In that case the uniquely defined central extension $\hat{\mathfrak{g}}$ of the loop algebra $\mathfrak{g}[T, T^{-1}]$ is called the *affine Kac–Moody algebra of* $\mathfrak{g}$.

- In a similar way the Virasoro algebra can be defined as a central extension of the Witt algebra (cf. Chap. 5).

Definition 4.2. An exact sequence of Lie algebra homomorphisms

$$0 \longrightarrow \mathfrak{a} \longrightarrow \mathfrak{h} \overset{\pi}{\longrightarrow} \mathfrak{g} \longrightarrow 0$$

splits if there is a Lie algebra homomorphism $\beta : \mathfrak{g} \to \mathfrak{h}$ with $\pi \circ \beta = \mathrm{id}_{\mathfrak{g}}$. The homomorphism β is called a *splitting map*. A central extension which splits is called a *trivial* extension, since it is equivalent to the exact sequence of Lie algebra homomorphisms

$$0 \longrightarrow \mathfrak{a} \longrightarrow \mathfrak{a} \oplus \mathfrak{g} \longrightarrow \mathfrak{g} \longrightarrow 0.$$

(Equivalence is defined in analogy to the group case, cf. Definition 3.18.)

If, in the preceding examples of central extensions of Lie groups, the exact sequence of Lie groups splits in the sense of Definition 3.19 with a differentiable homomorphism $S : G \to E$ as splitting map, then the corresponding sequence of Lie algebra homomorphisms also splits in the sense of Definition 4.2 with splitting map $\dot{S}$. In general, the reverse implication holds for connected and simply connected Lie groups G only. In this case, the sequence of Lie groups splits if and only if the associated sequence of Lie algebras splits. All this follows immediately from the properties stated at the beginning of this chapter.

Remark 4.3. For every central extension of Lie algebras

$$0 \longrightarrow \mathfrak{a} \longrightarrow \mathfrak{h} \overset{\pi}{\longrightarrow} \mathfrak{g} \longrightarrow 0,$$

there is a linear map $\beta : \mathfrak{g} \to \mathfrak{h}$ with $\pi \circ \beta = \mathrm{id}_{\mathfrak{g}}$ (β is in general not a Lie algebra homomorphism). Let

$$\Theta(X, Y) := [\beta(X), \beta(Y)] - \beta([X, Y]) \quad for \quad X, Y \in \mathfrak{g}.$$

Then β is a splitting map if and only if $\Theta = 0$.

It can easily be checked that the map $\Theta : \mathfrak{g} \times \mathfrak{g} \to \mathfrak{a}$ (depending on β) always has the following properties:

1° $\Theta : \mathfrak{g} \times \mathfrak{g} \to \mathfrak{a}$ is bilinear and alternating.
2° $\Theta(X,[Y,Z]) + \Theta(Y,[Z,X]) + \Theta(Z,[X,Y]) = 0.$

Moreover, $\mathfrak{h} \cong \mathfrak{g} \oplus \mathfrak{a}$ as vector spaces by the linear isomorphism

$$\psi : \mathfrak{g} \times \mathfrak{a} \to \mathfrak{h}, \quad X \oplus Y = (X,Y) \mapsto \beta(X) + Y.$$

Finally, with the Lie bracket on $\mathfrak{g} \oplus \mathfrak{a}$ given by

$$[X \oplus Z, Y \oplus Z']_{\mathfrak{h}} := [X,Y]_{\mathfrak{g}} + \Theta(X,Y)$$

for $X,Y \in \mathfrak{g}$ and $Z,Z' \in \mathfrak{a}$ the map ψ is a Lie algebra isomorphism.

The Lie bracket on $\mathfrak{h}$ can also be written as

$$[\beta(X) + Z, \beta(Y) + Z'] = \beta([X,Y]) + \Theta(X,Y).$$

Here, we treat $\mathfrak{a}$ as a subalgebra of $\mathfrak{h}$ again.

Definition 4.4. A map $\Theta : \mathfrak{g} \times \mathfrak{g} \to \mathfrak{a}$ with the properties $1°$ and $2°$ of Remark 4.3 will be called a *2-cocycle* on $\mathfrak{g}$ with values in $\mathfrak{a}$ or simply a *cocycle*.

The discussion in Remark 4.3 leads to the following classification.

Lemma 4.5. *With the notations just introduced we have*

1. *Every central extension $\mathfrak{h}$ of $\mathfrak{g}$ by $\mathfrak{a}$ comes from a cocycle $\Theta : \mathfrak{g} \times \mathfrak{g} \to \mathfrak{a}$ as in 4.3.*
2. *Every cocycle $\Theta : \mathfrak{g} \times \mathfrak{g} \to \mathfrak{a}$ generates a central extension $\mathfrak{h}$ of $\mathfrak{g}$ by $\mathfrak{a}$ as in 4.3.*
3. *Such a central extension splits (and this implies that it is trivial) if and only if there is a $\mu \in \mathrm{Hom}_{\mathbb{K}}(\mathfrak{g}, \mathfrak{a})$ with*

$$\Theta(X,Y) = \mu([X,Y])$$

for all $X,Y \in \mathfrak{g}$.

Proof.

1. is obvious from the preceding remark.
2. Let $\mathfrak{h}$ be the vector space $\mathfrak{h} := \mathfrak{g} \oplus \mathfrak{a}$. The bracket

$$[X \oplus Z, Y \oplus Z']_{\mathfrak{h}} := [X,Y]_{\mathfrak{g}} \oplus \Theta(X,Y)$$

for $X,Y \in \mathfrak{g}$ and $Z,Z' \in \mathfrak{a}$ is a Lie bracket if and only if Θ is a cocycle. Hence, $\mathfrak{h}$ with this Lie bracket defines a central extension of $\mathfrak{g}$ by $\mathfrak{a}$.
3. Let $\sigma : \mathfrak{g} \to \mathfrak{h} = \mathfrak{g} \oplus \mathfrak{a}$ a splitting map, that is a Lie algebra homomorphism with $\pi \circ \sigma = \mathrm{id}_{\mathfrak{g}}$. Then σ has to be of the form $\sigma(X) = X + \mu(X)$, $X \in \mathfrak{g}$, with a suitable $\mu \in \mathrm{Hom}_{\mathbb{K}}(\mathfrak{g}, \mathfrak{a})$. From the definition of the bracket on $\mathfrak{h}$, $[\sigma(X), \sigma(Y)] = [X,Y] + \Theta(X,Y)$ for $X,Y \in \mathfrak{g}$. Furthermore, since σ is a Lie algebra homomorphism, $[\sigma(X), \sigma(Y)] = \sigma([X,Y]) = [X,Y] + \mu([X,Y])$. It follows that $\Theta(X,Y) = \mu([X,Y])$. Conversely, if Θ has this form, it clearly satisfies $1°$

and 2°. The linear map $\sigma : \mathfrak{g} \to \mathfrak{h} = \mathfrak{g} \oplus \mathfrak{a}$ defined by $\sigma(X) := X + \mu(X)$, $X \in \mathfrak{g}$, turns out to be a Lie algebra homomorphism:

$$
\begin{aligned}
\sigma([X,Y]) &= [X,Y]_{\mathfrak{g}} + \mu([X,Y]) \\
&= [X,Y]_{\mathfrak{g}} + \Theta(X,Y) \\
&= [X + \mu(X), Y + \mu(Y)]_{\mathfrak{h}} \\
&= [\sigma(X), \sigma(Y)]_{\mathfrak{h}}.
\end{aligned}
$$

Hence, σ is a splitting map.

Examples of Lie algebras given by a suitable cocycle are the Heisenberg algebra and the Kac–Moody algebras, see above, and the Virasoro algebra, cf. Chap. 5.

As in the case of groups, the collection of all equivalence classes of central extensions for a Lie algebra is a cohomology group.

Definition 4.6.

$$
\begin{aligned}
\mathrm{Alt}^2(\mathfrak{g}, \mathfrak{a}) &:= \{\Theta : \mathfrak{g} \times \mathfrak{g} \to \mathfrak{a} | \Theta \text{ satisfies condition } 1°\}. \\
Z^2(\mathfrak{g}, \mathfrak{a}) &:= \{\Theta \in \mathrm{Alt}^2(\mathfrak{g}, \mathfrak{a}) | \Theta \text{ satisfies condition } 2°\}. \\
B^2(\mathfrak{g}, \mathfrak{a}) &:= \{\Theta : \mathfrak{g} \times \mathfrak{g} \to \mathfrak{a} | \exists \mu \in \mathrm{Hom}_{\mathbb{K}}(\mathfrak{g}, \mathfrak{a}) : \Theta = \tilde{\mu}\}. \\
H^2(\mathfrak{g}, \mathfrak{a}) &:= Z^2(\mathfrak{g}, \mathfrak{a}) / B^2(\mathfrak{g}, \mathfrak{a}).
\end{aligned}
$$

Here, $\tilde{\mu}$ is given by $\tilde{\mu}(X,Y) := \mu([X,Y])$ for $X, Y \in \mathfrak{g}$.

Z^2 and B^2 are linear subspaces of Alt^2 with $B^2 \subset Z^2$. The above vector spaces are, in particular, abelian groups. Z^2 is the space of 2-cocycles and $H^2(\mathfrak{g}, \mathfrak{a})$ is called the *second cohomology group* of $\mathfrak{g}$ with values in $\mathfrak{a}$. We have proven the following classification of central extensions of Lie algebras.

Remark 4.7. The cohomology group $H^2(\mathfrak{g}, \mathfrak{a})$ is in one-to-one correspondence with the set of equivalence classes of central extensions of $\mathfrak{g}$ by $\mathfrak{a}$.

Cf. Remark 3.27 for the case of group extensions.

4.2 Bargmann's Theorem

We now come back to the question of whether a projective representation can be lifted to a unitary representation.

Theorem 4.8 (Bargmann [Bar54]). *Let G be a connected and simply connected, finite-dimensional Lie group with*

$$
H^2(\mathrm{Lie}\ G, \mathbb{R}) = 0.
$$

Then every projective representation $T : G \to U(\mathbb{P})$ *has a* lift *as a unitary repre-sentation* $S : G \to U(\mathbb{H})$, *that is for every continuous homomorphism* $T : G \to U(\mathbb{P})$ *there is a continuous homomorphism* $S : G \to U(\mathbb{H})$ *with* $T = \widehat{\gamma} \circ S$.

Proof. By Theorem 3.10, there is a central extension E of G and a homomorphism $\widehat{T} : E \to U(\mathbb{H})$, such that the following diagram commutes:

$$
\begin{array}{ccccccccc}
1 & \longrightarrow & U(1) & \longrightarrow & E & \stackrel{\pi}{\longrightarrow} & G & \longrightarrow & 1 \\
& & \downarrow{\rm id} & & \downarrow{\widehat{T}} & & \downarrow{T} & & \\
1 & \longrightarrow & U(1) & \longrightarrow & U(\mathbb{H}) & \underset{\widehat{\gamma}}{\longrightarrow} & U(\mathbb{P}) & \longrightarrow & 1
\end{array}
$$

Here, $E = \{(U,g) \in U(\mathbb{H}) \times G \,|\, \widehat{\gamma}(U) = Tg\}$, $\pi = \mathrm{pr}_2$, and $\widehat{T} = \mathrm{pr}_1$. E is a topological group as a subgroup of the topological group $U(\mathbb{H}) \times G$ (cf. Proposition 3.11) and $\widehat{T}$ and π are continuous homomorphisms. The lower exact sequence has local contin-uous sections, as we will prove in Lemma 4.9: For every $A \in U(\mathbb{P})$ there is an open neighborhood $W \subset U(\mathbb{P})$ and a continuous map $v : W \to U(\mathbb{H})$ with $\widehat{\gamma} \circ v = \mathrm{id}_W$. Let now $V := T^{-1}(W)$. Then $\mu(g) := (v \circ T(g), g)$, $g \in V$, defines a local contin-uous section $\mu : V \to E$ of the upper sequence because $\widehat{\gamma}(v \circ T(g)) = Tg$, that is $(v \circ T(g), g) \in E$ for $g \in V$. μ is continuous because v and T are continuous. This implies that

$$
\psi : U(1) \times V \to \pi^{-1}(V) \subset E, \quad (\lambda, g) \mapsto (\lambda v \circ T(g), g),
$$

is a bijective map with a continuous inverse map

$$
\psi^{-1}(U, g) = (\lambda(U), g),
$$

where $\lambda(U) \in U(1)$ for $U \in \widehat{\gamma}^{-1}(W)$ is given by the equation $U = \lambda(U) v \circ \widehat{\gamma}(U)$. Hence, the continuity of ψ^{-1} is a consequence of the continuity of the multiplication

$$
U(1) \times U(\mathbb{H}) \to U(\mathbb{H}), \quad (\lambda, U) \mapsto \lambda U.
$$

We have shown that the open subset $\pi^{-1}(V) = (T \circ \pi)^{-1}(W) \subset E$ is homeomor-phic to $U(1) \times V$. Consequently, E is a topological manifold of dimension $1 + \dim G$. By using the theorem of Montgomory and Zippin mentioned above, the topological group E is even a $(1 + \dim G)$-dimensional Lie group and the upper sequence

$$
1 \longrightarrow U(1) \longrightarrow E \longrightarrow G \longrightarrow 1
$$

is a sequence of differentiable homomorphisms.

Now, according to Remark 4.7 the corresponding exact sequence of Lie algebras

$$
0 \longrightarrow \mathrm{Lie}\, U(1) \longrightarrow \mathrm{Lie}\, E \longrightarrow \mathrm{Lie}\, G \longrightarrow 0
$$

splits because of the condition $H^2(\text{Lie } G, \mathbb{R}) = 0$. Since G is connected and simply connected, the sequence

$$\cdot 1 \longrightarrow \mathrm{U}(1) \longrightarrow E \longrightarrow G \longrightarrow 1$$

splits with a differentiable homomorphism $\sigma : G \to E$ as splitting map: $\pi \circ \sigma = \mathrm{id}_G$. Finally, $S := \widehat{T} \circ \sigma$ is the postulated lift. S is a continuous homomorphism and $\widehat{\gamma} \circ \widehat{T} = T \circ \pi$ implies $\widehat{\gamma} \circ S = \widehat{\gamma} \circ \widehat{T} \circ \sigma = T \circ \pi \circ \sigma = T \circ \mathrm{id}_G = T$:

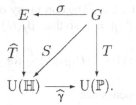

$\square$

Lemma 4.9. $\widehat{\gamma} : \mathrm{U}(\mathbb{H}) \to \mathrm{U}(\mathbb{P})$ *has local continuous sections and therefore can be regarded as a principal fiber bundle with structure group* $\mathrm{U}(1)$.

Proof. (cf. [Sim68, p. 10]) For $f \in \mathbb{H}$ let

$$V_f := \{U \in \mathrm{U}(\mathbb{H}) : \langle Uf, f \rangle \neq 0\}.$$

Then V_f is open in $\mathrm{U}(\mathbb{H})$, since $U \mapsto Uf$ is continuous in the strong topology. Hence, $U \mapsto \langle Uf, f \rangle$ is continuous as well. (For the strong topology all maps $U \mapsto Uf$ are continuous by definition.) The set

$$W_f := \widehat{\gamma}(V_f) = \{T \in \mathrm{U}(\mathbb{P}) : \delta(T\varphi, \varphi) \neq 0\}, \ \varphi = \widehat{\gamma}(f),$$

is open in $\mathrm{U}(\mathbb{P})$ since $\widehat{\gamma}^{-1}(W_f) = V_f$ is open. (The open subsets in $\mathrm{U}(\mathbb{P})$ are, by Definition 3.6, precisely the subsets $W \subset \mathrm{U}(\mathbb{P})$, such that $\widehat{\gamma}^{-1}(W) \subset \mathrm{U}(\mathbb{H})$ is open.) $(W_f)_{f \in \mathbb{H}}$ is, of course, an open cover of $\mathrm{U}(\mathbb{P})$. Let

$$\beta_f : V_f \to \mathrm{U}(1), \quad U \mapsto \frac{|\langle Uf, f \rangle|}{\langle Uf, f \rangle}.$$

β_f is continuous, since $U \mapsto \langle Uf, f \rangle$ is continuous. Furthermore, $\beta_f(e^{i\theta}U) = e^{-i\theta}\beta_f(U)$ for $U \in V_f$ and $\theta \in \mathbb{R}$, as one can see directly. One obtains a continuous section of $\widehat{\gamma}$ over W_f by

$$v_f : W_f \to \mathrm{U}(\mathbb{H}), \quad \widehat{\gamma}(U) \mapsto \beta_f(U)U.$$

v_f is well-defined, since $U' \in V_f$ with $\widehat{\gamma}(U') = \widehat{\gamma}(U)$, that is $U' = e^{i\theta}U$, implies

$$\beta_f(U')U' = \beta_f(e^{i\theta}U)e^{i\theta}U = \beta_f(U)U.$$

Now $\widehat{\gamma} \circ v_f = \mathrm{id}_{W_f}$, since

$$\widehat{\gamma} \circ v_f(\widehat{\gamma}(U)) = \widehat{\gamma}(\beta_f(U)U) = \widehat{\gamma}(U) \quad \text{for} \quad U \in V_f.$$

Eventually, v_f is continuous: let $V_1 \in W_f$ and $U_1 = v_f(V_1) \in v_f(W_f)$. Then $\beta_f(U_1) = 1$. Every open neighborhood of U_1 contains an open subset

$$B = \{U \in V_f : \|Ug_j - U_1 g_j\| < \varepsilon \text{ for } j = 1, \ldots, m\}$$

with $\varepsilon > 0$ and $g_j \in \mathbb{H}$, $j = 1, \ldots, m$. The continuity of β_f on W_f implies that there are further $g_{m+1}, \ldots, g_n \in \mathbb{H}$, $\|g_j\| = 1$, so that $|\beta_f(U) - 1| < \frac{\varepsilon}{2}$ for

$$U \in B' := \{U \in V_f : \|Ug_j - U_1 g_j\| < \frac{\varepsilon}{2} \text{ for } j = 1, \ldots, m, \ldots, n\}.$$

The image $D := \widehat{\gamma}(B')$ is open, since

$$\widehat{\gamma}^{-1}(D) = \bigcup_{\lambda \in U(1)} \{U \in V_f : \|Ug_j - \lambda U_1 g_j\| < \frac{\varepsilon}{2} \text{ for } j = 1, \ldots, n\}$$

is open. (We have shown that the map $\widehat{\gamma} : U(\mathbb{H}) \to U(\mathbb{P})$ is open.) Hence, D is an open neighborhood of V_1. v_f is continuous since $v_f(D) \subset B$: for $P \in D$ there is a $U \in B'$ with $P = \widehat{\gamma}(U)$, that is $v_f(P) = \beta_f(U)U$. This implies

$$\|v_f(P)g_j - U_1 g_j\| \le \|\beta_f(U)Ug_j - \beta_f(U)U_1 g_j\|$$
$$+ \|(\beta_f(U) - 1)U_1 g_j\|$$
$$< \frac{\varepsilon}{2} + \frac{\varepsilon}{2}$$

for $j = 1, \ldots, m$, that is $v_f(P) \in B$. Hence, the image $v_f(D)$ of the neighborhood D of V_1 is contained in B.

In spite of this nice result no reasonable differentiable structure seems to be known on the unitary group $U(\mathbb{H})$ and its quotient $U(\mathbb{P})$ with respect to the strong topology in order to prove a result which would state that $U(\mathbb{H}) \to U(\mathbb{P})$ is a differentiable principal fiber bundle. The difficulty in defining a Lie group structure on the unitary group lies in the fact that the corresponding Lie algebra should contain the (bounded and unbounded) self-adjoint operators on $\mathbb{H}$. In contrast to this situation, with respect to the operator norm topology the unitary group is a Lie group.

E is by construction the fiber product of $\widehat{\gamma}$ and T. Since $\widehat{\gamma}$ is locally trivial by Lemma 4.9 with general fiber $U(1)$, this must also hold for $E \to G$. Exactly this was needed in the proof of Theorem 4.8, to show that E actually is a Lie group.

Remark 4.10. For every finite-dimensional semi-simple Lie algebra $\mathfrak{g}$ over $\mathbb{K}$ one can show $H^2(\mathfrak{g}, \mathbb{K}) = 0$ (cf. [HN91]). As a consequence of the above discussion we thus have the following result which can be applied to the quantization of certain important symmetries: if G is a connected and simply connected finite-dimensional

Lie group with semi-simple Lie algebra $\text{Lie}(G) = \mathfrak{g}$, then every continuous representation $T : G \to U(\mathbb{P})$ has a lift to a unitary representation. In particular, to every continuous representation $T : SU(N) \to U(\mathbb{P})$ (resp. $T : SL(2, \mathbb{C}) \to U(\mathbb{P})$) there corresponds a unitary representation $S : SU(N) \to U(\mathbb{H})$ (resp. $SL(2, \mathbb{C}) \to U(\mathbb{H})$) with $\widehat{\gamma} \circ S = T$.

Note that $SL(2, \mathbb{C})$ is the universal covering group of the proper Lorentz group $SO(3, 1)$ and $SU(2)$ is the universal covering group of the rotation group $SO(3)$.

References

[Bar54] V. Bargmann. On unitary ray representations of continuous groups. *Ann. Math.* **59** (1954), 1–46.

[BR77] A.O. Barut and R. Raczka. *Theory of Group Representations and Applications.* PWN – Polish Scientific Publishers, Warsaw, 1977.

[HN91] J. Hilgert and K.-H. Neeb. *Lie Gruppen und Lie Algebren.* Vieweg, Braunschweig, 1991.

[MZ55] D. Montgomory and L. Zippin. *Topological Transformation Groups.* Interscience, New York, 1955.

[Sim68] D. Simms. *Lie Groups and Quantum Mechanics.* Lecture Notes in Mathematics **52**, Springer Verlag, Berlin, 1968.

Chapter 5
The Virasoro Algebra

In this chapter we describe how the Witt algebra and the Virasoro algebra as its essentially unique nontrivial central extension appear in the investigation of conformal symmetries. This result has been proven by Gelfand and Fuks in [GF68]. The last section discusses the question of whether there exists a Lie group whose Lie algebra is the Virasoro algebra.

5.1 Witt Algebra and Infinitesimal Conformal Transformations of the Minkowski Plane

The quantization of classical systems with symmetries yields representations of the classical symmetry group in $U(\mathbb{P})$ (with $\mathbb{P} = \mathbb{P}(\mathbb{H})$, the projective space of a Hilbert space $\mathbb{H}$, cf. Chap. 3), that is, the so-called *projective representations*. As we have explained in Corollary 2.15, the conformal group of $\mathbb{R}^{1,1}$ is isomorphic to $\mathrm{Diff}_+(\mathbb{S}) \times \mathrm{Diff}_+(\mathbb{S})$ (here and in the following $\mathbb{S} := \mathbb{S}^1$ is the unit circle). Hence, given a classical theory with this conformal group as symmetry group, one studies the group $\mathrm{Diff}_+(\mathbb{S})$ and its Lie algebra first. After quantization one is interested in the unitary representations of the central extensions of $\mathrm{Diff}_+(\mathbb{S})$ or $\mathrm{Lie}\,(\mathrm{Diff}_+(\mathbb{S}))$ in order to get representations in the Hilbert space as we have explained in the preceding two sections.

The group $\mathrm{Diff}_+(\mathbb{S})$ is in a canonical way an infinite dimensional Lie group modeled on the real vector space of smooth vector fields $\mathrm{Vect}(\mathbb{S})$. (We will discuss $\mathrm{Vect}(\mathbb{S})$ in more detail below.) $\mathrm{Diff}_+(\mathbb{S})$ is equipped with the topology of uniform convergence of the smooth mappings $\varphi : \mathbb{S} \to \mathbb{S}$ and all their derivatives. This topology is metrizable. Similarly, $\mathrm{Vect}\,(\mathbb{S})$ carries the topology of uniform convergence of the smooth vector fields $X : \mathbb{S} \to T\mathbb{S}$ and all their derivatives. With this topology, $\mathrm{Vect}(\mathbb{S})$ is a Fréchet space. In fact, $\mathrm{Vect}(\mathbb{S})$ is isomorphic to $C^\infty(\mathbb{S}, \mathbb{R})$, as we will see shortly. The proof that $\mathrm{Diff}_+(\mathbb{S})$ in this way actually becomes a differentiable manifold modeled on $\mathrm{Vect}(\mathbb{S})$ and that the group operation and the inversion are differentiable is elementary and can be carried out for arbitrary oriented, compact (finite-dimensional) manifolds M instead of $\mathbb{S}$ (cf. [Mil84]).

Schottenloher, M.: *The Virasoro Algebra*. Lect. Notes Phys. **759**, 75–85 (2008)
DOI 10.1007/978-3-540-68628-6_6   © Springer-Verlag Berlin Heidelberg 2008

Since $\mathrm{Diff}_+(\mathbb{S})$ is a manifold modeled on the vector space $\mathrm{Vect}(\mathbb{S})$, the tangent space $T_\varphi(\mathrm{Diff}_+(\mathbb{S}))$ at a point $\varphi \in \mathrm{Diff}_+(\mathbb{S})$ is isomorphic to the vector space $\mathrm{Vect}(\mathbb{S})$. Hence, $\mathrm{Vect}(\mathbb{S})$ is also the underlying vector space of the Lie algebra $\mathrm{Lie}(\mathrm{Diff}_+(\mathbb{S}))$. A careful investigation of the two Lie brackets on $\mathrm{Vect}(\mathbb{S})$ – one from $\mathrm{Vect}(\mathbb{S})$, the other from $\mathrm{Lie}(\mathrm{Diff}_+(\mathbb{S}))$ – shows that each Lie bracket is exactly the negative of the other (cf. [Mil84]). However, this subtle fact is not important for the representation theory of $\mathrm{Lie}(\mathrm{Diff}_+(\mathbb{S}))$. Consequently, it is usually ignored. So we set

$$\mathrm{Lie}(\mathrm{Diff}_+(\mathbb{S})) := \mathrm{Vect}(\mathbb{S}).$$

The vector space $\mathrm{Vect}(\mathbb{S})$ is – like the space $\mathrm{Vect}(M)$ of smooth vector fields on a smooth compact manifold M – an infinite dimensional Lie algebra over $\mathbb{R}$ with a natural Lie bracket: a smooth vector field X on M can be considered to be a derivation $X : C^\infty(M) \to C^\infty(M)$, that is a $\mathbb{R}$-linear map with

$$X(fg) = X(f)g + fX(g) \quad \text{for} \quad f, g \in C^\infty(M).$$

The Lie bracket of two vector fields X and Y is the *commutator*

$$[X,Y] := X \circ Y - Y \circ X,$$

which turns out to be a derivation again. Hence, $[X,Y]$ defines a smooth vector field on M. For $M = \mathbb{S}$ the space $C^\infty(\mathbb{S})$ can be described as the vector space $C^\infty_{2\pi}(\mathbb{R})$ of 2π-periodic functions $\mathbb{R} \to \mathbb{R}$. A general vector field $X \in \mathrm{Vect}(\mathbb{S})$ in this setting has the form $X = f\frac{d}{d\theta}$, where $f \in C^\infty_{2\pi}(\mathbb{R})$ and where the points z of $\mathbb{S}$ are represented as $z = e^{i\theta}$, θ being a variable in $\mathbb{R}$. For $X = f\frac{d}{d\theta}$ and $Y = g\frac{d}{d\theta}$ it is easy to see that

$$[X,Y] = (fg' - f'g)\frac{d}{d\theta} \quad \text{with} \quad g' = \frac{d}{d\theta}g \text{ and } f' = \frac{d}{d\theta}f. \tag{5.1}$$

The representation of f by a convergent Fourier series

$$f(\theta) = a_0 + \sum_{n=1}^{\infty} (a_n \cos(n\theta) + b_n \sin(n\theta))$$

leads to a natural (topological) generating system for $\mathrm{Vect}(\mathbb{S})$:

$$\frac{d}{d\theta}, \quad \cos(n\theta)\frac{d}{d\theta}, \quad \sin(n\theta)\frac{d}{d\theta}.$$

Of special interest is the complexification

$$\mathrm{Vect}^{\mathbb{C}}(\mathbb{S}) := \mathrm{Vect}(\mathbb{S}) \otimes \mathbb{C}$$

of $\mathrm{Vect}(\mathbb{S})$. To begin with, we discuss only the restricted Lie algebra $W \subset \mathrm{Vect}^{\mathbb{C}}(\mathbb{S})$ of polynomial vector fields on $\mathbb{S}$. Define

$$L_n := z^{1-n}\frac{d}{dz} = -iz^{-n}\frac{d}{d\theta} = -ie^{-in\theta}\frac{d}{d\theta} \in \text{Vect}^{\mathbb{C}}(\mathbb{S}),$$

for $n \in \mathbb{Z}$. $L_n : C^{\infty}(\mathbb{S},\mathbb{C}) \to C^{\infty}(\mathbb{S},\mathbb{C})$, $f \mapsto z^{1-n}f'$. The linear hull of the L_n over $\mathbb{C}$ is called the Witt algebra:

$$W := \mathbb{C}\{L_n : n \in \mathbb{Z}\}.$$

It has to be shown, of course, that W with the Lie bracket in $\text{Vect}^{\mathbb{C}}(\mathbb{S})$ actually becomes a Lie algebra over $\mathbb{C}$. For that, we determine the Lie bracket of the L_n, L_m, which can also be deduced from the above formula (5.1). For $n, m \in \mathbb{Z}$ and $f \in C^{\infty}(\mathbb{S},\mathbb{C})$,

$$L_n L_m f = z^{1-n}\frac{d}{dz}\left(z^{1-m}\frac{d}{dz}f\right)$$

$$= (1-m)z^{1-n-m}\frac{d}{dz}f - z^{1-n}z^{1-m}\frac{d^2}{dz^2}f.$$

This yields

$$[L_n, L_m]f = L_n L_m f - L_m L_n f$$

$$= ((1-m)-(1-n))z^{1-n-m}\frac{d}{dz}f$$

$$= (n-m)L_{n+m}f.$$

In a theory with conformal symmetry, the Witt algebra W is a part of the complexified Lie algebra $\text{Vect}^{\mathbb{C}}(\mathbb{S}) \times \text{Vect}^{\mathbb{C}}(\mathbb{S})$ belonging to the classical conformal symmetry. Hence, as we explained in the preceding chapter, the central extensions of W by $\mathbb{C}$ become important for the quantization process.

5.2 Witt Algebra and Infinitesimal Conformal Transformations of the Euclidean Plane

Before we focus on the central extensions of the Witt algebra in Theorem 5.1, another approach to the Witt algebra shall be described. This approach is connected with the discussion in Sect. 2.4 about the conformal group for the Euclidean plane. In fact, in the development of conformal field theory in the context of statistical mechanics mostly the Euclidean signature is used. This point of view is taken, for example, in the fundamental papers on conformal field theory in two dimensions (cf., e.g., [BPZ84], [Gin89], [GO89]).

The conformal transformations in domains $U \subset \mathbb{C} \cong \mathbb{R}^{2,0}$ are the holomorphic or antiholomorphic functions with nowhere-vanishing derivative (cf. Theorem 1.11). We will treat only the holomorphic case for the beginning. If one ignores the question of how these holomorphic transformations can form a group (cf. Sect. 2.4) and investigates infinitesimal holomorphic transformations, these can be written as

$$z \mapsto z + \sum_{n \in \mathbb{Z}} a_n z^n,$$

with convergent Laurent series $\sum_{n \in \mathbb{Z}} a_n z^n$. In the sense of the general relation between $\mathrm{Diff}_+(M)$ and $\mathrm{Vect}(M)$, the vector fields representing these infinitesimal transformations can be written as

$$\sum a_n z^{n+1} \frac{d}{dz}$$

in the fictional relation between the "conformal group" (see, however, Sect. 5.4) and the vector fields. The Lie algebra of all these vector fields has the sequence $(L_n)_{n \in \mathbb{Z}}$, $L_n = z^{1-n} \frac{d}{dz}$, as a (topological) basis with the Lie bracket derived above:

$$[L_n, L_m] = (n - m) L_{n+m}.$$

Hence, for the Euclidean case there are also good reasons to introduce the Witt algebra $\mathrm{W} = \mathbb{C}\{L_n : n \in \mathbb{Z}\}$ with this Lie bracket as the conformal symmetry algebra. The Witt algebra is a dense subalgebra of the Lie algebra of holomorphic vector fields on $\mathbb{C} \setminus \{0\}$. The same is true for an annulus $\{z \in \mathbb{C} : r < |z| < R\}$, $0 \leq r < R \leq \infty$. However, only the vector fields L_n with $n \leq 1$ can be continued holomorphically to a neighborhood of 0 in $\mathbb{C}$, the other L_n s are strictly singular at 0. As a consequence, contrary to what we have just stated the vector fields L_n, $n > 1$, cannot be considered to be infinitesimal conformal transformations on a suitable neighborhood of 0. Instead, these meromorphic vector fields correspond to proper deformations of the standard conformal structure on $\mathbb{R}^{2,0} \cong \mathbb{C}$.

Without having to speak of a specific "conformal group" one can require – as it is usually done in conformal field theory à la [BPZ84] – that the primary field operators of a conformal field theory transform infinitesimally according to the L_n (a condition which will be explained in detail in Sect. 9.3). This symmetry condition yields an infinite number of constraints. This viewpoint explains the claim of "infinite dimensionality" in the citations of Sect. 2.4.

Let us point out that there is no complex Lie group H with Lie $H = \mathrm{Vect}^{\mathbb{C}}(\mathbb{S})$ as is explained in Sect. 5.4.

The antiholomorphic transformations/vector fields yield a copy $\overline{\mathrm{W}}$ of W with basis $\overline{L}_n$, so that

$$[\overline{L}_n, \overline{L}_m] = (n - m) \overline{L}_{n+m} \quad \text{and} \quad [L_n, \overline{L}_m] = 0.$$

For the Minkowski plane one has a copy of the Witt algebra as well, which in this case originates from the second factor $\mathrm{Diff}_+(\mathbb{S})$ in the characterization

$$\mathrm{Conf}(\mathbb{R}^{1,1}) \cong \mathrm{Diff}_+(\mathbb{S}) \times \mathrm{Diff}_+(\mathbb{S}).$$

In both cases there is a natural isomorphism $t : \mathrm{W} \to \mathrm{W}$ of the Witt algebra, defined by $t(L_n) := -L_{-n}$ on the basis. t is a linear isomorphism and respects the Lie bracket:

$$[t(L_n), t(L_m)] = [L_{-n}, L_{-m}] = -(n-m)L_{-(n+m)} = (n-m)t(L_{n+m}).$$

Hence, t is a Lie algebra isomorphism. Since $t^2 = \mathrm{id}_W$, t is an involution. These facts explain that in many texts on conformal field theory the basis

$$L_n^{\sim} = -z^{n+1}\frac{d}{dz} = t\left(z^{1-n}\frac{d}{dz}\right)$$

instead of $L_n = z^{1-n}\frac{d}{dz}$ is used. Incidentally, the involution t induced on W by the biholomorphic coordinate change $z \mapsto w = \frac{1}{z}$ of the punctured plane $\mathbb{C} \setminus \{0\}$: $dz = -w^{-2}dw$ implies

$$z^{1-n}\frac{d}{dz} = w^{n-1}(-w^2)\frac{d}{dw} = -w^{n+1}\frac{d}{dw}.$$

5.3 The Virasoro Algebra as a Central Extension of the Witt Algebra

After these two approaches to the Witt algebra W we now come to the Virasoro algebra, which is a proper central extension of W. For existence and uniqueness we need

Theorem 5.1. [GF68] $H^2(W, \mathbb{C}) \cong \mathbb{C}$.

Proof. In the following we show: the linear map $\omega : W \times W \to \mathbb{C}$ given by

$$\omega(L_n, L_m) := \delta_{n+m}\frac{n}{12}(n^2 - 1), \delta_k := \begin{cases} 1 & \text{for } k = 0 \\ 0 & \text{for } k \neq 0 \end{cases}$$

defines a nontrivial central extension of W by $\mathbb{C}$ and up to equivalence this is the only nontrivial extension of W by $\mathbb{C}$. In order to do this we prove

1. $\omega \in Z^2(W, \mathbb{C})$.

2. $\omega \notin B^2(W, \mathbb{C})$.

3. $\Theta \in Z^2(W, \mathbb{C}) \Rightarrow \exists \lambda \in \mathbb{C} : \Theta \sim \lambda\omega$.

Remark: The choice of the factor $\frac{1}{12}$ in the definition of ω is in accordance with the zeta function regularization using the Riemann zeta function, cf. [GSW87, p. 96].

1. Evidently, ω is bilinear and alternating. In order to show $\omega \in Z^2(W, \mathbb{C})$, that is $2°$ of Remark 4.3, we have to check that

$$\omega(L_k, [L_m, L_n]) + \omega(L_m, [L_n, L_k]) + \omega(L_n, [L_k, L_m]) = 0$$

for $k, m, n \in \mathbb{Z}$. This can be calculated easily:

$$12(\omega(L_k, [L_m, L_n]) + \omega(L_m, [L_n, L_k])$$
$$+ \omega(L_n, [L_k, L_m]))$$
$$= \delta_{k+m+n}((m-n)k(k^2-1) + (n-k)m(m^2-1)$$
$$+ (k-m)n(n^2-1))$$
$$= -(m-n)(m+n)((m+n)^2-1)$$
$$+ (2n+m)m(m^2-1)$$
$$- (2m+n)n(n^2-1)$$
$$= 0.$$

2. Assume that there exists $\mu \in \mathrm{Hom}_{\mathbb{C}}(W, \mathbb{C})$ with $\omega(X, Y) = \mu([X, Y])$ for all $X, Y \in W$. Then for every $n \in \mathbb{N}$ we have

$$\omega(L_n, L_{-n}) = \tilde{\mu}(L_n, L_{-n})$$
$$\Rightarrow \tfrac{n}{12}(n^2-1) = \mu([L_n, L_{-n}])$$
$$\Rightarrow \tfrac{n}{12}(n^2-1) = 2n\mu(L_0)$$
$$\Rightarrow \mu(L_0) = \tfrac{1}{24}(n^2-1).$$

The last equation cannot hold for every $n \in \mathbb{N}$. So the assumption was wrong, which implies $\omega \notin B^2(W, \mathbb{C})$.

3. Let $\Theta \in Z^2(W, \mathbb{C})$. Then for $k, m, n \in \mathbb{Z}$ we have

$$0 = \Theta(L_k, [L_m, L_n]) + \Theta(L_m, [L_n, L_k]) + \Theta(L_n, [L_k, L_m])$$
$$= (m-n)\Theta(L_k, L_{m+n}) + (n-k)\Theta(L_m, L_{n+k})$$
$$+ (k-m)\Theta(L_n, L_{k+m}).$$

For $k = 0$ we get

$$(m-n)\Theta(L_0, L_{m+n}) + n\Theta(L_m, L_n) - m\Theta(L_n, L_m) = 0.$$

Hence

$$\Theta(L_n, L_m) = \frac{m-n}{m+n}\Theta(L_0, L_{m+n}) \quad \text{for} \quad m, n \in \mathbb{Z}; \ m \neq -n.$$

We define a homomorphism $\mu \in \mathrm{Hom}_{\mathbb{C}}(W, \mathbb{C})$ by

$$\mu(L_n) := \frac{1}{n}\Theta(L_0, L_n) \quad \text{for} \quad n \in \mathbb{Z} \setminus \{0\},$$
$$\mu(L_0) := -\frac{1}{2}\Theta(L_1, L_{-1}),$$

and let $\Theta' := \Theta + \tilde{\mu}$. Then $\Theta'(L_n, L_m) = 0$ for $m, n \in \mathbb{Z}, m \neq -n$, since

$$\Theta'(L_n, L_m) = \Theta(L_n, L_m) + \mu([L_n, L_m])$$
$$= \frac{m-n}{m+n}\Theta(L_0, L_{n+m}) + \mu((n-m)L_{n+m})$$
$$= \frac{m-n}{m+n}\Theta(L_0, L_{n+m}) + \frac{n-m}{m+n}\Theta(L_0, L_{n+m})$$
$$= 0.$$

So there is a map $h : \mathbb{Z} \to \mathbb{C}$ with

$$\Theta'(L_n, L_m) = \delta_{n+m}h(n) \quad \text{for} \quad n, m \in \mathbb{Z}.$$

Since Θ' is alternating, it follows:

$$h(0) = 0 \quad \text{and} \quad h(-k) = -h(k) \quad \text{for all} \quad k \in \mathbb{Z}.$$

By definition of μ we have

$$h(1) = \Theta'(L_1, L_{-1})$$
$$= \Theta(L_1, L_{-1}) + \mu([L_1, L_{-1}])$$
$$= \Theta(L_1, L_{-1}) + \mu(2L_0)$$
$$= \Theta(L_1, L_{-1}) - \Theta(L_1, L_{-1})$$
$$= 0.$$

It remains to be shown that there is a $\lambda \in \mathbb{C}$ with $\Theta' = \lambda\omega$, that is

$$h(n) = \frac{\lambda}{12}n(n^2 - 1) \quad \text{for} \quad n \in \mathbb{N}. \tag{5.2}$$

Since $\Theta' \in Z^2(W, \mathbb{C})$, we have for $k, m, n \in \mathbb{N}$,

$$0 = \Theta'(L_k, [L_m, L_n]) + \Theta'(L_m, [L_n, L_k])$$
$$\quad + \Theta'(L_n, [L_k, L_m])$$
$$= (m-n)\Theta'(L_k, L_{m+n}) + (n-k)\Theta'(L_m, L_{n+k})$$
$$\quad + (k-m)\Theta'(L_n, L_{k+m}).$$

For $k + m + n = 0$ we get

$$0 = (m-n)h(k) + (n-k)h(m) + (k-m)h(n)$$
$$= -(m-n)h(m+n) + (2n+m)h(m)$$
$$\quad - (2m+n)h(n).$$

The substitution $n = 1$ yields the equation

$$-(m-1)h(m+1) + (2+m)h(m) - (2m+1)h(1) = 0,$$

for $m \in \mathbb{N}$. Combined with $h(1) = 0$ this implies the recursion formula

$$h(m+1) = \frac{m+2}{m-1}h(m) \quad \text{for} \quad m \in \mathbb{N}\setminus\{1\}.$$

Consequently, the map h is completely determined by $h(2) \in \mathbb{C}$. We now show by induction $n \in \mathbb{N}$ that for $\lambda := 2h(2)$ the relation (5.2) holds. The cases $n = 1$ and $n = 2$ are obvious. So let $m \in \mathbb{N}$, $n > 1$, and $h(m) = \frac{\lambda}{12}m(m^2 - 1)$. Then

$$
\begin{aligned}
h(m+1) &= \frac{m+2}{m-1}\, h(m) \\
&= \frac{m+2}{m-1}\frac{\lambda}{12}m(m^2 - 1) \\
&= \frac{\lambda}{12}m(m+1)(m+2) \\
&= \frac{\lambda}{12}(m+1)((m+1)^2 - 1).
\end{aligned}
$$
□

Definition 5.2. The *Virasoro algebra* Vir is the central extension of the Witt algebra W by $\mathbb{C}$ defined by ω, that is

$$\text{Vir} = \text{W} \oplus \mathbb{C}Z \quad \text{as a complex vector space,}$$

$$[L_n, L_m] = (n-m)L_{n+m} + \delta_{n+m}\frac{n}{12}(n^2 - 1)Z,$$

$$[L_n, Z] = 0 \quad \text{for} \quad n, m \in \mathbb{Z}.$$

5.4 Does There Exist a Complex Virasoro Group?

In Sect. 2.3 we have shown that the conformal group $\text{Conf}(\mathbb{R}^{2,0})$ of the Euclidean plane is not infinite dimensional. Instead, it is isomorphic to the familiar finite-dimensional group Mb of Möbius transformations which in turn is isomorphic to the Lorentz group $SO(3, 1)$. Here, the conformal group is defined to be the group of global conformal transformations defined on open dense subsets $M \subset \mathbb{R}^{2,0}$.

It is, however, a fact and an essential feature that in conformal field theory the infinite dimensional Lie algebra Vir is used as the fundamental set of (infinitesimal) symmetries. Even if it is impossible to interpret these symmetries as generators of conformal transformations on open subsets of the euclidean plane (cf. Sect. 2.3) it is in principle not excluded that there exists an infinite dimensional complex Lie group $\mathscr{G}$ such that the Virasoro algebra Vir is essentially the Lie algebra of $\mathscr{G}$. Such a Lie group would be called a *Virasoro group*. Such a group would play the role of an abstract infinite dimensional conformal group related to the Euclidean plane embodying all conformal symmetries.

We are thus led to discuss the following questions:

1. **Question**: Does there exist a complex Lie group $\mathscr{G}$ with the Virasoro algebra Vir as its Lie algebra?
   Closely related to this question are the following two questions.
2. **Question**: Does there exist a complex Lie group $\mathscr{H}$ with the Witt algebra W as its Lie algebra?
3. **Question**: Does there exist a real Lie group $\mathscr{F}$ such that the Lie algebra of $\mathscr{F}$ is the central extension $\mathrm{Vir}^{\mathbb{R}}$ of the real version $\mathrm{W}^{\mathbb{R}}$ of the Witt algebra given by the same cocycle ω as in Theorem 5.1?

The questions have to be formulated in a more precise manner, but the answer to the first question in its most natural setting is no, as we report in the following.

The questions are not clearly stated in the infinite dimensional setting because answering them requires to specify a topology on Vir since there is no natural topology on an infinite dimensional complex vector space in contrast to the finite-dimensional case. Since Vir can be equipped with many different topologies compatible with its structure of a complex Lie algebra we obtain a series of questions depending on the topologies considered. The topology to be chosen should be at least a locally convex topology since there exists a reasonable theory of Lie groups and Lie algebras (cf. [Mil84]) with models in locally convex spaces. However, only for Banach Lie groups one has an exponential mapping which is a local embedding and thus gives coordinates. In fact, the nonexistence of a Virasoro group is closely related to deficiencies of the exponential mapping.

If one considers locally convex topologies on Vir, it is quite natural to require that the corresponding Lie group has its models in the completion $\widehat{\mathrm{Vir}}$ of Vir. Consequently, the questions 1–3 have to be refined by asking for Lie groups such that their Lie algebras are isomorphic as topological Lie algebras to the completions $\widehat{\mathrm{Vir}}, \widehat{\mathrm{W}}$ resp. $\widehat{\mathrm{Vir}^{\mathbb{R}}}$.

What is the right topology on Vir and on the other two related Lie algebras? Regarding the definition of Vir as the central extension of the Witt algebra W and taking into account the origin of W as a Lie algebra of complex vector fields on $\mathbb{S}$ it is natural to start with the topology on W which is induced from $\mathrm{Vect}(\mathbb{S})^{\mathbb{C}}$ where on $\mathrm{Vect}(\mathbb{S})$ the natural Fréchet topology on compact convergence of the vector fields and all its derivatives is considered. The completion $\widehat{\mathrm{W}}$ of W is $\mathrm{Vect}(\mathbb{S})^{\mathbb{C}}$, and the second question reduces to the existence of a complexification of the real Lie group $\mathrm{Diff}_+(\mathbb{S})$. By a result of Lempert [Lem97*],

Theorem 5.3. $\mathrm{Diff}_+(\mathbb{S})$ *has no complexification. In particular, there even does not exist a real Lie group* $\mathscr{H}$ *with Lie* $\mathscr{H} = \widehat{\mathrm{W}} = \mathrm{Vect}(\mathbb{S})^{\mathbb{C}}$.

Of course, the notion of a complexification has to be made precise, in particular, since in the literature different concepts are used. A (universal) complexification of a real Lie group G is a complex Lie group $G^{\mathbb{C}}$ together with a homomorphism $j : G \to G^{\mathbb{C}}$ such that any homomorphism $\psi : G \to H$ into a complex Lie group

H factors uniquely through j, that is there exists a unique complex analytic morphism $\hat{\psi} : G^{\mathbb{C}} \to H$ with $\psi = \hat{\psi} \circ j$. Finite-dimensional Lie groups always have a complexification although the homomorphism need not be injective.

Note that Theorem 5.3 would follow from the conjecture that every homomorphism ψ into a complex Lie group H is necessarily trivial. This conjecture is stated in [PS86*] (3.2.3) using the fact that $\mathrm{Diff}_+(\mathbb{S})$ is simple according to [Her71]. But in [PS86*] it is implicitly used that H has a reasonable exponential mapping which is not true in general.

Therefore, the proof of Theorem 5.3 in [Lem97*] is based on completely different methods and the result holds for arbitrary compact and connected manifolds M of finite dimension ≥ 1 instead of $\mathbb{S}$.

With the same arguments as in [Lem97*] it can be shown that there is no Virasoro group with respect to the natural topology on Vir induced by the embedding $\mathrm{Vir} \to \mathrm{Vect}(\mathbb{S})^{\mathbb{C}} \oplus \mathbb{C}$ as vector spaces over $\mathbb{C}$ (cf. [Nit06*]):

Theorem 5.4. *There does not exist a complex Lie group $\mathscr{G}$ with Lie $\mathscr{G} = \widehat{\mathrm{Vir}}$.*

In other words, there does not exist an abstract Virasoro group. On the other hand, the third question can be answered in the affirmative. There is a real Lie group $\mathscr{F}$ whose Lie algebra is the (real) nontrivial central extension of $\mathrm{Vect}(\mathbb{S})$. $\mathscr{F}$ is a nontrivial central extension of $\mathrm{Diff}_+(\mathbb{S})$ by $\mathbb{S}^1$.

To construct the extension group $\mathscr{F}$ we can use the restricted unitary group $\mathrm{U}_{\mathrm{res}}(\mathbb{H}_+)$ introduced in Definition 3.16. With a suitable choice of $\mathbb{H}_+ \subset \mathbb{H} = L^2(\mathbb{S})$ (the space of functions $f \in L^2(\mathbb{S})$ without negative Fourier coefficients) one obtains a natural embedding of $\mathrm{Diff}_+(\mathbb{S})$ into $\mathrm{U}_{\mathrm{res}}(\mathbb{H}_+)$ (cf. [PS86*]) and differentiating this sequence yields a nontrivial central extension

$$0 \longrightarrow \mathbb{R} \longrightarrow \mathrm{Vect}(\mathbb{S})^{\sim} \longrightarrow \mathrm{Vect}(\mathbb{S}) \longrightarrow 0$$

of $\mathrm{Vect}(\mathbb{S}) \cong \widehat{\mathrm{W}^{\mathbb{R}}}$.

References

[BPZ84]   A.A. Belavin, A.M. Polyakov, and A.B. Zamolodchikov. In- finite conformal symmetry in two-dimensional quantum field theory. *Nucl. Phys.* **B 241** (1984), 333–380.

[GF68]   I.M. Gelfand and D.B. Fuks. Cohomology of the Lie algebra of vector fields of a circle. *Funct. Anal. Appl.* **2** (1968), 342–343.

[Gin89]   P. Ginsparg. *Introduction to Conformal Field Theory. Fields, Strings and Critical Phenomena*, Les Houches 1988, Elsevier, Amsterdam, 1989.

[GO89]   P. Goddard and D. Olive. Kac-Moody and Virasoro algebras in relation to quantum mechanics. *Int. J. Mod. Phys.* **A1** (1989), 303–414.

[GSW87]   M.B. Green, J.H. Schwarz, and E. Witten. *Superstring Theory, Vol. 1*. Cambridge University Press, Cambridge, 1987.

[Her71]   M.-R. Herman. Simplicité du groupe des Difféomorphismes de classe C^∞, isotope à l'identité, du tore de dimension n. *C.R. Acad. Sci. Paris* **273** (1971), 232–234.

[Lem97*]   L. Lempert. The problem of complexifying a Lie group. In: *Multidimensional Complex Analysis and Partial Differential Equations*, P.D. Cordaro et al. (Eds.), Contemporary Mathematics **205**, 169–176. AMS, Providence, RI, 1997.

[Mil84]   J. Milnor. Remarks on infinite dimensional Lie groups. In: *Relativity, Groups and Topology II*, Les Houches 1983, 1007–1058. North-Holland, Amsterdam, 1984.

[Nit06*]   T. Nitschke. *Komplexifizierung unendlichdimensionaler Lie-Gruppen*. Diplomarbeit, LMU München, 2006.

[PS86*]   A. Pressley and G. Segal. *Loop Groups*. Oxford University Press, Oxford, 1986.

Part II
First Steps Toward Conformal Field Theory

The term "conformal field theory" stands for a variety of different formulations and slightly different structures. The aim of the second part of these notes is to describe some of these formulations and structures and thereby contribute to answering the question of what conformal field theory is.

Conformal field theories are best described either by the way they appear and are constructed or by properties and axioms which provide classes of conformal field theories. The most common theories by examples are

- free bosons or fermions (σ-models on a torus),
- WZW-models[1] for compact Lie groups and gauged WZW-models,
- coset and orbifold constructions of WZW-models.

Systematic descriptions of conformal field theory emphasizing the fundamental structures and properties comprise

- various combinatorical approaches like the axioms of Moore–Seiberg [MS89], Friedan–Shenker [FS87], or Segal [Seg88a].
- the Osterwalder–Schrader axioms with conformal invariance [FFK89],
- the vertex algebras or chiral algebras [BD04*] as their generalizations,

A common feature and essential point of all these approaches to conformal field theory is the appearance of representations of the Virasoro algebra which play a central role. The simple reason for this major role of the Virasoro is based on the fact that the elements of the Virasoro algebra are symmetries of the quantum system and these elements are regarded as the most important observables in conformal field theory. In this context the generators L_n can be compared in their physical significance to the momentum or angular momentum in conventional one-particle quantum mechanics.

Since the Witt algebra W is a generating subalgebra of the infinitesimal classical conformal transformations of the Minkowski plane in each of the two light cone variables (cf. Corollary 2.15 and Sect. 5.1), the set of all observables of conformal field theory contains the direct product $\text{Vir} \times \overline{\text{Vir}}$ of two copies of the Virasoro algebra. (Note that after quantization, the Witt algebra has to be replaced by its nontrivial central extension, the Virasoro algebra Vir, cf. Chaps. 3 and 4.) In general, one assumes the full set $\mathscr{A}_{\text{tot}}$ of observables to form an algebra which decomposes into a direct product of algebras $\mathscr{A} \times \mathscr{A}'$ containing the Virasoro algebras $\text{Vir} \subset \mathscr{A}, \overline{\text{Vir}} \subset \mathscr{A}'$. The two components of the full algebra of observables are called chiral halves or holomorphic/antiholomorphic or similar.

As a consequence of the product structure, for many purposes one can restrict the investigations to one "chiral half" of the theory in such a way that only $\text{Vir} \subset \mathscr{A}$ resp. $\overline{\text{Vir}} \subset \mathscr{A}'$ is studied. The restrictions to one chiral half requires among other things to regard the light cone variables t_+ and t_- as completely independent variables, and, in the same way, the complex variables z and $\bar{z}$ as completely independent. The identification of $\bar{z}$ with the complex conjugate only takes place when the two chiral halves of the conformal field theory are combined.

---

[1] WZW = Wess–Zumino–Witten

Restricting now to one chiral half $\mathscr{A}$ and, furthermore, restricting to the subalgebra Vir we are led, first of all, to study the representations of the Virasoro algebra.

In a certain way one could claim now that conformal field theory is the representation theory of the Virasoro algebra and of certain algebras (namely chiral algebras) containing the Virasoro algebra. Therefore, in this second part of the notes we first describe the representations of the Virasoro algebra (Chap. 6) and explain as an example how the quantization of strings leads to a representation of the Virasoro algebra (Chap. 7). Next we discuss the axiomatic approach to quantum field theory according to Wightman as well as the Euclidean version according to Osterwalder–Schrader (Chap. 8) and treat the case of two-dimensional conformal field theory in a separate chapter (Chap. 9). In Chap. 10 we connect all these with the theory of vertex algebras, and in Chap. 11 we present as an example of an application of conformal field theory to complex algebraic geometry the Verlinde formula in the context of holomorphic vector bundles and moduli spaces.

Chapter 6
Representation Theory of the Virasoro Algebra

Most of the results in this chapter can be found in [Kac80]. A general treatment of the Virasoro algebra and its significance in geometry and algebra is given in [GR05*].

6.1 Unitary and Highest-Weight Representations

Let V be a vector space over $\mathbb{C}$.

Definition 6.1 (Unitary Representation). A representation $\rho : \mathrm{Vir} \to \mathrm{End}_{\mathbb{C}}V$ (that is a Lie algebra homomorphism ρ) is called *unitary* if there is a positive semi-definite hermitian form $H : V \times V \to \mathbb{C}$, so that for all $v, w \in V$ and $n \in \mathbb{Z}$ one has

$$H(\rho(L_n)v, w) = H(v, \rho(L_{-n})w),$$

$$H(\rho(Z)v, w) = H(v, \rho(Z)w).$$

Note that this notion of a unitary representation differs from that introduced in Definition 3.7 where a unitary representation of a topological group G was defined to be a continuous homomorphism $G \to \mathrm{U}(\mathbb{H})$ into the unitary group of a Hilbert space. This is so, because we do not consider any topological structure in Vir.

One requires that $\rho(L_n)$ is formally adjoint to $\rho(L_{-n})$, to ensure that ρ maps the generators $\frac{d}{d\theta}$, $\cos(n\theta)\frac{d}{d\theta}$, $\sin(n\theta)\frac{d}{d\theta}$ (cf. Chap. 5) of the real Lie algebra $\mathrm{Vect}(\mathbb{S})$ to skew-symmetric operators. Since

$$\frac{d}{d\theta} = iL_0, \quad \cos(n\theta)\frac{d}{d\theta} = -\frac{i}{2}(L_n + L_{-n}), \quad \text{and}$$

$$\sin(n\theta)\frac{d}{d\theta} = -\frac{1}{2}(L_n - L_{-n}),$$

it follows from $H(\rho(L_n)v, w) = H(v, \rho(L_{-n})w)$ that

$$H(\rho(D)v, w) + H(v, \rho(D)w) = 0$$

Schottenloher, M.: *Representation Theory of the Virasoro Algebra*. Lect. Notes Phys. **759**, 91–102 (2008)
DOI 10.1007/978-3-540-68628-6_7   © Springer-Verlag Berlin Heidelberg 2008

for all

$$D \in \left\{ \frac{d}{d\theta}, \cos(n\theta)\frac{d}{d\theta}, \sin(n\theta)\frac{d}{d\theta} \right\}.$$

So, in principle, these unitary representations of Vir can be integrated to pro-jective representations $\mathrm{Diff}_+(\mathbb{S}) \to \mathrm{U}(\mathbb{P}(\mathbb{H}))$ (cf. Sect. 6.5), where $\mathbb{H}$ is the Hilbert space given by (V, H).

Definition 6.2. A vector $v \in V$ is called a *cyclic vector* for a representation $\rho : \mathrm{Vir} \to \mathrm{End}(V)$ if the set

$$\{\rho(X_1)\ldots\rho(X_m)v : X_j \in \mathrm{Vir} \quad \text{for} \quad j = 1,\ldots m, m \in \mathbb{N}\}$$

spans the vector space V.

Definition 6.3. A representation $\rho : \mathrm{Vir} \to \mathrm{End}(V)$ is called a *highest-weight repre-sentation* if there are complex numbers $h, c \in \mathbb{C}$ and a cyclic vector $v_0 \in V$, so that

$$\begin{aligned}
\rho(Z)v_0 &= cv_0, \\
\rho(L_0)v_0 &= hv_0, \text{and} \\
\rho(L_n)v_0 &= 0 \quad \text{for } n \in \mathbb{Z}, n \geq 1.
\end{aligned}$$

The vector v_0 is then called the *highest-weight vector* (or *vacuum vector*) and V is called a *Virasoro module* (via ρ) with *highest weight* (c, h), or simply a *Virasoro module for* (c, h).

Such a representation is also called a *positive energy representation* if $h \geq 0$. The reason of this terminology is the fact that L_0 often has the interpretation of the energy operator which is assumed to be diagonalizable with spectrum bounded from below. With this assumption any representation ρ respecting this property sat-isfies $\rho(L_n)v_0 = 0$ for all $n \in \mathbb{Z}, n > 0$, if v_0 is an eigenvector of $\rho(L_0)$ with lowest eigenvalue $h \in \mathbb{R}$. This follows from the fact that $w = \rho(L_n)(v_0)$ is an eigenvec-tor of $\rho(L_0)$ with eigenvalue $h - n$ or $w = 0$ as can be seen by using the relation $L_0 L_n = L_n L_0 - nL_n$:

$$\rho(L_0)(w) = \rho(L_n)\rho(L_0)v_0 - n\rho(L_n)v_0 = \rho(L_n)(hv_0) - nw = (h - n)w.$$

Now, since h is the lowest eigenvalue of $\rho(L_0)$, w has to vanish for $n > 0$.

The notation often used by physicists is $|h\rangle$ instead of v_0 and $L_n|h\rangle$ instead of $\rho(L_n)v_0$ so that, in particular, $L_0|h\rangle = h|h\rangle$.

6.2 Verma Modules

Definition 6.4. A *Verma module* for $c, h \in \mathbb{C}$ is a complex vector space $M(c, h)$ with a highest-weight representation

$$\rho : \text{Vir} \to \text{End}_{\mathbb{C}}(M(c,h))$$

and a highest-weight vector $v_0 \in M(c,h)$, so that

$$\{\rho(L_{-n_1})\dots\rho(L_{-n_k})v_0 : n_1 \geq \dots \geq n_k > 0 \,,\, k \in \mathbb{N}\} \cup \{v_0\}$$

is a vector space basis of $M(c,h)$.

Every Verma module $M(c,h)$ yields a highest-weight representation with highest weight (c,h). For fixed $c,h \in \mathbb{C}$ the Verma module $M(c,h)$ is unique up to isomorphism. For every Virasoro module V with highest weight (c,h) there is a surjective homomorphism $M(c,h) \to V$, which respects the representation. This holds, since

Lemma 6.5. *For every $h,c \in \mathbb{C}$ there exists a Verma module $M(c,h)$.*

Proof. Let

$$M(c,h) := \mathbb{C}v_0 \oplus \bigoplus \mathbb{C}\{v_{n_1\dots n_k} : n_1 \geq \dots \geq n_k > 0 \,,\, k \in \mathbb{Z}, \, k > 0\}$$

be the complex vector space spanned by v_0 and $v_{n_1,\dots,n_k}$, $n_1 \geq \dots \geq n_k > 0$. We define a representation

$$\rho : \text{Vir} \to \text{End}_{\mathbb{C}}(M(c,h))$$

by

$$\rho(Z) := c\,\text{id}_{M(c,h)},$$
$$\rho(L_n)v_0 := 0 \quad \text{for} \quad n \in \mathbb{Z}, n \geq 1,$$
$$\rho(L_0)v_0 := hv_0,$$
$$\rho(L_0)v_{n_1\dots n_k} := \left(\sum_{j=1}^{k} n_j + h\right) v_{n_1\dots n_k},$$
$$\rho(L_{-n})v_0 := v_n \quad \text{for} \quad n \in \mathbb{Z}, n \geq 1,$$
$$\rho(L_{-n})v_{n_1\dots n_k} := v_{nn_1\dots n_k} \quad \text{for} \quad n \geq n_1.$$

For all other $v_{n_1\dots n_k}$ with $1 \leq n < n_1$ one obtains $\rho(L_{-n})v_{n_1\dots n_k}$ by permutation, taking into account the commutation relations $[L_n, L_m] = (n-m)L_{n+m}$ for $n \neq m$, e.g., for $n_1 > n \geq n_2$:

$$\rho(L_{-n})v_{n_1\dots n_k}$$
$$= \rho(L_{-n})\rho(L_{-n_1})v_{n_2\dots n_k}$$
$$= (\rho(L_{-n_1})\rho(L_{-n}) + (-n+n_1)\rho(L_{-(n+n_1)}))v_{n_2\dots n_k}$$
$$= v_{n_1 n n_2\dots n_k} + (n_1 - n)v_{(n_1+n)n_2\dots n_k}.$$

So

$$\rho(L_{-n})v_{n_1\dots n_k} := v_{n_1 n n_2\dots n_k} + (n_1 - n)v_{(n_1+n)n_2\dots n_k}.$$

Similarly one defines $\rho(L_n)v_{n_1\dots n_k}$ for $n \in \mathbb{N}$ taking into account the commutation relations, e.g.,

$$\rho(L_n)v_{n_1} := \begin{cases} 0 & \text{for} \quad n > n_1 \\ (2nh + \frac{n}{12}(n^2-1)c)v_0 & \text{for} \quad n = n_1 \\ (n+n_1)v_{n_1-n} & \text{for} \quad 0 < n < n_1. \end{cases}$$

Hence, ρ is well-defined and $\mathbb{C}$-linear. It remains to be shown that ρ is a representation, that is

$$[\rho(L_n), \rho(L_m)] = \rho([L_n, L_m]).$$

For instance, for $n \geq n_1$ we have

$$\begin{aligned} & [\rho(L_0), \rho(L_{-n})]v_{n_1 \dots n_k} \\ &= \rho(L_0)v_{nn_1 \dots n_k} - \rho(L_{-n})\left(\textstyle\sum n_j + h\right)v_{n_1 \dots n_k} \\ &= \left(\textstyle\sum n_j + n + h\right)v_{nn_1 \dots n_k} - \left(\textstyle\sum n_j + h\right)v_{nn_1 \dots n_k} \\ &= nv_{nn_1 \dots n_k} \\ &= n\rho(L_{-n})v_{n_1 \dots n_k} \\ &= \rho([L_0, L_{-n}])v_{n_1 \dots n_k} \end{aligned}$$

and for $n \geq m \geq n_1$

$$\begin{aligned} & [\rho(L_{-m}), \rho(L_{-n})]v_{n_1 \dots n_k} \\ &= \rho(L_{-m})v_{nn_1 \dots n_k} - v_{nmn_1 \dots n_k} \\ &= v_{nmn_1 \dots n_k} + (n-m)v_{(n+m)n_1 \dots n_k} - v_{nmn_1 \dots n_k} \quad \text{(s.o.)} \\ &= (n-m)v_{(n+m)n_1 \dots n_k} \\ &= (n-m)\rho(L_{-(m+n)})v_{n_1 \dots n_k} \\ &= \rho([L_{-m}, L_{-n}])v_{n_1 \dots n_k}. \end{aligned}$$

The other identities follow along the same lines from the respective definitions. $\square$

$M(c,h)$ can also be described as an induced representation, a concept which is explained in detail in Sect. 10.49. To show this, let

$$B^+ := \mathbb{C}\{L_n : n \in \mathbb{Z}, n \geq 0\} \oplus \mathbb{C}Z.$$

B^+ is a Lie subalgebra of Vir. Let $\sigma : B^+ \to \mathrm{End}_{\mathbb{C}}(\mathbb{C})$ be the one-dimensional representation with $\sigma(Z) := c$, $\sigma(L_0) := h$, and $\sigma(L_n) = 0$ for $n \geq 1$. Then the representation ρ described explicitly above is induced by σ on Vir with representation module

$$\mathrm{U}(\mathrm{Vir}) \otimes_{U(B^+)} \mathbb{C} \cong M(c,h).$$

($\mathrm{U}(\mathfrak{g})$ is the universal enveloping algebra of a Lie algebra $\mathfrak{g}$, see Definition 10.45.)

Remark 6.6. Let V be a Virasoro module for $c, h \in \mathbb{C}$. Then we have the direct sum decomposition $V = \bigoplus_{N \in \mathbb{N}} V_N$, where $V_0 := \mathbb{C}v_0$ and V_N for $N \in \mathbb{N}$ is, $N > 0$, the complex vector space generated by

$$\rho(L_{-n_1})\ldots\rho(L_{-n_k})v_0$$

$$\text{with}\quad n_1 \geq \ldots \geq n_k > 0,\ \sum_{j=1}^{k} n_j = N,\ k \in \mathbb{N},\ k > 0.$$

The V_N are eigenspaces of $\rho(L_0)$ for the eigenvalue $(N+h)$, that is

$$\rho(L_0)\,|_{V_N} = (N+h)\mathrm{id}_{V_N}.$$

This follows from the definition of a Virasoro module and from the commutation relations of the L_m.

Lemma 6.7. *Let V be a Virasoro module for $c, h \in \mathbb{C}$ and U a submodule of V. Then*

$$U = \bigoplus_{N \in \mathbb{N}_0} (V_N \cap U).$$

A submodule of V is an *invariant linear subspace* of V, that is a complex-linear subspace U of V with $\rho(D)U \subset U$ for $D \in \mathrm{Vir}$.

Proof. Let $w = w_0 \oplus \ldots \oplus w_s \in U$, where $w_j \in V_j$ for $j \in \{1,\ldots,s\}$. Then

$$\begin{aligned}
w &= w_0 + \ldots + w_s,\\
\rho(L_0)w &= hw_0 + \ldots + (s+h)w_s,\\
&\ \vdots\\
\rho(L_0)^{s-1}w &= h^{s-1}w_0 + \ldots + (s+h)^{s-1}w_s.
\end{aligned}$$

This is a system of linear equations for $w_0,\ldots,w_s$ with regular coefficient matrix. Hence, the $w_0,\ldots,w_s$ are linear combinations of the $w,\ldots,\rho(L_0)^{s-1}w \in U$. So $w_j \in V_j \cap U$. □

6.3 The Kac Determinant

We are mainly interested in unitary representations of the Virasoro algebra, since the representations of Vir appearing in conformal field theory shall be unitary. To find a suitable hermitian form on a Verma module $M(c,h)$, we need to define the notion of the expectation value $\langle w \rangle$ of a vector $w \in M(c,h)$: with respect to the decomposition $M(c,h) = \bigoplus V_N$ according to Lemma 6.7, w has a unique component $w' \in V_0$. The expectation value is simply the coefficient $\langle w \rangle \in \mathbb{C}$ of this component w' for the basis $\{v_0\}$, that is $w' = \langle w \rangle v_0$. ($\langle w \rangle$ makes sense for general Virasoro modules as well.)

Let $M = M(c,h)$, $c, h \in \mathbb{R}$, be the Verma module with highest-weight representation $\rho : \mathrm{Vir} \to \mathrm{End}_{\mathbb{C}}(M(c,h))$ and let v_0 be the respective highest-weight vector. Instead of $\rho(L_n)$ we mostly write L_n in the following. We define a hermitian form $H : M \times M \to \mathbb{C}$ on the basis $\{v_{n_1\ldots n_k}\} \cup \{v_0\}$:

$$H(v_{n_1 \ldots n_k}, v_{m_1 \ldots m_j}) := \langle L_{n_k} \ldots L_{n_1} v_{m_1 \ldots m_j} \rangle$$
$$= \langle L_{n_k} \ldots L_{n_1} L_{-m_1} \ldots L_{-m_j} v_0 \rangle.$$

In particular, this definition includes

$$H(v_0, v_0) := 1 \quad \text{and} \quad H(v_0, v_{n_1 \ldots n_k}) := 0 =: H(v_{n_1 \ldots n_k}, v_0).$$

The condition $c, h \in \mathbb{R}$ implies $H(v, v') = H(v', v)$ for all basis vectors

$$v, v' \in B := \{v_{n_1 \ldots n_k} : n_1 \geq \ldots \geq n_k > 0\} \cup \{v_0\}.$$

The elementary but lengthy proof of this statement consists in a repeated use of the commutation relations of the L_ns. Now, the map $H : B \times B \to \mathbb{R}$ has an $\mathbb{R}$-bilinear continuation to $M \times M$, which is $\mathbb{C}$-antilinear in the first and $\mathbb{C}$-linear in the second variable:

For $w, w' \in M$ with unique representations $w = \sum \lambda_j w_j$, $w' = \sum \mu_k w'_k$ relative to basis vectors $w_j, w'_k \in B$, one defines

$$H(w, w') := \sum \sum \overline{\lambda}_j \mu_k H(w_j, w'_k).$$

$H : M \times M \to \mathbb{C}$ is a hermitian form. However, it is not positive definite or positive semi-definite in general. Just in order to decide this, the Kac determinant is used. H has the following properties:

Theorem 6.8. *Let $h, c \in \mathbb{R}$ and $M = M(c, h)$.*

1. *$H : M \times M \to \mathbb{C}$ is the unique hermitian form satisfying $H(v_0, v_0) = 1$, as well as $H(L_n v, w) = H(v, L_{-n} w)$ and $H(Zv, w) = H(v, Zw)$ for all $v, w \in M$ and $n \in \mathbb{Z}$.*
2. *$H(v, w) = 0$ for $v \in V_N$, $w \in V_M$ with $N \neq M$, that is the eigenspaces of L_0 are pairwise orthogonal.*
3. *$\ker H$ is the maximal proper submodule of M.*

Proof.

1. That the identity

$$H(L_n v, w) = H(v, L_{-n} w)$$

   holds for the hermitian form introduced above can again be seen using the commutation relations. The uniqueness of such a hermitian form follows immediately from

$$H(v_{n_1 \ldots n_k}, v_{m_1 \ldots m_j}) = H(v_0, L_{n_k} \ldots L_{n_1} v_{m_1 \ldots m_j}).$$

2. For $n_1 + \ldots + n_k > m_1 + \ldots + m_j$ the commutation relations of the L_n imply that $L_{n_k} \ldots L_{n_1} L_{-m_1} \ldots L_{-m_j} v_0$ can be written as a sum $\sum P_l v_0$, where the operator P_l begins with an L_s, $s \in \mathbb{Z}$, $s \geq 1$, that is $P_l = Q_l L_s$. Consequently, $H(v_{n_1 \ldots n_k}, v_{m_1 \ldots m_j}) = 0$.
3. $\ker H := \{v \in M : H(w, v) = 0 \ \forall w \in M\}$ is a submodule, because $v \in \ker H$ implies $L_n v \in \ker H$ since $H(w, L_n v) = H(L_{-n} w, v) = 0$. Naturally, $M \neq \ker H$ because $v_0 \notin \ker H$. Let $U \subset M$ be an arbitrary proper submodule. To show $U \subset$

$\ker H$, let $w \in U$. For $n_1 \geq \ldots \geq n_k > 0$ one has $H(v_{n_1 \ldots n_k}, w) = H(v_0, L_{n_k} \ldots L_{n_1} w)$. Assume $H(v_{n_1 \ldots n_k}, w) \neq 0$. Then $\langle L_{n_k} \ldots L_{n_1} w \rangle \neq 0$. By Lemma 6.7 this implies $v_0 \in U$ (because $L_{n_k} \ldots L_{n_1} w \in U$), and also $v_{m_1 \ldots m_j} \in U$, in contradiction to $M \neq U$. Similarly we get $H(v_0, w) = 0$, so $w \in \ker H$. $\qquad\square$

Remark 6.9. $M(c,h)/\ker H$ is a Virasoro module with a nondegenerate hermitian form H. However, H is not definite, in general.

Corollary 6.10. *If H is positive semi-definite then $c \geq 0$ and $h \geq 0$.*

Proof. For $n \in \mathbb{N}$, $n > 0$, we have

$$
\begin{aligned}
H(v_n, v_n) &= H(v_0, L_n L_{-n} v_0) \\
&= H(v_0, \rho([L_n, L_{-n}]) v_0) \\
&= 2nh + \frac{n}{12}(n^2 - 1)c.
\end{aligned}
$$

$H(v_1, v_1) \geq 0$ implies $h \geq 0$. Then, from $H(v_n, v_n) \geq 0$ we get $2nh + \frac{n}{12}(n^2 - 1)c \geq 0$ for all $n \in \mathbb{N}$, hence $c \geq 0$. $\qquad\square$

Definition 6.11. Let $P(N) := \dim_{\mathbb{C}} V_N$ and $\{b_1, \ldots, b_{P(N)}\}$ be a basis of V_N. We define matrices A^N by $A_{ij}^N := H(b_i, b_j)$ for $i, j \in \{1, \ldots, P(N)\}$.

Obviously, H is positive semi-definite if all these matrices A^N are positive semi-definite. For $N = 0$ and $N = 1$ one has $A^0 = (1)$ and $A^1 = (h)$ relative to the bases $\{v_0\}$ and $\{v_1\}$, respectively. V_2 has $\{v_2, v_{1,1}\}$ ($v_2 = L_{-2} v_0$ and $v_{1,1} = L_{-1} L_{-1} v_0$) as basis. For instance,

$$
\begin{aligned}
H(v_2, v_2) &= \langle L_2 L_{-2} v_0 \rangle = \langle L_{-2} L_2 v_0 + 4 L_0 v_0 + \frac{2}{12} 3 c v_0 \rangle \\
&= 4h + \frac{1}{2}c, \\
H(v_{1,1}, v_{1,1}) &= 8h^2 + 4h, \\
H(v_2, v_{1,1}) &= 6h.
\end{aligned}
$$

Hence, the matrix A^2 relative to $\{v_2, v_{1,1}\}$ is

$$
A^2 = \begin{pmatrix} 4h + \frac{1}{2}c & 6h \\ 6h & 8h^2 + 4h. \end{pmatrix}
$$

A^2 is (for $c \geq 0$ and $h \geq 0$) positive semi-definite if and only if

$$
\det A^2 = 2h(16h^2 - 10h + 2hc + c) \geq 0.
$$

This condition restricts the choice of $h \geq 0$ and $c \geq 0$ even more if H has to be positive semi-definite. In the case $c = \frac{1}{2}$, for instance, h must be outside the interval $]\frac{1}{16}, \frac{1}{2}[$. (Taking into account the other A^N, h can only have the values $0, \frac{1}{16}, \frac{1}{2}$; for these values H is in fact unitary, see below.)

Theorem 6.12. [Kac80] *The* Kac determinant det A^N *depends on* (c,h) *as follows:*

$$\det A^N(c,h) = K_N \prod_{\substack{p,q \in \mathbb{N} \\ pq \leq N}} (h - h_{p,q}(c))^{P(N-pq)},$$

where $K_N \geq 0$ *is a constant which does not depend on* (c,h), *the* $P(M)$ *is an in Definition 6.11, and*

$$h_{p,q}(c) := \frac{1}{48}((13-c)(p^2+q^2) + \sqrt{(c-1)(c-25)}(p^2-q^2))$$
$$-24pq - 2 + 2c).$$

A proof can be found in [KR87] or [CdG94], for example.

To derive $\det A^N(c,h) > 0$ for all $c > 1$ and $h > 0$ from Theorem 6.12, it makes sense to define

$$\varphi_{q,q} := h - h_{q,q}(c),$$
$$\varphi_{p,q} := (h - h_{p,q}(c))(h - h_{q,p}(c)), \quad p \neq q.$$

Then by Theorem 6.12 we have

$$\det A^N(c,h) = K_N \prod_{\substack{p,q \in \mathbb{N} \\ pq \leq N, p \leq q}} (\varphi_{p,q})^{P(N-pq)}.$$

For $1 \leq p,q \leq N$ and $c > 1, h > 0$ one has

$$\varphi_{q,q}(c) = h + \frac{1}{24}(c-1)(q^2-1) > 0,$$

$$\varphi_{p,q}(c) = \left(h - \left(\frac{p-q}{2} \right)^2 \right)^2 + \frac{1}{24}h(p^2+q^2-2)(c-1)$$

$$+ \frac{1}{576}(p^2-1)(q^2-1)(c-1)^2$$

$$+ \frac{1}{48}(c-1)(p-q)^2(pq+1) > 0.$$

Hence, $\det A^N(c,h) > 0$ for all $c > 1, h > 0$.

So the hermitian form H is positive definite for the entire region $c > 1, h > 0$ if there is just one example $M(c,h)$ with $c > 1, h > 0$, such that H is positive definite. We will find such an example in the context of string theory (cf. Theorem 7.11).

The investigation of the region $0 \leq c < 1, h \geq 0$ is much more difficult. The following theorem contains a complete description:

Theorem 6.13. *Let* $c,h \in \mathbb{R}$.

1. $M(c,h)$ *is unitary (positive definite) for* $c > 1, h > 0$.
1a. $M(c,h)$ *is unitary (positive semi-definite) for* $c \geq 1, h \geq 0$.

2. $M(c,h)$ *is unitary for* $0 \leq c < 1, h > 0$ *if and only if there exists some* $m \in \mathbb{N}, m > 0$, *so that* $c = c(m)$ *and* $h = h_{p,q}(m)$ *for* $1 \leq p \leq q < m$ *with*

$$h_{p,q}(m) := \frac{((m+1)p - mq)^2 - 1}{4m(m+1)}, \ m \in \mathbb{N},$$

$$c(m) := 1 - \frac{6}{m(m+1)}, \ m \in \mathbb{N} \setminus \{1\}.$$

For the proof of 2: Using the Kac determinant, Friedan, Qiu, and Shenker have shown in [FQS86] that in the region $0 \leq c < 1$ the hermitian form H can be unitary only for the values of $c = c(m)$ and $h = h_{p,q}(m)$ stated in 2. Goddard, Kent, and Olive have later proven in [GKO86], using Kac–Moody algebras, that $M(c,h)$ actually gives a unitary representation in all these cases.

If $M(c,h)$ is unitary and positive semi-definite, but not positive definite, we let

$$W(c,h) := M(c,h)/\ker H.$$

Now $W(c,h)$ is a unitary highest-weight representation (positive definite).

Remark 6.14. Up to isomorphism, for every $c,h \in \mathbb{R}$ there is at most one positive definite unitary highest-weight representation, which must be $W(c,h)$. If $\rho : \mathrm{Vir} \to \mathrm{End}_{\mathbb{C}}(V)$ is a positive definite unitary highest-weight representation with vacuum vector $v_0' \in V$ and hermitian form H', the map

$$v_0 \mapsto v_0', \quad v_{n_1 \ldots n_k} \mapsto \rho(L_{-n_1} \ldots L_{-n_k})v_0,$$

defines a surjective linear homomorphism $\varphi : M(c,h) \to V$, which respects the hermitian forms H and H':

$$H'(\varphi(v), \varphi(w)) = H(v,w).$$

Therefore, H is positive semi-definite and φ factorizes over $W(c,h)$ as a homomorphism $\overline{\varphi} : W(c,h) \to V$.

6.4 Indecomposability and Irreducibility of Representations

Definition 6.15. M is *indecomposable* if there are no invariant proper subspaces V, W of M, so that $M = V \oplus W$. Otherwise M is *decomposable*.

Definition 6.16. M is called *irreducible* if there is no invariant proper subspace V of M. Otherwise M is called *reducible*.

Theorem 6.17. *For each weight (c,h) we have the following:*

1. *The Verma module $M(c,h)$ is indecomposable.*
2. *If $M(c,h)$ is reducible, then there is a maximal invariant subspace $I(c,h)$, so that $M(c,h)/I(c,h)$ is an irreducible highest-weight representation.*
3. *Any positive definite unitary highest-weight representation (that is $W(c,h)$, see above) is irreducible.*

Proof.

1. Let V,W be invariant subspaces of $M = M(c,h)$, and $M = V \oplus W$. By Remark 6.7, we have the direct sum decompositions

$$V = \bigoplus (M_j \cap V) \quad \text{and} \quad W = \bigoplus (M_j \cap W).$$

   Since $\dim M_0 = 1$, this implies $(M_0 \cap V) = 0$ or $(M_0 \cap W) = 0$. So the highest-weight vector v_0 is contained either in V or in W. From the invariance of V and W it follows that $V = M$ or $W = M$.

2. Let $I(c,h)$ be the sum of the invariant proper subspaces of M. Then $I(c,h)$ is an invariant proper subspace of M and $M(c,h)/I(c,h)$ is an irreducible highest-weight representation.

3. Let V be a positive definite unitary highest-weight representation and $U \subsetneq V$ be an invariant subspace. Then

$$U^\perp = \{v \in V : H(u,v) = 0 \; \forall u \in U\}$$

   is an invariant subspace as well, since

$$H(u, L_n v) = H(L_{-n} u, v) = 0$$

   and $U \oplus U^\perp = V$. So 3 follows from 1. $\square$

6.5 Projective Representations of $\mathrm{Diff}_+(\mathbb{S})$

We know the unitary representations $\rho_{c,h} : \mathrm{Vir} \to \mathrm{End}(W_{c,h})$ for $c \geq 1, h \geq 0$ or $c = c(m)$, $h = h_{p,q}(m)$ from the discrete series, where $W_{c,h} := W(c,h)$ is the unique unitary highest-weight representation of the Virasoro algebra Vir described in the preceding section. Let $\mathbb{H} := \widehat{W}_{c,h}$ be the completion of $W_{c,h}$ with respect to its hermitean form. It can be shown that there is a linear subspace $\widetilde{W}_{c,h} \subset \mathbb{H}$, $W_{c,h} \subset \widetilde{W}_{c,h}$, so that $\rho_{c,h}(\xi)$ has a linear continuation $\overline{\rho}_{c,h}(\xi)$ on $\widetilde{W}_{c,h}$ for all $\xi \in \mathrm{Vir} \cap (\mathrm{Vect}(\mathbb{S}))$, where $\overline{\rho}_{c,h}(\xi)$ is an essentially self-adjoint operator. The representation $\rho_{c,h}$ is integrable in the following sense:

Theorem 6.18. [GW85] *There is a projective unitary representation $U_{c,h} : \mathrm{Diff}_+(\mathbb{S}) \to U(\mathbb{P}(\mathbb{H}))$, so that*

$$\widehat{\gamma}(\exp(\overline{\rho}_{c,h}(\xi))) = U_{c,h}(\exp(\xi))$$

for all $\xi \in$ Vect(S), that is for all real vector fields ξ in S. Furthermore, for $X \in$ Vect(S) $\otimes \mathbb{C}$ and $\varphi \in$ Diff$_+$(S) one has

$$U_{c,h}(\varphi)\rho_{c,h}(X) = (\rho_{c,h}(T\varphi X) + c\alpha(X, \varphi))U_{c,h}(\varphi)$$

with a map α on Vect(S) $\times$ Diff$_+$(S). *Here, the $U_{c,h}(\varphi)$ are suitable lifts to $\mathbb{H}$ of the original $U_{c,h}(\varphi)$ (cf. Chap. 3).*

Further investigations in the setting of conformal field theory lead to representations of

- "chiral" algebras $\mathscr{A} \times \overline{\mathscr{A}}$ with Vir $\subset \mathscr{A}$, $\overline{\text{Vir}} \subset \overline{\mathscr{A}}$ (here $\overline{\text{Vir}}$ is an isomorphic copy of Vir and $\mathscr{A}$ as well as $\overline{\mathscr{A}}$ are further algebras), e.g., $\mathscr{A} = U(\widehat{\mathfrak{g}})$ (universal enveloping algebra of a Kac–Moody algebra), but also algebras, which are neither Lie algebras nor enveloping algebras of Lie algebras. (Cf., e.g., [BPZ84], [MS89], [FFK89], [Gin89], [GO89].)
- Semi-groups $\mathscr{E} \times \overline{\mathscr{E}}$ with Diff$_+$(S) $\subset \mathscr{E}$, Diff$_+$(S) $\subset \overline{\mathscr{E}}$. One discusses semi-group extensions Diff$_+$(S), because there is no complex Lie group with Vect$^{\mathbb{C}}$(S) as the associated Lie algebra (cf. 5.4). Interesting cases in this context are the semi-group of Shtan and the semi-group of Neretin which are considered, for instance, in [GR05*].

We just present a first example of such a semi-group here (for a survey cf. [Gaw89]):

Example 6.19. Let $q \in \mathbb{C}$, $\tau \in \mathbb{C}$, $q = \exp(2\pi i\tau)$, $|q| < 1$, and $\Sigma_q = \{z \in \mathbb{C} \mid |q| \leq |z| \leq 1\}$ be the closed annulus with outer radius 1 and inner radius $|q|$. Let $g_1, g_2 \in$ Diff$_+$(S) be real analytic diffeomorphisms on the circle S. Then one gets the following parameterizations of the boundary curves of Σ_q:

$$p_1(e^{i\theta}) := qg_1(e^{i\theta}), \qquad p_2(e^{i\theta}) := g_2(e^{i\theta}).$$

The mentioned semi-group $\mathscr{E}$ is the quotient of $\mathscr{E}_0$, where $\mathscr{E}_0$ is the set of pairs (Σ, p') of Riemann surfaces Σ with exactly two boundary curves parameterized by $p' = (p'_1, p'_2)$, for which there is a $q \in \mathbb{C}$ and a biholomorphic map $\varphi : \Sigma_q \to \Sigma$ (where p_1, p_2 is a parameterization of $\partial\Sigma_q$ as above), so that $\varphi \circ p_j = p'_j$. As a set one has $\mathscr{E} = \mathscr{E}_0/\sim$, where $\sim$ means biholomorphic equivalence preserving the parameterization. The product of two equivalence classes $[(\Sigma, p')], [(\Sigma', p'')] \in \mathscr{E}$ is defined by "gluing" Σ and Σ', where we identify the outer boundary curve of Σ with the inner boundary curve of Σ' taking into account the parameterizations. The ansatz

$$A_{c,h}([\Sigma_q, p]) := const\, U_{c,h}(g_2^{-1})q\exp(\overline{\rho}_{c,h}(L_0))U_{c,h}(g_1)$$

leads to a projective representation of $\mathscr{E}$ using Theorem 6.18.

More general semi-groups can be obtained by looking at more general Riemann surfaces, that is compact Riemann surfaces with finitely many boundary curves, which are parameterized and divided into incoming ("in") and outgoing ("out") boundary curves. The semi-groups defined in this manner have unitary representations as well (cf. [Seg91], [Seg88b], and [GW85]). Starting with these observations, Segal has suggested an interesting set of axioms to describe conformal field theory (cf. [Seg88a]).

References

[BD04*]   A. Beilinson and V. Drinfeld. *Chiral Algebras*. AMS Colloquium Publications **51**, AMS, Providence, RI, 2004.

[BPZ84]   A.A. Belavin, A.M. Polyakov, and A.B. Zamolodchikov. In- finite conformal symmetry in two-dimensional quantum field theory. *Nucl. Phys.* **B 241** (1984), 333–380.

[CdG94]   F. Constantinescu and H.F. de Groote. *Geometrische und Algebraische Methoden der Physik: Supermannigfaltigkeiten und Virasoro-Algebren*. Teubner, Stuttgart, 1994.

[FFK89]   G. Felder, J. Fröhlich, and J. Keller. On the structure of unitary conformal field theory, I. Existence of conformal blocks. *Comm. Math. Phys.* **124** (1989), 417–463.

[FQS86]   D. Friedan, Z. Qiu, and S. Shenker. Details of the nonunitary proof for highest weight representations of the Virasoro algebra. *Comm. Math. Phys.* **107** (1986), 535–542.

[FS87]   D. Friedan and S. Shenker. The analytic geometry of two-dimensional conformal field theory. *Nucl. Phys.* **B 281** (1987), 509–545.

[Gaw89]   K. Gawedski. Conformal field theory. *Sém. Bourbaki 1988–89*, Astérisque **177–178** (no 704) (1989) 95–126.

[Gin89]   P. Ginsparg. *Introduction to Conformal Field Theory. Fields, Strings and Critical Phenomena*, Les Houches 1988, Elsevier, Amsterdam, 1989.

[GKO86]   P. Goddard, A. Kent, and D. Olive. Unitary representations of the Virasoro and Super-Virasoro algebras. *Comm. Math. Phys.* **103** (1986), 105–119.

[GO89]   P. Goddard and D. Olive. Kac-Moody and Virasoro algebras in relation to quantum mechanics. *Int. J. Mod. Phys.* **A1** (1989), 303–414.

[GR05*]   L. Guieu and C. Roger. *L'algèbre et le groupe de Virasoro: aspects géometriques et algébriques, généralisations*. Preprint, 2005.

[GW85]   R. Goodman and N.R. Wallach. Projective unitary positive-energy representations of Diff($\mathbb{S}$). *Funct. Anal.* **63** (1985), 299–321.

[Kac80]   V. Kac. Highest weight representations of infinite dimensional Lie algebras. In: *Proc. Intern. Congress Helsinki*, Acad. Sci. Fenn., 299–304, 1980.

[KR87]   V. Kac and A.K. Raina. *Highest Weight Representations of Infinite Dimensional Lie Algebras*. World Scientific, Singapore, 1987.

[MS89]   G. Moore and N. Seiberg. Classical and conformal field theory. *Comm. Math. Phys.* **123** (1989), 177–254.

[Seg88a] G. Segal. The definition of conformal field theory. Unpublished Manuscript, 1988. Reprinted in *Topology, Geometry and Quantum Field Theory*, U. Tillmann (Ed.), 432–574, Cambridge University Press, Cambridge, 2004.

[Seg88b] G. Segal. Two dimensional conformal field theories and modular functors. In: *Proc. IXth Intern. Congress Math. Phys.* Swansea, 22–37, 1988.

[Seg91]   G. Segal. Geometric aspects of quantum field theory. *Proc. Intern. Congress Kyoto 1990, Math. Soc.* Japan, 1387–1396, 1991.

Chapter 7
String Theory as a Conformal Field Theory

We give an exposition of the classical system of a bosonic string and its quantization.

In bosonic string theory as a classical field theory we have the flat semi-Riemannian manifold

$$(\mathbb{R}^D, \eta) \text{ with } \eta = \mathrm{diag}(-1, 1, \ldots, 1)$$

as background space and a *world sheet* in this space, that is a C^∞-parameterization

$$x : Q \to \mathbb{R}^D$$

of a surface $W = x(Q) \subset \mathbb{R}^D$, where $Q \subset \mathbb{R}^2$ is an open or closed rectangle. This corresponds to the idea of a one-dimensional object, the *string*, which moves in the space $\mathbb{R}^D$ and wipes out the two-dimensional surface $W = x(Q)$. The classical fields (that is the kinematic variables of the theory) are the components $x^\mu : Q \to \mathbb{R}$ of the parameterization $x = (x^0, x^1, \ldots, x^{D-1}) : Q \to \mathbb{R}^D$ of the surface $W = x(Q) \subset \mathbb{R}^D$,

7.1 Classical Action Functionals and Equations of Motion for Strings

In classical string theory the admissible parameterizations, that is the dynamic variables of the world sheet, are those for which a given action functional is stationary. A natural action of the classical field theory uses the "area" of the world sheet. One defines the so-called *Nambu–Goto action*:

$$S_{NG}(x) := -\kappa \int_Q \sqrt{-\det g}\, dq^0 dq^1,$$

with a constant $\kappa \in \mathbb{R}$ (the "string tension", cf. [GSW87]). Here,

$$g := x^* \eta, (x^* \eta)_{\mu\nu} = \eta_{ij} \partial_\mu x^i \partial_\nu x^j,$$

is the metric on Q induced by $x : Q \to \mathbb{R}^D$ and the variation is taken only over those parameterizations x, for which g is a *Lorentz metric* (at least in the interior of Q), that is

Schottenloher, M.: *String Theory as a Conformal Field Theory*. Lect. Notes Phys. **759**, 103–120 (2008)
DOI 10.1007/978-3-540-68628-6_8

$$\det(g_{\mu\nu}) < 0.$$

Hence, (Q, g) is a two-dimensional *Lorentz manifold*, that is a two-dimensional semi-Riemannian manifold with a Lorentz metric g.

From the action principle

$$\frac{d}{d\varepsilon} S_{NG}(x + \varepsilon y)|_{\varepsilon=0} = 0$$

with suitable boundary conditions, one derives the equations of motion. Since it is quite difficult to make calculations with respect to the action S_{NG}, one also uses a different action, which leads to the same equations of motion. The *Polyakov action*

$$S_P(x, h) := -\frac{\kappa}{2} \int_Q \sqrt{-\det h}\, h^{ij}\, g_{ij}\, dq^0\, dq^1$$

depends, in addition, on a (Lorentz) metric h on Q. A separate variation of S_P with respect to h only leads to the former action S_{NG}:

Lemma 7.1.

$$\frac{d}{d\varepsilon} S_P(x, h + \varepsilon f)|_{\varepsilon=0} = 0$$

holds precisely for those Lorentzian metrics h on Q which satisfy $g = \lambda h$, where $\lambda : Q \to \mathbb{R}_+$ is a smooth function. Substitution of $h = \frac{1}{\lambda} g$ into S_P yields the original action S_{NG}.

Proof. In order to show the first statement let $(\widetilde{h}^{ij})$ be the matrix satisfying

$$2 \det h = \widetilde{h}^{ij} h_{ij}, \quad h^{ij} = (\det h)^{-1} \widetilde{h}^{ij}.$$

Then $\widetilde{h}^{00} = h_{11}$, $\widetilde{h}^{11} = h_{00}$, and $\widetilde{h}^{01} = -h_{10}$. Hence,

$$\sqrt{-\det(h + \varepsilon f)}(h + \varepsilon f)^{ij} = -(\sqrt{-\det(h + \varepsilon f)})^{-1} \widetilde{(h + \varepsilon f)}^{ij}$$

for symmetric $f = (f_{ij})$ with $\det(h + \varepsilon f) < 0$, and it follows

$$S_P(x, h + \varepsilon f) = \frac{\kappa}{2} \int_Q (\sqrt{-\det(h + \varepsilon f)})^{-1} (\widetilde{h}^{ij} + \varepsilon \widetilde{f}^{ij}) g_{ij} dq^0 dq^1.$$

Since $h^{ij} = -(-\det h)^{-1} \widetilde{h}^{ij}$ and $\widetilde{h}^{\alpha\beta} f_{\alpha\beta} = \widetilde{f}^{\alpha\beta} h_{\alpha\beta}$, we have

$$\frac{\partial}{\partial \varepsilon} S_P(x, h + \varepsilon f) \bigg|_{\varepsilon=0}$$

$$= \frac{\kappa}{2} \int_Q \left(\frac{\widetilde{f}^{ij}}{\sqrt{-\det h}} + \frac{\widetilde{h}^{ij} \widetilde{f}^{\alpha\beta} h_{\alpha\beta}}{2\sqrt{-\det h}^3} \right) g_{ij} dq^0 dq^1$$

$$= \frac{\kappa}{2} \int_Q \frac{\widetilde{f}^{ij}}{\sqrt{-\det h}} \left(g_{ij} - \frac{1}{2} h^{\alpha\beta} g_{\alpha\beta} h_{ij} \right) dq^0 dq^1.$$

This implies that $\delta S_P(x,h) = 0$ for fixed x leads to the "equation of motion"

$$g_{ij} - \frac{1}{2}h^{\alpha\beta}g_{\alpha\beta}h_{ij} = 0 \tag{7.1}$$

for h. Equivalently, the *energy–momentum tensor*

$$T_{ij} := g_{ij} - \frac{1}{2}h^{\alpha\beta}g_{\alpha\beta}h_{ij} \tag{7.2}$$

has to vanish. The solution h of (7.1) is $g = \lambda h$ with

$$\lambda = \frac{1}{2}h^{\alpha\beta}g_{\alpha\beta} > 0$$

($\lambda > 0$ follows from $\det g < 0$ and $\det h < 0$).

Substitution of the solution $h = \frac{1}{\lambda}g$ of the equation $T = 0$ in the action $S_P(x,h)$ yields the original action $S_{NG}(x)$. □

Invariance of the Action. It is easy to show that the action S_P is invariant with respect to

- Poincaré transformations,
- Reparameterizations of the world sheet, and
- Weyl rescalings: $h \mapsto h' := \Omega^2 h$.

S_{NG} is invariant with respect to Poincaré transformations and reparameterizations only.

Because of the invariance with respect to reparameterizations, the action S_P can be simplified by a suitable choice of parameterization. To achieve this, we need the following theorem:

Theorem 7.2. *Every two-dimensional Lorentz manifold (M, g) is conformally flat, that is there are local parameterizations ψ, such that for the induced metric g one has*

$$\psi^* g = \Omega^2 \eta = \Omega^2 \begin{pmatrix} -1 & 0 \\ 0 & 1 \end{pmatrix} \tag{7.3}$$

with a smooth function Ω. Coordinates for which the metric tensor is of this form are called isothermal coordinates.

For a positive definite metric g (on a surface) the existence of isothermal coordinates can be derived from the solution of the Beltrami equation (cf. [DFN84, p. 110]). In the Lorentzian case the existence of isothermal coordinates is much easier to prove. Since the issue of existence of isothermal coordinates has been neglected in the respective literature and since it seems to have no relation to the Beltrami equation, a proof shall be provided in the sequel. A proof can also be found in [Dic89].

Proof. [1] Let $x \in M$ and let $\psi : \mathbb{R}^2 \supset U \to M$ be a chart for M with $x \in \psi(U)$. We denote the matrix representing $\psi^* g$ by $g_{\mu\nu} \in C^\infty(U, \mathbb{R})$. If we choose a suitable linear map $A \in GL(\mathbb{R}^2)$ and replace ψ with $\psi \circ A : A^{-1}(U) \to M$, we can assume that

---

[1] By A. Jochens

$$(g_{\mu\nu}(\xi)) = \eta = \begin{pmatrix} -1 & 0 \\ 0 & 1 \end{pmatrix},$$

where $\xi := \psi^{-1}(x)$. We also have

$$\det(g_{\mu\nu}) = g_{11}g_{22} - g_{12}^2 < 0$$

since g is a Lorentz metric. We define

$$a := \sqrt{g_{12}^2 - g_{11}g_{22}} \in C^\infty(U, \mathbb{R}).$$

By our choice of the chart ψ we have $g_{22}(\xi) = 1$. The continuity of g_{22} implies that there is an open neighborhood $V \subset U$ of ξ with $g_{22}(\xi') > 0$ for $\xi' \in V$.

Now, there are two positive integrating factors $\lambda, \mu \in C^\infty(V', \mathbb{R}^+)$ and two functions $F, G \in C^\infty(V', \mathbb{R})$ on an open neighborhood $V' \subset V$ of ξ, so that

$$\partial_1 F = \lambda \sqrt{g_{22}}, \qquad \partial_2 F = \lambda \frac{g_{12} + a}{\sqrt{g_{22}}},$$

$$\partial_1 G = \mu \sqrt{g_{22}}, \qquad \partial_2 G = \mu \frac{g_{12} - a}{\sqrt{g_{22}}}.$$

The existence of F and λ can be shown as follows: we apply to the function $f \in C^\infty(V, \mathbb{R})$ defined by

$$f(t, x) := (g_{12}(x, t) + a(x, t))/g_{22}(x, t)$$

a theorem of the theory of ordinary differential equations, which guarantees the existence of a family of solutions depending differentiably on the initial conditions (cf. [Die69, 10.8.1 and 10.8.2]). By this theorem, we get an open interval $J \subset \mathbb{R}$ and open subsets $U_0, U \subset \mathbb{R}$ with $\xi \in U_0 \times J \subset U \times J \subset V$, as well as a map $\phi \in C^\infty(J \times J \times U_0, U)$, so that for all $t, s \in J$ and $x \in U_0$ we have

$$\frac{d}{dt}\phi(t, s, x) = f(t, \phi(t, s, x)) \qquad \text{and} \qquad \phi(t, t, x) = x. \tag{7.4}$$

Using the uniqueness theorem for ordinary differential equations, it can be shown that $\partial_3 \phi$ is positive and that

$$\phi(\tau, t, x) \in U_0 \Rightarrow \phi(s, \tau, \phi(\tau, t, x)) = \phi(s, t, x)$$

for $t, s, \tau \in J$ and $x \in U_0$. Defining

$$F(x, t) := \phi(t_0, t, x) \qquad \text{and} \qquad \lambda(x, t) := \frac{\partial_1 F(x, t)}{\sqrt{g_{22}(x, t)}}$$

for $(x,t) \in U_0 \times J$ and a fixed $t_0 \in J$ we obtain functions $F, \lambda \in C^\infty(U_0 \times J, \mathbb{R})$ with the required properties. By the same argument we also obtain the functions G and μ. The open subset $V' \subset V$ is the intersection of the domains of F and G.

For the map $\varphi = \begin{pmatrix} \varphi^1 \\ \varphi^2 \end{pmatrix} := \begin{pmatrix} F - G \\ F + G \end{pmatrix} \in C^\infty(V', \mathbb{R}^2)$ we have

$$\partial_1 \varphi^1 = (\lambda - \mu)\sqrt{g_{22}}, \qquad \partial_2 \varphi^1 = \lambda \frac{g_{12} + a}{\sqrt{g_{22}}} - \mu \frac{g_{12} - a}{\sqrt{g_{22}}},$$

$$\partial_1 \varphi^2 = (\lambda + \mu)\sqrt{g_{22}}, \qquad \partial_2 \varphi^2 = \lambda \frac{g_{12} + a}{\sqrt{g_{22}}} + \mu \frac{g_{12} - a}{\sqrt{g_{22}}}.$$

After a short calculation we get

$$\partial_\mu \varphi^\rho \partial_\nu \varphi^\sigma \eta_{\rho\sigma} = \partial_\mu \varphi^1 \partial_\nu \varphi^1 - \partial_\mu \varphi^2 \partial_\nu \varphi^2 = 4\lambda \mu g_{\mu\nu},$$

that is $\varphi^* \eta = 4\lambda \mu \psi^* g$. Furthermore,

$$\det D\varphi = \partial_1 \varphi^1 \partial_2 \varphi^2 - \partial_1 \varphi^2 \partial_2 \varphi^1 = -4\lambda \mu a \neq 0.$$

Hence, by the inverse mapping theorem there exists an open neighborhood $W \subset V'$ of ξ, so that $\widetilde{\varphi} := \varphi|_W : W \to \varphi(W)$ is a C^∞ diffeomorphism. $\varphi^* \eta = 4\lambda \mu \psi^* g$ implies

$$\eta = \left(\widetilde{\varphi}^{-1}\right)^* \varphi^* \eta = 4\lambda \mu \left(\widetilde{\varphi}^{-1}\right)^* \psi^* g = 4\lambda \mu \left(\psi \circ \widetilde{\varphi}^{-1}\right)^* g.$$

Now $\widetilde{\psi} := \psi \circ \widetilde{\varphi}^{-1} : \varphi(W) \to M$ is a chart for M with $x \in \widetilde{\psi}(\varphi(W))$ and we have

$$\widetilde{\psi}^* g - \Omega^2 \eta$$

with $\Omega := 1/(2\sqrt{\lambda \mu})$. $\qquad \qquad \square$

By Theorem 7.2 one can choose a local parameterization of the world sheet in such a way that

$$h = \Omega^2 \eta = \Omega^2 \begin{pmatrix} -1 & 0 \\ 0 & 1 \end{pmatrix}.$$

This fixing of h is called *conformal gauge*. Even after conformal gauge fixing a residual symmetry remains: it is easy to see that $S_P(x)$ in conformal gauge is invariant with respect to conformal transformations on the world sheet. In this manner, the conformal group $\text{Conf}(\mathbb{R}^{1,1}) \cong \text{Diff}_+(\mathbb{S}) \times \text{Diff}_+(\mathbb{S})$ turns out to be a symmetry group of the system, even if this holds only on the level of "constraints". In any case, the classical field theory of the bosonic string can be viewed as a conformally invariant field theory.

To simplify the equations of motion and, furthermore, to present solutions as certain Fourier series, we need a generalization of Theorem 7.2, stating that (in the case of closed strings, to which we restrict our discussion here) there exists a conformal gauge not only in a neighborhood of any given point, but also in a neighborhood of a closed injective curve (as a starting curve for the "time $\tau = 0$"). The existence of such isothermal coordinates can be shown by the same argumentation as Theorem 7.2. Finally, for the variation in the conformal gauge, it can be assumed that isothermal coordinates exist on the rectangle

$$Q = [0, 2\pi] \times [0, 2\pi]$$

and that $\sigma \mapsto x(0, \sigma)$, $\sigma \in [0, 2\pi]$ describes a simple closed curve. This is possible at least up to an irrelevant distortion factor (cf. [Dic89]).

Theorem 7.3. *The variation of S_{NG} or S_P in the conformal gauge leads to the equations of motion on $Q = [0, 2\pi] \times [0, 2\pi]$: These are the two-dimensional wave equations*

$$\partial_0^2 x - \partial_1^2 x = 0 \quad resp. \quad x_{\tau\tau} - x_{\sigma\sigma} = 0$$

with the constraints

$$\langle x_\sigma, x_\tau \rangle = 0 = \langle x_\sigma, x_\sigma \rangle + \langle x_\tau, x_\tau \rangle, \quad \langle x_\tau, x_\tau \rangle < 0,$$

imposed by the conformal gauge.

By x_σ we denote the partial derivative of $x = x(\tau, \sigma)$ with respect to σ (that is $\tau := q^0$, $\sigma := q^1$), and $\langle v, w \rangle$ is the inner product $\langle v, w \rangle = v^\mu w^\nu \eta_{\mu\nu}$ for $v, w \in \mathbb{R}^D$.

Proof. To derive the equations of motion and the constraints we start by writing S_P in the conformal gauge $h = \Omega^2 \eta$ with $\sqrt{-\det h} = \Omega^2$ and $h^{ij} g_{ij} = \Omega^2(-g_{00} + g_{11})$:

$$S_P(x) = S_P(x, \Omega^2 \eta) = \frac{\kappa}{2} \int_Q ((\partial_0 x, \partial_0 x) - \langle \partial_1 x, \partial_1 x \rangle) dq^0 dq^1.$$

For $y : Q \to \mathbb{R}^D$ and suitable boundary conditions $y|_{\partial Q} = 0$ we have

$$\frac{\partial}{\partial \varepsilon} S_P(x + \varepsilon y) \Big|_{\varepsilon=0} = \kappa \int_Q (\langle \partial_0 x, \partial_0 y \rangle - \langle \partial_1 x, \partial_1 y \rangle) dq^0 dq^1$$

$$= \kappa \int_Q \langle \partial_{11} x - \partial_{00} x, y \rangle dq^0 dq^1$$

(integration by parts). This yields

$$\partial_{11} x - \partial_{00} x = 0$$

as the equations of motion in the conformal gauge.

Because of the description of the metric h by $h = \frac{1}{\lambda} g$ with $\lambda > 0$, that is

$$\lambda h = \lambda(h_{ij}) = \begin{pmatrix} \langle x_\tau, x_\tau \rangle & \langle x_\sigma, x_\tau \rangle \\ \langle x_\tau, x_\sigma \rangle & \langle x_\sigma, x_\sigma \rangle \end{pmatrix},$$

the gauge fixing $h = \Omega^2 \eta$ implies the conditions

$$\langle x_\sigma, x_\tau \rangle = 0, \quad \langle x_\sigma, x_\sigma \rangle = -\langle x_\tau, x_\tau \rangle > 0. \qquad \square$$

The constraints are equivalent to the vanishing of the energy–momentum T, which is given by

$$T_{ij} = \langle x_i, x_j \rangle - \frac{1}{2} h_{ij} h^{kl} \langle x_k, x_l \rangle, \quad i,j,k,l \in \{\tau, \sigma\}$$

(see (7.2) and cf. [GSW87, p. 62ff]).

The solutions of the two-dimensional wave equations are

$$x(\tau, \sigma) = x_R(\tau - \sigma) + x_L(\tau + \sigma)$$

with two arbitrary differentiable maps x_R and x_L on Q with values in $\mathbb{R}^D$. For the closed string we get on $Q := [0, 2\pi] \times [0, 2\pi]$ (that is $x(\tau, \sigma) = x(\tau, \sigma + 2\pi)$) the following Fourier series expansion:

$$x_R^\mu(\tau - \sigma) = \frac{1}{2} x_0^\mu + \frac{1}{4\pi\kappa} p_0^\mu(\tau - \sigma) + \frac{i}{\sqrt{4\pi\kappa}} \sum_{n \neq 0} \frac{1}{n} \alpha_n^\mu e^{-in(\tau - \sigma)},$$

$$x_L^\mu(\tau + \sigma) = \frac{1}{2} x_0^\mu + \frac{1}{4\pi\kappa} p_0^\mu(\tau + \sigma) + \frac{i}{\sqrt{4\pi\kappa}} \sum_{n \neq 0} \frac{1}{n} \overline{\alpha}_n^\mu e^{-in(\tau + \sigma)}. \quad (7.5)$$

x_0 and p_0 can be interpreted as the center of mass and the center of momentum, respectively, while $\alpha_n^\mu, \overline{\alpha}_n^\nu$ are the oscillator modes of the string. x_L and x_R are viewed as "left movers" and "right movers". We have $x_0^\mu, p_0^\mu \in \mathbb{R}$ and $\alpha_n^\mu, \overline{\alpha}_m^\nu \in \mathbb{C}$. $\overline{\alpha}_m^\nu$ is not the complex conjugate of α_m^ν, but completely independent of α_m^ν. For x_R and x_L to be real, it is necessary that

$$(\alpha_n^\mu)^* = (\alpha_{-n}^\mu) \quad \text{and} \quad (\overline{\alpha}_n^\mu)^* = (\overline{\alpha}_{-n}^\mu) \quad (7.6)$$

hold for all $\mu \in \{0, \ldots, D-1\}$ and $n \in \mathbb{Z} \setminus \{0\}$, where $c \mapsto c^*$ denotes the complex conjugation. We let $\alpha_0^\mu := \overline{\alpha}_0^\mu := \frac{1}{\sqrt{4\pi\kappa}} p_0^\mu$. The $x = x_L + x_R$ with (7.5) can be written as

$$x(\sigma, \tau) = x_0 + \frac{2}{\sqrt{4\pi\kappa}} \alpha_0 \tau + \frac{i}{\sqrt{4\pi\kappa}} \sum_{n \neq 0} \frac{1}{n} \left(\alpha_n e^{-in(\tau - \sigma)} + \overline{\alpha}_n e^{-in(\tau + \sigma)} \right).$$

Hence, arbitrary $\alpha_n, \overline{\alpha}_n, x_0, p_0$ with (7.6) yield solutions of the one-dimensional wave equation. In order that these solutions are, in fact, solutions of the equations of motion for the actions S_{NG} or S_P, they must, in addition, respect the conformal gauge. Using

$$L_n := \frac{1}{2} \sum_{k \in \mathbb{Z}} \langle \alpha_k, \alpha_{n-k} \rangle \quad \text{and} \quad \overline{L}_n := \frac{1}{2} \sum_{k \in \mathbb{Z}} \langle \overline{\alpha}_k, \overline{\alpha}_{n-k} \rangle \quad \text{for } n \in \mathbb{Z}, \quad (7.7)$$

the gauge condition can be expressed as follows:

Lemma 7.4. *A parameterization* $x(\tau,\sigma) = x_L(\tau - \sigma) + x_R(\tau + \sigma)$ *of the world sheet with* x_R, x_L *as in (7.5) and (7.6) gives isothermal coordinates if and only if* $L_n = \bar{L}_n = 0$ *for all* $n \in \mathbb{Z}$.

Proof. We have isothermal coordinates if and only if

$$\langle x_\tau + x_\sigma, x_\tau + x_\sigma \rangle = \langle x_\tau - x_\sigma, x_\tau - x_\sigma \rangle = 0.$$

Using the identities

$$x_\tau - x_\sigma = \frac{2}{\sqrt{4\pi\kappa}} \sum_{n \in \mathbb{Z}} \alpha_n e^{-in(\tau-\sigma)} \quad \text{and}$$

$$x_\tau + x_\sigma = \frac{2}{\sqrt{4\pi\kappa}} \sum_{n \in \mathbb{Z}} \bar{\alpha}_n e^{-in(\tau+\sigma)},$$

we get

$$\langle x_\tau - x_\sigma, x_\tau - x_\sigma \rangle = 0$$

$$\iff 0 = \left\langle \sum_{n \in \mathbb{Z}} \alpha_n e^{-in(\tau-\sigma)}, \sum_{n \in \mathbb{Z}} \alpha_n e^{-in(\tau-\sigma)} \right\rangle$$

$$\iff 0 = \sum_{n \in \mathbb{Z}} \sum_{k \in \mathbb{Z}} e^{-i(n+k)(\tau-\sigma)} \langle \alpha_n, \alpha_k \rangle$$

$$\iff 0 = \sum_{m \in \mathbb{Z}} \sum_{n+k=m} e^{-im(\tau-\sigma)} \langle \alpha_n, \alpha_k \rangle$$

$$\iff \forall m \in \mathbb{Z}: \sum_{n+k=m} \langle \alpha_n, \alpha_k \rangle = 0$$

$$\iff \forall m \in \mathbb{Z}: \sum_{k \in \mathbb{Z}} \langle \alpha_{m-k}, \alpha_k \rangle = 0$$

$$\iff \forall m \in \mathbb{Z}: L_m = 0.$$

The same argument holds for $x_\tau + x_\sigma$ and $\bar{L}_m$. $\square$

Altogether, we have the following:

Theorem 7.5. *The solutions of the string equations of motion are the functions*

$$x(\tau,\sigma) = x_0 + \frac{2}{\sqrt{4\pi\kappa}} \alpha_0 \tau + \frac{i}{\sqrt{4\pi\kappa}} \sum_{n \neq 0} \frac{1}{n} \left(\alpha_n e^{-in(\tau-\sigma)} + \bar{\alpha}_n e^{-in(\tau+\sigma)} \right),$$

for which the conditions (7.6) and $L_n = \bar{L}_n = 0$ *hold.*

For a connection of the energy–momentum tensor T of a conformal field theory with the Virasoro generators L_n and $\bar{L}_n$ we refer to (9.3) and to Sect. 10.5 in the context of conformal vertex operators.

The oscillator modes α_n^μ and $\bar{\alpha}_m^\nu$ are observables of the classical system. Obviously, they are constants of motion. Hence, one should try to quantize the $\alpha_n^\mu, \bar{\alpha}_m^\nu$.

In order to quantize the classical field theory of the bosonic string one needs the Poisson brackets of the classical system:

$$\{\alpha_m^\mu, \alpha_n^\nu\} = im\eta^{\mu\nu}\delta_{m+n} = \{\overline{\alpha}_m^\mu, \overline{\alpha}_n^\nu\}, \tag{7.8}$$

$$\{\alpha_m^\mu, \overline{\alpha}_n^\nu\} = 0, \tag{7.9}$$

$$\{p_0^\mu, x_0^\nu\} = \eta^{\mu\nu}, \tag{7.10}$$

$$\{x_0^\mu, x_0^\nu\} = \{x_0^\mu, \alpha_m^\nu\} = \{x_0^\mu, \overline{\alpha}_m^\nu\} = 0, \tag{7.11}$$

for all $\mu, \nu \in \{0, \ldots, D-1\}$ and $m, n \in \mathbb{Z}$ (here and in the following we set $4\pi\kappa = 1$).

Observe that for each single index ν the collection of the observables $\alpha_n^\nu, n \in \mathbb{Z}$, define a Lie algebra with respect to the Poisson bracket which is isomorphic to the Heisenberg algebra.

Lemma 7.6. *For $n, m \in \mathbb{Z}$ one has*

$$\{L_m, L_n\} = i(n-m)L_{m+n}, \quad \{\overline{L}_m, \overline{L}_n\} = i(n-m)\overline{L}_{m+n},$$

$$and \quad \{\overline{L}_m, L_n\} = 0.$$

This follows from the general formula

$$\{AB, C\} = A\{B, C\} + \{A, C\}B$$

for the Poisson bracket.

7.2 Canonical Quantization

In general, quantization of a classical system shall provide quantum models reflecting the basic properties of the original classical system. A common quantization procedure is *canonical quantization*. In canonical quantization a complex Hilbert space $\mathbb{H}$ has to be constructed in order to represent the quantum mechanical states as one-dimensional subspaces of $\mathbb{H}$ and to represent the observables as self-adjoint operators in $\mathbb{H}$. (The notion of a self-adjoint operator is briefly recalled on p. 130.) Thereby the relevant classical observables $f, g, \ldots$ have to be replaced with operators $\widehat{f}, \widehat{g}$ such that the Poisson bracket is preserved in the sense that it is replaced with the commutator of operators in $\mathbb{H}$

$$\{\cdot, \cdot\} \longmapsto -i[\cdot, \cdot].$$

Hence, for the relevant $f, g, \ldots$ the following relations should be satisfied on a common domain of definitions of the operators

$$[\widehat{f}, \widehat{g}] = -i\widehat{\{f, g\}}.$$

In addition, some natural identities have to be satisfied. For example, in the situation of the classical phase space $\mathbb{R}^{2n}$ with its Poisson structure on the space of observables $f : \mathbb{R}^{2n} \to \mathbb{C}$ induced by the natural symplectic structure on $\mathbb{R}^{2n}$ it is natural to require the *Dirac conditions*:

1. $\widehat{1} = \mathrm{id}_{\mathbb{H}}$,
2. $[\widehat{q^\mu}, \widehat{p_\nu}] = i\delta^\mu_\nu, [\widehat{q^\mu}, \widehat{q^\nu}] = [\widehat{p_\mu}, \widehat{p_\nu}] = 0$,

with respect to the standard canonical coordinates (q^μ, p_ν) of $\mathbb{R}^{2n}$.

In general, one cannot quantize all classical observables (due to a result of van Hove) and one chooses a suitable subset $\mathscr{A}$ which can be assumed to be a Lie algebra with respect to the Poisson bracket. The canonical quantization of this subalgebra $\mathscr{A}$ of the Poisson algebra of all observables means essentially to find a representation of $\mathscr{A}$ in the Hilbert space $\mathbb{H}$.

The Harmonic Oscillator. Let us present as an elementary example a canonical quantization of the one-dimensional harmonic oscillator. The classical phase space is $\mathbb{R}^2$ with coordinates (q, p). The Poisson bracket of two classical observables f, g, that is smooth functions $f, g : \mathbb{R}^2 \to \mathbb{C}$, is

$$\{f, g\} = \frac{\partial f}{\partial q}\frac{\partial g}{\partial p} - \frac{\partial f}{\partial p}\frac{\partial g}{\partial q}.$$

The hamiltonian function (that is the energy) of the harmonic oscillator is $h(q, p) = \frac{1}{2}(q^2 + p^2)$. The set of observables one wants to quantize contains at least four functions $1, p, q, h$. Because of $\{1, f\} = 0, \{q, p\} = 1, \{h, p\} = q$, and $\{h, q\} = -p$ the vector space $\mathscr{A}$ generated by $1, q, p, h$ is a Lie algebra with respect to the Poisson bracket.

As the Hilbert space of states one typically takes the space of square integrable functions $\mathbb{H} := L^2(\mathbb{R})$ in the variable q. The quantization of 1 is prescribed by the first Dirac condition. As the quantization of q one then chooses the *position operator* $\widehat{q} = Q$ defined by $\varphi(q) \mapsto q\varphi(q)$ with domain of definition $D_Q = \{\varphi \in \mathbb{H} : \int_{\mathbb{R}} |q\varphi(q)|^2 dq < \infty\}$. Q is an unbounded self-adjoint operator. This holds also for the *momentum operator* P which is the quantization of p: $P = \widehat{p}$. P is defined as $P(\varphi) = -i\frac{\partial\varphi(q)}{\partial q}$ for φ in the space D of all smooth functions on $\mathbb{R}$ with compact support and can be continued to D_P such that the continuation is self-adjoint. Observe that D is dense in $\mathbb{H}$. The second Dirac condition is satisfied on D, i.e

$$[Q, P]\varphi = i\varphi, \varphi \in D.$$

Finally, the quantization $\widehat{h}$ of the hamiltonian function h is the hamiltonian operator H, given by

$$H(\varphi) = \frac{1}{2}\left(\frac{\partial^2 \varphi}{\partial q^2}(q) + q^2\varphi(q)\right)$$

on D with domain D_H such that H is self-adjoint. It is easy to verify $[H, Q] = -iP, [H, P] = iQ$ on D from which we deduce $[\widehat{a}, \widehat{b}] = -i\widehat{\{a, b\}}$ for all $a, b \in \mathscr{A}$ on D.

Note that $\rho(a) := i\hat{a}$ defines a representation of $\mathscr{A}$ in $\mathbb{H}$.

A different realization of a canonical quantization of the harmonic oscillator is the following. The Hilbert space is the space $\mathbb{H} = \ell^2$ of complex sequence $z = (z_\nu)_{\nu \in \mathbb{N}}$ which are square summable $\|z\|^2 = \sum_{\nu=0}^{\infty} |z_\nu|^2 < \infty$. Let $(e_n)_{\nu \in \mathbb{N}}$ be the standard (Schauder) basis of ℓ^2, that is $e_n = (\delta_n^k)$. By

$$H(e_n) := (n + \frac{1}{2})e_n,$$
$$A^*(e_n) := \sqrt{2n+2}\,e_{n+1},$$
$$A(e_0) := 0, A(e_{n+1}) := \sqrt{2n+2}\,e_n,$$

we define operators H, A, A^* on the subspace $D \subset \mathbb{H}$ of finite sequences, that is finite linear combinations of the e_ns. H is an essentially self-adjoint operator and A^* is the adjoint of A as the notation already suggests. (More precisely, A and A^* are the restrictions to D of operators which are adjoint to each other.)

With $Q := \frac{1}{2}(A + A^*)$ and $P := \frac{1}{2}(A - A^*)$ the operators $\mathrm{id}_{\mathbb{H}}, Q, P, H$ satisfy in D the same commutation relations

$$[Q, P] = i \, \mathrm{id}_{\mathbb{H}}, [H, Q] = -iP, [H, P] = iQ$$

as before, and therefore constitute another canonical quantization of $\mathscr{A}$. The two quantizations are equivalent.

Note that D can be identified with the space of complex-valued polynomials $\mathbb{C}[T]$ by $e_n \mapsto T^n$. This opens the possibility to purely algebraic methods in quantum field theory by restricting all operations to the vector space $D = \mathbb{C}[T]$ as, e.g., in the quantization of strings (see below), in the representation of the Virasoro algebra (cf. Sect. 6.5), or in the theory of vertex operators (cf. Chap. 10).

For obvious reasons, A is called the *annihilation operator* and A^* is called *creation operator*.

Returning to the question of quantizing a string one observes immediately that for any fixed index μ the Poisson brackets of the (α_m^μ) are those of an infinite sequence of one-dimensional harmonic oscillators (up to a constant). The corresponding *oscillator algebra* $\mathscr{A}$ generated by (α_m^μ) (with fixed μ) can therefore be interpreted as the algebra of an infinite dimensional harmonic oscillator. For a fixed index $\mu > 0$ (which we omit for the rest of this section) the relevant Poisson brackets of the oscillator algebra $\mathscr{A}$ are, according to (7.8),

$$\{\alpha_m, \alpha_n\} = im\delta_{n+m}, \{1, \alpha_n\} = 0.$$

After quantization the operators $a_n := \widehat{\alpha_n}$ generate a Lie algebra which is the complex vector space generated by $a_n, n \in \mathbb{Z}$, and Z (sometimes denoted $Z = 1$) with the Lie bracket given by

$$[a_m, a_n] = m\delta_{n+m}Z, \quad [Z, a_m] = 0.$$

We see that this Lie algebra is nothing else than the Heisenberg algebra H (cf. (4.1)).

We conclude that constructing a canonical quantization of the infinite dimensional harmonic oscillator is the same as finding a representation $\rho : H \to \text{End } D$ of the Heisenberg algebra H in a suitable dense subspace $D \subset \mathbb{H}$ of a Hilbert space $\mathbb{H}$ with $\rho(Z) = \text{id}_{\mathbb{H}}$.

Fock Space Representation. As the appropriate Fock space (that is representation space) we choose the complex vector space

$$S := \mathbb{C}[T_1, T_2, \ldots] \tag{7.12}$$

of polynomials in an infinite number of variables. We have to find a representation of the Heisenberg algebra in $\text{End}_{\mathbb{C}} S$. Define

$$\rho(a_n) := \frac{\partial}{\partial T_n} \quad \text{for } n > 0,$$
$$\rho(a_0) := \mu \text{id}_S \quad \text{where } \mu \in \mathbb{C},$$
$$\rho(a_{-n}) := nT_n \quad \text{for } n > 0, \quad \text{and}$$
$$\rho(Z) := \text{id}_S.$$

Then the commutation relations obviously hold and the representation is irreducible. Moreover, it is a unitary representation in the following sense:

Lemma 7.7. *For each $\mu \in \mathbb{R}$ there is a unique positive definite hermitian form on S, so that $H(1,1) = 1$ (1 stands for the vacuum vector) and*

$$H(\rho(a_n)f, g) = H(f, \rho(a_{-n})g)$$

for all $f, g \in S$ and $n \in \mathbb{Z}$, $n \neq 0$.

Proof. First of all one sees that distinct monomials $f, g \in S$ have to be orthogonal for such a hermitian form H on S. (The *monomials* are the polynomials of the form $T_{n_1}^{k_1} T_{n_2}^{k_2} \ldots T_{n_r}^{k_r}$ with $n_j, k_j \in \mathbb{N}$ for $j = 1, 2, \ldots, r$.) Given two distinct monomials f, g there exist an index $n \in \mathbb{N}$ and exponents $k \neq l$, $k, l \geq 0$, such that $f = T_n^k f_1, g = T_n^l g_1$ for suitable monomials f_1, g_1 which are independent of T_n. Without loss of generality let $k < l$. Then

$$H((\rho(a_n))^{k+1}f, T_n^{l-k-1}g_1) = H((\frac{\partial}{\partial T_n})^{k+1} T_n^k f_1, T_n^{l-k-1}g_1)$$
$$= H(0, T_n^{l-k-1}g_1)$$
$$= 0$$

and

$$H((\rho(a_n))^{k+1}f, T_n^{l-k-1}g_1) = H(f, (\rho(a_{-n})^{k+1} T_n^{l-k-1}g_1))$$
$$= H(f, n^{k+1} T_n^l g_1)$$
$$= H(f, g)$$

imply $H(f,g) = 0$. Moreover,

$$
\begin{aligned}
H(f,f) &= H(f, n^{-k}(\rho(a_n))^k f_1) \\
&= n^{-k} H(\rho(a_n)^k T_n^k f_1, f_1) \\
&= \frac{k!}{n^k} H(f_1, f_1).
\end{aligned}
$$

Using $H(1,1) = 1$, it follows for monomials $f = T_{n_1}^{k_1} T_{n_2}^{k_2} \ldots T_{n_r}^{k_r}$ with $n_1 < n_2 < \ldots < n_r$

$$
H(f,f) = \frac{k_1! k_2! \ldots k_r!}{n^{k_1} n^{k_2} \ldots n^{k_r}}. \tag{7.13}
$$

Since the monomials constitute a (Hamel) basis of S, H is uniquely determined as a positive definite hermitian form by (7.13) and the orthogonality condition. Reversing the arguments, by using (7.13) and the orthogonality condition $H(f,g) = 0$ for distinct monomials $f, g \in$ S as a definition for H, one obtains a hermitian form H on S with the required properties.   □

Note that $\rho(a_n)^* = \rho(a_{-n})$ by the last result and for each $n > 0$ the operator $\rho(a_n)$ is an annihilation operator while $\rho(a_n)^*$ is a creation operator.

7.3 Fock Space Representation of the Virasoro Algebra

In order to obtain a representation of the Virasoro algebra Vir on the basis of the Fock space representation $\rho : $ H $\to$ End(S) of the Heisenberg algebra described in the last section it seems to be natural to use the definition of the Virasoro observables L_n in classical string theory, cf. (7.7),

$$
L_n = \frac{1}{2} \sum_{k \in \mathbb{Z}} \alpha_k \alpha_{n-k} = \frac{1}{2} \sum_{k \in \mathbb{Z}} \alpha_{n-k} \alpha_k,
$$

which satisfy the Witt relations (up to the constant i, see Lemma 7.6).

In a first naive attempt one could try to define the operators $L_n : $ S $\to$ S by $L_n = \frac{1}{2} \sum_{k \in \mathbb{Z}} a_k a_{n-k}$ resp. $L_n = \frac{1}{2} \sum_{k \in \mathbb{Z}} \rho(a_k) \rho(a_{n-k})$. But this procedure is not well-defined on S, since

$$
\rho(a_k) \rho(a_{n-k}) \neq \rho(a_{n-k}) \rho(a_k),
$$

in general.

However, the *normal ordering*

$$
:\rho(a_i)\rho(a_j): := \begin{cases} \rho(a_i)\rho(a_j) & \text{for } i \leq j \\ \rho(a_j)\rho(a_i) & \text{for } i > j \end{cases}
$$

defines operators

$$
\rho(L_n) : \text{S} \to \text{S}, \quad \rho(L_n) := \frac{1}{2} \sum_{k \in \mathbb{Z}} :\rho(a_k)\rho(a_{n-k}):.
$$

The $\rho(L_m)$ are well-defined operators, since the application to an arbitrary polynomial $P \in S = \mathbb{C}[T_1, T_2, \ldots]$ yields only a finite number of nonzero terms. The normal ordering constitutes a difference compared to the classical summation for the case $n = 0$ only. This follows from

$$\rho(a_i)\rho(a_j) = \rho(a_j)\rho(a_i) \quad \text{for} \quad i+j \neq 0,$$

$$:\rho(a_k)\rho(a_{-k}): = \rho(a_{-k})\rho(a_k) \quad \text{for} \quad k \in \mathbb{N}.$$

Consequently, the operators $\rho(L_n)$ can be represented as

$$\rho(L_0) = \frac{1}{2}\rho(a_0)^2 + \sum_{k \in \mathbb{N}_1} \rho(a_{-k})\rho(a_k),$$

$$\rho(L_{2m}) = \frac{1}{2}(\rho(a_m))^2 + \sum_{k \in \mathbb{N}_1} \rho(a_{m-k})\rho(a_{m+k}),$$

$$\rho(L_{2m+1}) = \sum_{k \in \mathbb{N}_0} \rho(a_{m-k})\rho(a_{m+k+1}),$$

for $m \in \mathbb{N}_0$ (here $\mathbb{N}_k = \{n \in \mathbb{Z} : n \geq k\}$).

We encounter normal ordering as an important tool in a more general context in Chap. 10 on vertex algebras.

Theorem 7.8. *In the Fock space representation we have*

$$[L_n, L_m] = (n-m)L_{n+m} + \frac{n}{12}(n^2 - 1)\delta_{n+m}\mathrm{id}$$

(with L_n instead of $\rho(L_n)$). Hence, it is a representation of the Virasoro algebra.

Proof. First of all we show

$$[L_n, a_m] = -m a_{m+n}, \tag{7.14}$$

where $m, n \in \mathbb{Z}$, using the commutation relations for the a_ns. (Here and in the following we write L_n instead of $\rho(L_n)$ and a_n instead of $\rho(a_n)$.) Let $n \neq 0$.

$$L_n a_m = \frac{1}{2} \sum_{k \in \mathbb{Z}} a_{n-k} a_k a_m$$

$$= \frac{1}{2} \sum_{k \in \mathbb{Z}} a_{n-k}(a_m a_k + k\delta_{k+m})$$

$$= \frac{1}{2} \sum_{k \in \mathbb{Z}} ((a_m a_{n-k} + (n-k)\delta_{n+m-k})a_k + k\delta_{k+m}a_{n-k})$$

$$= a_m L_n + \frac{1}{2}(-m a_{n+m} - m a_{n+m})$$

$$= a_m L_n - m a_{n+m}.$$

The case $n = 0$ is similar. From $[L_n, a_m] = -ma_{n+m}$ one can deduce

$$[[L_n, L_m], a_k] = -k(n-m)a_{n+m+k}. \tag{7.15}$$

In fact,

$$
\begin{aligned}
L_n L_m a_k &= L_n(a_k L_m - ka_{m+k}) \\
&= a_k L_n L_m - ka_{n+k} L_m - kL_n a_{m+k}.
\end{aligned}
$$

Hence,

$$
\begin{aligned}
[L_n, L_m] a_k &= a_k[L_n, L_m] + k[L_m, a_{n+k}] - k[L_n, a_{m+k}] \\
&= a_k[L_n, L_m] - k(n+k)a_{m+n+k} + k(m+k)a_{m+n+k} \\
&= a_k[L_n, L_m] - k(n-m)a_{n+m+k}.
\end{aligned}
$$

It is now easy to deduce from (7.14) and (7.15) that for every $f \in S$ with

$$[L_n, L_m]f = (n-m)L_{n+m}f + \frac{n}{12}(n^2 - 1)\delta_{n+m}f$$

and every $k \in \mathbb{Z}$ we have

$$[L_n, L_m](a_k f) = (n-m)L_{n+m}(a_k f) + \frac{n}{12}(n^2 - 1)\delta_{n+m}(a_k f).$$

As a consequence, the commutation relation we want to prove has only to be checked on the vacuum vector $\Omega = 1 \in S$. The interesting case is to calculate $[L_n, L_{-n}]\Omega$. Let $n > 0$. Then $L_n \Omega = 0$. Hence $[L_n, L_{-n}]\Omega = L_n L_{-n}\Omega$. In case of $n = 2m + 1$ we obtain

$$
\begin{aligned}
L_{-n}\Omega &= \frac{1}{2}\sum_{k \in \mathbb{Z}} a_{-n-k}a_k \Omega \\
&= \frac{1}{2}\sum_{k \in \mathbb{Z}} a_{-n+k}a_{-k} \Omega \\
&= \frac{1}{2}\sum_{k=0}^{n} a_{-n+k}a_{-k} \Omega \\
&= \mu n T_n + \frac{1}{2}\sum_{k=1}^{n-1} k(n-k)T_k T_{n-k} \\
&= \mu n T_n + \sum_{k=1}^{m} k(n-k)T_k T_{n-k} =: P_n.
\end{aligned}
$$

Now, $a_l a_{n-l} P_n \neq 0$ holds for $l \in \{0, 1, \ldots n\}$ only and we infer $a_l a_{n-l} P_n = l(n-l), 1 \leq l \leq n-1$, and $a_l a_{n-l} P_n = \mu^2 n$ for $l = 0, l = n$. It follows that

$$[L_n, L_{-n}]\Omega = \mu^2 n + \sum_{k=1}^{m} k(n-k)$$

$$= 2nL_0\Omega + n\sum_{k=1}^{m} k - \sum_{k=1}^{m} k^2$$

$$= 2nL_0\Omega + n\frac{m}{2}(m+1) - \frac{1}{6}m(m+1)(2m+1)$$

$$= 2nL_0\Omega + \frac{n}{3}m(m+1)$$

$$= 2nL_0\Omega + \frac{n}{12}(n^2 - 1).$$

The case $n = 2m$ can be treated in the same manner. Similarly, one checks that $[L_n, L_m]\Omega = (n-m)L_{n+m}$ for the relatively simple case $n + m \neq 0$.   □

Another proof can be found, for instance, in [KR87, p. 15ff]. Here, we wanted to demonstrate the impact of the commutation relations of the Heisenberg algebra respectively the oscillator algebra $\mathscr{A}$.

Corollary 7.9. *The representation of Theorem 7.8 yields a positive definite unitary highest-weight representation of the Virasoro algebra with the highest weight $c = 1, h = \frac{1}{2}\mu^2$ (cf. Chap. 6).*

Proof. For the highest-weight vector $v_0 := 1$ let

$$V := \mathrm{span}_\mathbb{C}\{L_n v_0 : n \in \mathbb{Z}\}.$$

Then the restrictions of $\rho(L_n)$ to the subspace $V \subset \mathsf{S}$ of S define a highest-weight representation of Vir with highest weight $(1, \frac{1}{2}\mu^2)$ and Virasoro module V.   □

Remark 7.10. In most cases one has $\mathsf{S} = V$. But this does not hold for $\mu = 0$, for instance.

More unitary highest-weight representations can be found by taking tensor products: for $f \otimes g \in V \otimes V$ let

$$(\rho \otimes \rho)(L_n)(f \otimes g) := (\rho(L_n)f) \otimes g + f \otimes (\rho(L_n)g).$$

As a simple consequence one gets

Theorem 7.11. $\rho \otimes \rho : \mathrm{Vir} \to \mathrm{End}_\mathbb{C}(V \otimes V)$ *is a positive definite unitary highest-weight representation for the highest weight $c = 2, h = \mu^2$. By iteration of this procedure one gets unitary highest-weight representations for every weight (c, h) with $c \in \mathbb{N}_1$ and $h \in \mathbb{R}_+$.*

For the physics of strings, these representations resp. quantizations are not sufficient, since only some of the important observables are represented. It is our aim in this section, however, to present a straightforward construction of a unitary Verma module with $c > 1$ and $h \geq 0$ for the discussion in Chap. 6 based on quantization. Indeed, the starting point was the attempt of quantizing string theory. But for the construction of the Verma module only the Fock space representation of the Heisenberg algebra as the algebra of the infinite dimensional harmonic oscillator was used by restricting to one single coordinate.

We now come back to strings in taking care of all coordinates $x^\mu, \mu \in \{0, 1, \ldots d - 1\}$.

7.4 Quantization of Strings

In (non-compactified bosonic) string theory, the Poisson algebra

$$\mathscr{A} := \mathbb{C}1 \oplus \bigoplus_{\mu=0}^{D-1}(\mathbb{C}x_0^\mu \oplus \mathbb{C}p_0^\mu) \oplus \bigoplus_{\mu=0}^{D-1}\bigoplus_{m \neq 0}(\mathbb{C}\alpha_m^\mu)$$

of the classical oscillator modes and of the coordinates x_0^μ, p_0^ν has to be quantized. (See (7.8) for their Poisson brackets.) Equivalently, one has to find a representation of the string algebra

$$\mathscr{L} := \mathbb{C}1 \oplus \bigoplus_{\mu=0}^{D-1}(\widehat{\mathbb{C}x_0^\mu} \oplus \widehat{\mathbb{C}p_0^\mu}) \oplus \bigoplus_{\mu=0}^{D-1}\bigoplus_{m \neq 0}(\mathbb{C}\alpha_m^\mu)$$

with the following Lie brackets

$$\{a_m^\mu, a_n^\nu\} = m\eta^{\mu\nu}\delta_{m+n},$$

$$\{\widehat{p_0^\mu}, \widehat{x_0^\nu}\} = -i\eta^{\mu\nu},$$

$$\{\widehat{x_0^\mu}, \widehat{x_0^\nu}\} = \{\widehat{x_0^\mu}, a_m^\nu\} = 0,$$

according to (7.8).

The corresponding Fock space is

$$\mathsf{S} := \mathbb{C}[T_n^\mu : n \in \mathbb{N}_0, \mu = 0, \ldots, D-1]$$

and the respective representation is given by

$$\rho(a_m) := \eta^{\mu\nu}\frac{\partial}{\partial T_m^\nu} \qquad \text{for } m > 0,$$

$$P^\mu := \rho(a_0^\mu) := i\eta^{\mu\nu}\frac{\partial}{\partial T_0^\mu} \qquad (\alpha_0^\mu = p_0^\mu \text{ if } 4\pi\kappa = 1),$$

$$\rho(a_{-m}^\mu) := mT_m^\mu \qquad \text{for } m > 0,$$

$$Q^\mu := \rho(\widehat{x_0^\mu}) := T_0^\mu.$$

The natural hermitian form on S with $H(1,1) = 1$ and

$$H(\rho(\alpha_m^\mu)f, g) = H(f, \rho(\alpha_{-m}^\mu)g)$$

is no longer positive semi-definite. For instance,

$$
\begin{aligned}
H(T_1^0, T_1^0) &= H(\alpha_{-1}^0 1, \alpha_{-1}^0 1) = H(1, \alpha_1^0 \alpha_{-1}^0 1) \\
&= H(1, [\alpha_1^0, \alpha_{-1}^0]1) = H(1, -1) \\
&= -1.
\end{aligned}
$$

Moreover, this representation does not respect the gauge conditions $L_n = 0$. A solution of both problems is provided by the so-called "no-ghost theorem" (cf. [GSW87]). It essentially states that taking into account the gauge conditions $L_n = 0$, $n > 0$, the representation becomes unitary for the dimension $D = 26$. This means that the restriction of the hermitian form to the space of "physical states"

$$\mathscr{P} := \{f \in S : L_n f = 0 \text{ for all } n > 0, L_0 f = f\}$$

is positive semi-definite ($D = 26$). A proof of the no-ghost theorem using the Kac determinant can be found in [Tho84].

References

[DFN84]   B.A. Dubrovin, A.T. Fomenko, and S.P. Novikov. *Modern Geometry – Methods and Applications I*. Springer-Verlag, Berlin, 1984.

[Dic89]   R. Dick. *Conformal Gauge Fixing in Minkowski Space*. Letters in Mathematical Physics **18**, Springer, Dordrecht (1989), 67–76.

[Die69]   J. Dieudonné. *Foundations of Modern Analysis, Volume 10-1*. Academic Press, New York-London, 1969.

[GSW87]   M.B. Green, J.H. Schwarz, and E. Witten. *Superstring Theory, Volume 1*. Cambridge University Press, Cambridge, 1987.

[KR87]   V. Kac and A.K. Raina. *Highest Weight Representations of Infinite Dimensional Lie Algebras*. World Scientific, Singapore, 1987.

[Tho84]   C.B. Thorn. A proof of the no-ghost theorem using the Kac determinant. In: *Vertex Operators in Mathematics and Physics*, Lepowsky et al. (Eds.), 411–417. Springer Verlag, Berlin, 1984.

Chapter 8
Axioms of Relativistic Quantum Field Theory

Although quantum field theories have been developed and used for more than 70 years a generally accepted and rigorous description of the structure of quantum field theories does not exist. In many instances quantum field theory is approached by quantizing classical field theories as for example elaborated in the last chapter on strings. A more systematic specification uses axioms. We present in Sect. 8.3 the system of axioms which has been formulated by Arthur Wightman in the early 1950s. This chapter follows partly the thorough exposition of the subject in [SW64*]. In addition, we have used [Simo74*], [BLT75*], [Haa93*], as well as [OS73] and [OS75].

The presentation of axiomatic quantum field theory in this chapter serves several purposes:

- It gives a general motivation for the axioms of two-dimensional conformal field theory in the Euclidean setting which we introduce in the next chapter.
- It explains in particular the transition from Minkowski spacetime to Euclidean spacetime (Wick rotation) and thereby the transition from relativistic quantum field theory to Euclidean quantum field theory (cf. Sect. 8.5).
- It explains the equivalence of the two descriptions of a quantum field theory using either the fields (as operator-valued distributions) or the correlation functions (resp. correlation distributions) as the main objects of the respective system (cf. Sect. 8.4).
- It motivates how the requirement of conformal invariance in addition to the Poincaré invariance leads to the concept of a vertex algebra.
- It points out important work which is known already for about 50 years and still leads to many basic open problems like one out of the seven millennium problems (cf. the article of Jaffe and Witten [JW06*]).
- It gives the opportunity to describe the general framework of quantum field theory and to introduce some concepts and results on distributions and functional analysis (cf. Sect. 8.1).

The results from functional analysis and distributions needed in this chapter can be found in most of the corresponding textbooks, e.g., in [Rud73*] or [RS80*].

First of all, we recall some aspects of distribution theory in order to present a precise concept of a quantum field.

Schottenloher, M.: *Axioms of Relativistic Quantum Field Theory*. Lect. Notes Phys. **759**, 121–152 (2008)
DOI 10.1007/978-3-540-68628-6_9 © Springer-Verlag Berlin Heidelberg 2008

8.1 Distributions

A quantum field theory consists of quantum states and quantum fields with various properties. The quantum states are represented by the lines through 0 (resp. by the rays) of a separable complex Hilbert space $\mathbb{H}$, that is by points in the associated projective space $\mathbb{P} = \mathbb{P}(\mathbb{H})$ and the observables of the quantum theory are the self-adjoint operators in $\mathbb{H}$.

In a direct analogy to classical fields one is tempted to understand quantum fields as maps on the configuration space $\mathbb{R}^{1,3}$ or on more general spacetime manifolds M with values in the set of self-adjoint operators in $\mathbb{H}$. However, one needs more general objects, the quantum fields have to be operator-valued distributions. We therefore recall in this section the concept of a distribution with a couple of results in order to introduce the concept of a quantum field or field operator in the next section.

Distributions. Let $\mathscr{S}(\mathbb{R}^n)$ be the *Schwartz space* of *rapidly decreasing smooth functions*, that is the complex vector space of all functions $f : \mathbb{R}^n \to \mathbb{C}$ with continuous partial derivatives of any order for which

$$|f|_{p,k} := \sup_{|\alpha| \leq p} \sup_{x \in \mathbb{R}^n} |\partial^\alpha f(x)| (1 + |x|^2)^k < \infty, \qquad (8.1)$$

for all $p, k \in \mathbb{N}$. (∂^α is the partial derivative for the multi-index $\alpha = (\alpha_1, \ldots, \alpha_n) \in \mathbb{N}^n$ with respect to the usual cartesian coordinates $x = (x^1, x^2, \ldots, x^n)$ in $\mathbb{R}^n$.)

The elements of $\mathscr{S} = \mathscr{S}(\mathbb{R}^n)$ are the *test functions* and the dual space contains the (tempered) distributions.

Observe that (8.1) defines seminorms $f \mapsto |f|_{p,k}$ on $\mathscr{S}$.

Definition 8.1. A *tempered distribution* T is a linear functional $T : \mathscr{S} \to \mathbb{C}$ which is continuous with respect to all the seminorms $|\ |_{p,k}$ defined in (8.1), $p, k \in \mathbb{N}$.

Consequently, a linear $T : \mathscr{S} \to \mathbb{C}$ is a tempered distribution if for each sequence (f_j) of test functions which converges to $f \in \mathscr{S}$ in the sense that

$$\lim_{j \to \infty} |f_j - f|_{p,k} = 0 \quad \text{for all } p, k \in \mathbb{N},$$

the corresponding sequence $(T(f_j))$ of complex numbers converges to $T(f)$. Equivalently, a linear $T : \mathscr{S} \to \mathbb{C}$ is continuous if it is bounded, that is there are $p, k \in \mathbb{N}$ and $C \in \mathbb{R}$ such that

$$|T(f)| \leq C|f|_{p,k}$$

for all $f \in \mathscr{S}$.

The vector space of tempered distributions is denoted by $\mathscr{S}' = \mathscr{S}'(\mathbb{R}^n)$. $\mathscr{S}'$ will be endowed with the topology of uniform convergence on all the compact subsets of $\mathscr{S}$. Since we only consider tempered distributions in these notes we often call a tempered distribution simply a distribution in the sequel.

Some distributions are represented by functions, for example for an arbitrary measurable and bounded function g on $\mathbb{R}^n$ the functional

$$T_g(f) := \int_{\mathbb{R}^n} g(x)f(x)dx, f \in \mathscr{S},$$

defines a distribution. A well-known distribution which cannot be represented as a distribution of the form T_g for a function g on $\mathbb{R}^n$ is the *delta distribution*

$$\delta_y : \mathscr{S} \to \mathbb{C}, f \mapsto f(y),$$

the evaluation at $y \in \mathbb{R}^n$. Nevertheless, δ_y is called frequently the delta function at y and one writes $\delta_y = \delta(x-y)$ in order to use the formal integral

$$\delta_y(f) = f(y) = \int_{\mathbb{R}^n} \delta(x-y)f(x)dx.$$

Here, the right-hand side of the equation is defined by the left-hand side. Distributions T have derivatives. For example

$$\frac{\partial}{\partial q^j}T(f) := -T(\frac{\partial}{\partial q^j}f),$$

and $\partial^\alpha T$ is defined by

$$\partial^\alpha T(f) := (-1)^{|\alpha|}T(\partial^\alpha f), f \in \mathscr{S}.$$

By using partial integration one obtains $\partial^\alpha T_g = T_{\partial^\alpha g}$ if g is differentiable and suitably bounded.

An important example in the case of $n = 1$ is $T_H(f) := \int_0^\infty f(x)dx$, $f \in \mathscr{S}$, with

$$\frac{d}{dt}T(f) = -\int_0^\infty f'(x)dx = f(0) = \delta_0(f).$$

We observe that the delta distribution δ_0 has a representation as the derivative of a function (the *Heaviside function* $H(x) = \chi_{[0,\infty[}$) although δ_0 is not a true function. This fact has the following generalization:

Proposition 8.2. *Every tempered distribution* $T \in \mathscr{S}'$ *has a representation as a finite sum of derivatives of continuous functions of polynomial growth, that is there exist* $g_\alpha : \mathbb{R}^n \to \mathbb{C}$ *such that*

$$T = \sum_{0 \le |\alpha| \le k} \partial^\alpha T_{g_\alpha}.$$

Partial Differential Equations. Since a distribution possesses partial derivatives of arbitrary order it is possible to regard partial differential equations as equations for distributions and not only for differentiable functions. Distributional solutions in general lead to results for true functions. This idea works especially well in the case of partial differential equations with constant coefficients.

For a polynomial $P(X) = c_\alpha X^\alpha \in \mathbb{C}[X_1, \ldots, X_n]$ in n variables with complex coefficients $c_\alpha \in \mathbb{C}$ one obtains the partial differential operator

$$P(-i\partial) = c_\alpha(-i\partial)^\alpha = \sum c_{(\alpha_1,\dots,\alpha_n)}\partial_1^{\alpha_1}\dots\partial_n^{\alpha_n},$$

and the corresponding inhomogeneous partial differential equation

$$P(-i\partial)u = v,$$

which is meaningful for functions as well as for distributions. As an example, the basic partial differential operator determined by the geometry of the Euclidean space $\mathbb{R}^n = \mathbb{R}^{n,0}$ is the Laplace operator

$$\Delta = \partial_1^2 + \dots + \partial_n^2,$$

with $\Delta = P(-i\partial)$ for $P = -(X_1^2 + \dots + X_n^2)$.

In the same way, the basic partial differential operator determined by the geometry of the Minkowski space $\mathbb{R}^{1,D-1}$ is the wave operator (the Laplace–Beltrami operator with respect to the Minkowski-metric, cf. 1.6)

$$\Box = \partial_0{}^2 - (\partial_1^2 + \dots + \partial_{D-1}^2) = \partial_0^2 - \Delta,$$

and $\Box = P(-i\partial)$ for $P = -X_0^2 + X_1^2 + \dots + X_{D-1}^2$.

A *fundamental solution* of the partial differential equation $P(-i\partial)u = v$ is any distribution G satisfying

$$P(-i\partial)G = \delta.$$

Proposition 8.3. *Such a fundamental solution provides solutions of the inhomogeneous partial differential equation $P(-i\partial)u = v$ by convolution of G with v:*

$$P(-i\partial)(G * v) = v.$$

Proof. Here, the *convolution* of two rapidly decreasing smooth functions $u, v \in \mathscr{S}$, is defined by

$$u * v(x) := \int_{\mathbb{R}^n} u(y)v(x-y)dy = \int_{\mathbb{R}^n} u(x-y)v(y)dy.$$

The identity $\partial_j(u*v) = (\partial_j u) * v = u * \partial_j v$ holds. The convolution is extended to the case of a distribution $T \in \mathscr{S}'$ by $T * v(u) := T(v * u)$. This extension again satisfies

$$\partial_j(T*v) = (\partial_j T) * v = T * \partial_j v.$$

Furthermore, we see that

$$\delta * v(u) = \delta(v * u) = \int_{\mathbb{R}^n} v(y)u(y)dy,$$

thus $\delta * v = v$. Now, the defining identity $P(-i\partial)G = \delta$ for the fundamental solution implies $P(-i\partial)(G * v) = \delta * v = v$. $\Box$

Fundamental solutions are not unique, the difference u of two fundamental solutions is evidently a solution of the homogeneous equation $P(-i\partial)u = 0$.

Fundamental solutions are not easy to obtain directly. They often can be derived using Fourier transform.

Fourier Transform. The Fourier transform of a suitably bounded measurable function $u : \mathbb{R}^n \to \mathbb{C}$ is

$$\widehat{u}(p) := \int_{\mathbb{R}^n} u(x) e^{ix \cdot p} dx$$

for $p = (p_1, \dots, p_n) \in (\mathbb{R}^n)' \cong \mathbb{R}^n$ whenever this integral is well-defined. Here, $x \cdot p$ stands for a nondegenerate bilinear form appropriate for the problem one wants to consider. For example, it might be the Euclidean scalar product or the Minkowski scalar product in $\mathbb{R}^n = \mathbb{R}^{1,D-1}$ with $x \cdot p = x^\mu \eta_\mu^\nu p_\nu = x^0 p_0 - x^1 p_1 - \dots - x^{D-1} p_{D-1}$.

The Fourier transform is, in particular, well-defined for a rapidly decreasing smooth function $u \in \mathscr{S}(\mathbb{R}^n) = \mathscr{S}$ and, moreover, the transformed function $\mathscr{F}(u) = \widehat{u}$ is again a rapidly decreasing smooth function $\mathscr{F}(u) \in \mathscr{S}$. The inverse Fourier transform of a function $v = v(p)$ is

$$\mathscr{F}^{-1} v(x) := (2\pi)^{-n} \int_{\mathbb{R}^n} v(p) e^{-ix \cdot p} dp.$$

Proposition 8.4. *The Fourier transform is a linear continuous map*

$$\mathscr{F} : \mathscr{S} \to \mathscr{S}$$

whose inverse is $\mathscr{F}^{-1}$. *As a consequence,* $\mathscr{F}$ *has an adjoint*

$$\mathscr{F}' : \mathscr{S}' \to \mathscr{S}', T \mapsto T \circ \mathscr{F}.$$

On the basis of this result we can define the Fourier transform $\mathscr{F}(T)$ of a tempered distribution T as the adjoint

$$\mathscr{F}(T)(v) := T(\mathscr{F}(v)) = \mathscr{F}'(T)(v), v \in \mathscr{S},$$

and we obtain a map $\mathscr{F} : \mathscr{S}' \to \mathscr{S}'$ which is linear, continuous, and invertible. Note that for a function $g \in \mathscr{S}$ the Fourier transforms of the corresponding distribution T_g and that of g are the same:

$$\mathscr{F}(T_g)(v) = T_g(\widehat{v}) = \int_{\mathbb{R}^n} \int_{(\mathbb{R}^n)'} g(x) v(p) e^{ix \cdot p} dp dx = T_{\mathscr{F}(g)}(v).$$

Typical examples of Fourier transforms of distributions are

$$\mathscr{F}(H)(\omega) = \int_0^\infty e^{it\omega} dt = \frac{i}{\omega + i0},$$

$$\mathscr{F}(\delta_0) = \int_{\mathbb{R}^D} \delta_0(x) e^{ix \cdot p} dx = 1,$$

$$\mathscr{F}^{-1}(e^{ip \cdot y}) = (2\pi)^{-D} \int_{\mathbb{R}^D} e^{ip \cdot (y-x)} dp = \delta(x-y).$$

The fundamental importance of the Fourier transform is that it relates partial derivatives in the x^k with multiplication by the appropriate coordinate functions p_k after Fourier transformation:

$$\mathscr{F}(\partial_k u) = -ip_k \mathscr{F}(u)$$

by partial integration

$$\mathscr{F}(\partial_k u)(p) = \int \partial_k u(x)e^{ix\cdot p}dx = -\int u(x)ip_k e^{ix\cdot p}dx = -ip_k \mathscr{F}(u)(p),$$

and consequently,

$$\mathscr{F}(\partial^\alpha u) = (-ip)^\alpha \mathscr{F}(u).$$

This has direct applications to partial differential equations of the type

$$P(-i\partial)u = v.$$

The general differential equation $P(-i\partial)u = v$ will be transformed by $\mathscr{F}$ into the equation

$$P(p)\widehat{u} = \widehat{v}.$$

Now, trying to solve the original partial differential equation leads to a division problem for distributions. Of course, the multiplication of a polynomial $P = P(p)$ and a distribution $T \in \mathscr{S}'$ given by $PT(u) := T(Pu)$ is well-defined because $Pu(p) = P(p)u(p)$ is a function $Pu \in \mathscr{S}$ for each $u \in \mathscr{S}$. Solving the division problem, that is determining a distribution T with $PT = f$ for a given polynomial P and function f, is in general a difficult task.

For a polynomial P let us denote $G = G_P$ the inverse Fourier transform $\mathscr{F}^{-1}(T)$ of a solution of the division problem $PT = 1$, that is $P\widehat{G} = 1$. Then G is a *fundamental solution* of $P(-i\partial)u = v$, that is

$$P(-i\partial)G = \delta$$

since $\mathscr{F}(P(-i\partial)G) = P(p)\widehat{G} = 1$ and $\mathscr{F}^{-1}(1) = \delta$.

Klein–Gordon Equation. We study as an explicit example the fundamental solution of the Klein–Gordon equation. The results will be used later in the description of the free boson within the framework of Wightman's axioms, cf. p. 135, in order to construct a model satisfying all the axioms of quantum field theory.

The dynamics of a free bosonic classical particle is governed by the Klein–Gordon equation. The Klein–Gordon equation with mass $m > 0$ is

$$(\square + m^2)u = v,$$

where $\square$ is the wave operator for the Minkowski space $\mathbb{R}^{1,D-1}$ as before. A fundamental solution can be determined by solving the division problem

$$(-p^2 + m^2)\, T = 1:$$

A suitable

$$T \text{ is } (m^2 - p^2)^{-1}$$

as a distribution given by

$$T(v) = \int_{\mathbb{R}^{D-1}} \left(PV \int_{\mathbb{R}} \frac{v(p)}{\omega(p) - p_0^2} dp_0 \right) dp,$$

where $PV \int$ is the principal value of the integral. The corresponding fundamental solution (the propagator) is

$$G(x) = (2\pi)^{-D} \int_{\mathbb{R}^D} (m^2 - p^2)^{-1} e^{-ix \cdot p} dp.$$

G can be expressed more concretely by Bessel, Hankel, etc., functions.

We restrict our considerations to the free fields which are the solutions of the homogeneous equation

$$(\Box - m^2)\phi = 0.$$

The Fourier transform $\widehat{\phi}$ satisfies

$$(p^2 - m^2)\widehat{\phi} = 0,$$

where $p^2 = \langle p, p \rangle = p_0^2 - (p_1^2 + \ldots + p_{D-1}^2)$. Therefore, $\widehat{\phi}$ has its support in the mass-shell $\{p \in (\mathbb{R}^{1,D-1})' : p^2 = m^2\}$. Consequently, $\widehat{\phi}$ is proportional to $\delta(p^2 - m^2)$ as a distribution, that is $\widehat{\phi} = g(p)\delta(p^2 - m^2)$, and we get ϕ by the inverse Fourier transform

$$\phi(x) = (2\pi)^{-D} \int_{\mathbb{R}^D} g(p)\delta(p^2 - m^2) e^{-ip \cdot x} dp.$$

Definition 8.5. The distribution

$$D_m(x) := 2\pi i \mathscr{F}^{-1}((\operatorname{sgn}(p_0)\delta(p^2 \quad m^2))(x)$$

is called the *Pauli–Jordan function.*

($\operatorname{sgn}(t)$ is the sign of t, $\operatorname{sgn}(t) = H(t) - H(-t)$.) D_m generates all solutions of the homogeneous Klein–Gordon equation. In order to describe D_m in detail and to use the integration

$$D_m(x) = 2\pi i (2\pi)^{-D} \int_{\mathbb{R}^D} \operatorname{sgn}(p_0)\delta(p^2 - m^2) e^{-ip \cdot x} dp$$

for further calculations we observe that for a general g the distribution

$$\widehat{\phi} = g(p)\delta(p^2 - m^2)$$

can also be written as

$$\widehat{\phi} = H(p_0)g_+(p)\delta(p^2 - m^2) - H(-p_0)g_-(p)\delta(p^2 - m^2)$$

taking into account the two components of the hyperboloid $\{p \in (\mathbb{R}^{1,D-1})' : p^2 = m^2\}$: the upper hyperboloid

$$\Gamma_m := \{p \in (\mathbb{R}^{1,D-1})' : p^2 = m^2, p_0 > 0\}$$

and the lower hyperboloid

$$-\Gamma_m = \{p \in ((\mathbb{R}^{1,D-1}))' : p^2 = m^2, p_0 < 0\}.$$

Here, the g_+, g_- are distributions on the upper resp. lower hyperboloid, which in our situation can be assumed to be functions which simply depend on $\mathsf{p} \in \mathbb{R}^{D-1}$ via the global charts

$$\xi_\pm : \mathbb{R}^{D-1} \to \pm\Gamma_m, \mathsf{p} \mapsto (\pm\omega(\mathsf{p}), \mathsf{p}),$$

where $\omega(\mathsf{p}) := \sqrt{\mathsf{p}^2 + m^2}$ and $\mathsf{p} = (p_1, \ldots, p_{D-1})$, hence $\mathsf{p}^2 = p_1^2 + \ldots + p_{D-1}^2$. Let λ_m be the invariant measure on Γ_m given by the integral

$$\int_{\Gamma_m} h(\xi)d\lambda_m(\xi) := \int_{\mathbb{R}^{D-1}} h(\xi_+(\mathsf{p}))(2\omega(\mathsf{p}))^{-1}d\mathsf{p}$$

for functions h defined on Γ_m and analogously on $-\Gamma_m$. Then for $v \in \mathscr{S}(\mathbb{R}^D)$ the value of $\delta(p^2 - m^2)$ is

$$\delta(p^2 - m^2)(v) = \int_{\Gamma_m} v(\omega(\mathsf{p}), \mathsf{p})d\lambda_m + \int_{-\Gamma_m} v(-\omega(\mathsf{p}), \mathsf{p})d\lambda_m.$$

Here, we use the identity $\delta(t^2 - b^2) = (2b)^{-1}(\delta(t-b) + \delta(t+b))$ in one variable t with respect to a constant $b > 0$.

These considerations lead to the following ansatz which is in close connection to the formulas in the physics literature. We separate the coordinates $x \in \mathbb{R}^{1,D-1}$ into $x = (t, \mathsf{x})$ with $t = x^0$ and $\mathsf{x} = (x^1, \ldots, x^{D-1})$. Let

$$\phi(t, \mathsf{x}) := (2\pi)^{-D} \int_{\mathbb{R}^{D-1}} (a(\mathsf{p})e^{i(\mathsf{p}\cdot\mathsf{x} - \omega(\mathsf{p})t)} + a^*(\mathsf{p})e^{-i(\mathsf{p}\cdot\mathsf{x} - \omega(\mathsf{p})t)})d\lambda_m(\mathsf{p})$$

for arbitrary functions $a, a^* \in \mathscr{S}(\mathbb{R}^{D-1})$ in $D-1$ variables. Then $\phi(t, \mathsf{x})$ satisfies $(\square + m^2)\phi = 0$ which is clear from the above derivation (because of $a(\mathsf{p}) = g_+(\omega(\mathsf{p}), \mathsf{p}), a^*(\mathsf{p}) = g_-(-\omega(\mathsf{p}), \mathsf{p})$ up to a constant). That $\phi(t, \mathsf{x})$ satisfies $(\square + m^2)\phi = 0$ is in fact very easy to show directly: With the abbreviation

$$k(t, \mathsf{x}, \mathsf{p}) := (2\pi)^{-D}(a(\mathsf{p})e^{i(\mathsf{p}\cdot\mathsf{x} - \omega(\mathsf{p})t)} + a^*(\mathsf{p})e^{-i(\mathsf{p}\cdot\mathsf{x} - \omega(\mathsf{p})t)})$$

we have

$$\partial_0^2 \phi(t, \mathsf{p}) = \int_{\gamma_m} i^2 \omega(\mathsf{p})^2 k(t, \mathsf{x}, \mathsf{p})d\lambda_m \text{ and}$$

$$\partial_j^2 \phi(t, \mathsf{p}) = \int_{\gamma_m} i^2 p_j^2 k(t, \mathsf{x}, \mathsf{p})d\lambda_m \text{ for } j > 0.$$

Hence,

$$\Box \phi(t,\mathsf{p}) = -\int_{\gamma_m} (\omega(\mathsf{p})^2 - \mathsf{p}^2)k(t,\mathsf{x},\mathsf{p})d\lambda_m = -m^2\phi(t,\mathsf{x}).$$

We have shown the following result:

Proposition 8.6. *Each solution* $\phi \in \mathscr{S}$ *of* $(\Box + m^2)\phi = 0$ *can be represented uniquely as*

$$\phi(t,\mathsf{x}) := (2\pi)^D \int_{\mathbb{R}^{D-1}} (a(\mathsf{p})e^{i(\mathsf{p}\cdot\mathsf{x}-\omega(\mathsf{p})t)} + a^*(\mathsf{p})e^{-i(\mathsf{p}\cdot\mathsf{x}-\omega(\mathsf{p})t)})d\lambda_m(\mathsf{p})$$

with $a, a^* \in \mathscr{S}((\mathbb{R}^{D-1})')$. *The real solutions correspond to the case* $a^* = \bar{a}$.

8.2 Field Operators

Operators and Self-Adjoint Operators. Let $\mathscr{S}\mathscr{O} = \mathscr{S}\mathscr{O}(\mathbb{H})$ denote the set of self-adjoint operators in $\mathbb{H}$ and $\mathscr{O} = \mathscr{O}(\mathbb{H})$ the set of all densely defined operators in $\mathbb{H}$. (A general reference for operator theory is [RS80*].) Here, an *operator* in $\mathbb{H}$ is a pair (A, D) consisting of a subspace $D = D_A \subset \mathbb{H}$ and a $\mathbb{C}$-linear mapping $A : D \to \mathbb{H}$, and A is densely defined whenever D_A is dense in $\mathbb{H}$. In the following we are interested only in densely defined operators. Recall that such an operator can be unbounded, that is $\sup\{\|Af\| : f \in D, \|f\| \leq 1\} = \infty$, and many relevant operators in quantum theory are in fact unbounded. As an example, the position and momentum operators mentioned in Sect. 7.2 in the context of quantization of the harmonic oscillator are unbounded.

If a densely defined operator A is bounded (that is $\sup\{\|Af\| : f \in D_A, \|f\| \leq 1\} < \infty$), then A is continuous and possesses a unique linear and continuous continuation to all of $\mathbb{H}$.

Let us also recall the notion of a self-adjoint operator. Every densely defined operator A in $\mathbb{H}$ has an adjoint operator A^* which is given by

$$D_{A^*} := \{f \in \mathbb{H} | \exists h \in \mathbb{H} \; \forall g \in D_A : \langle h,g \rangle = \langle f, Ag \rangle\},$$

$$\langle A^*f, g \rangle = \langle f, Ag \rangle, f \in D_{A^*}, g \in D_A.$$

A^*f for $f \in D_{A^*}$ is thus the uniquely determined $h = A^*f \in \mathbb{H}$ with $\langle h,g \rangle = \langle f, Ag \rangle$ for all $g \in D_A$.

It is easy to show that the adjoint A^* of a densely defined operator A is a *closed operator*. A closed operator B in $\mathbb{H}$ is defined by the property that the graph of B, that is the subspace

$$\Gamma(B) = \{(f, B(f)) : f \in D_B\} \subset \mathbb{H} \times \mathbb{H}$$

of $\mathbb{H} \times \mathbb{H}$, is closed, where the Hilbert space structure on $\mathbb{H} \times \mathbb{H} \cong \mathbb{H} \oplus \mathbb{H}$ is defined by the inner product

$$\langle (f,f'),(g,g')\rangle := \langle f,g\rangle + \langle f',g'\rangle.$$

Hence, an operator B is closed if for all sequences (f_n) in D_B such that $f_n \to f \in \mathbb{H}$ and $Bf_n \to g \in \mathbb{H}$ it follows that $f \in D_B$ and $Bf = g$. Of course, every continuous operator defined on all of $\mathbb{H}$ is closed. Conversely, every closed operator B defined on all of $\mathbb{H}$ is continuous by the closed graph theorem. Note that a closed densely defined operator which is continuous satisfies $D_B = \mathbb{H}$.

Self-adjoint operators are sometimes mixed up with symmetric operators. For operators with domain of definition $D_B = \mathbb{H}$ the two notions agree and this holds more generally for closed operators also. A *symmetric operator* is a densely defined operator A such that

$$\langle Af,g\rangle = \langle f,Ag\rangle, f,g \in D_A.$$

By definition, a *self-adjoint operator* A is an operator which agrees with its adjoint A^* in the sense of $D_A = D_{A^*}$ and $A^* f = Af$ for all $f \in D_A$. Clearly, a self-adjoint operator is symmetric and it is closed since adjoint operators are closed in general. Conversely, it can be shown that a symmetric operator is self-adjoint if it is closed. An operator B is called *essentially self-adjoint* when it has a unique continuation to a self-adjoint operator, that is there is a self-adjoint operator A with $D_B \subset D_A$ and $B = A|_{D_B}$.

For a closed operator A, the spectrum $\sigma(A) = \{\lambda \in \mathbb{C} : (A - \lambda \,\mathrm{id}_{\mathbb{H}})^{-1}$ does not exist as a bounded operator$\}$ is a closed subset of $\mathbb{C}$. Whenever A is self-adjoint, the spectrum $\sigma(A)$ is completely contained in $\mathbb{R}$.

For a self-adjoint operator A there exists a unique representation $U : \mathbb{R} \to U(\mathbb{H})$ satisfying

$$\lim_{t\to 0} \frac{U(t)f - f}{t} = -iAf$$

for each $f \in D_A$ according to the spectral theorem. U is denoted $U(t) = e^{-itA}$ and A (or sometimes $-iA$) is called the *infinitesimal generator* of $U(t)$. Conversely (cf. [RS80*]),

Theorem 8.7 (Theorem of Stone). *Let $U(t)$ be a one parameter group of unitary operators in the complex Hilbert space $\mathbb{H}$, that is U is a unitary representation of $\mathbb{R}$. Then the operator A, defined by*

$$Af := \lim_{t\to 0} i\frac{U(t)f - f}{t}$$

in the domain in which this limit exists with respect to the norm of $\mathbb{H}$, is self-adjoint and generates $U(t) : U(t) = e^{-itA}$, $t \in \mathbb{R}$.

With the aid of (tempered) distributions and (self-adjoint) operators we are now in the position to explain what quantum fields are.

Field Operators. The central objects of quantum field theory are the quantum fields or field operators. A field operator is the analogue of a classical field but now in the quantum model. Therefore, in a first attempt, one might try to consider a field operator Φ to be a map from M to $\mathscr{SO}$ assigning to a point $x \in M = \mathbb{R}^{1,D-1}$ a self-adjoint operator $\Phi(x)$ in a suitable way. However, for various reasons such a map is not sufficient to describe quantum fields (see also Proposition 8.15). For example, in some classical field theories the Poisson bracket of a field ϕ at points $x, y \in M$ with $x^0 = y^0$ (at equal time) is of the form

$$\{\phi(x), \phi(y)\} = \delta(\mathsf{x} - \mathsf{y}),$$

where $\mathsf{x} := (x^1, \ldots, x^{D-1})$, the space part of $x = (x^0, x^1, \ldots, x^{D-1})$. This equation has a rigorous interpretation in the context of the theory of distributions.

As a consequence, a quantum field will be an operator-valued distribution.

Definition 8.8. A *field operator* or *quantum field* is now by definition an *operator-valued distribution* (on $\mathbb{R}^n$), that is a map

$$\Phi : \mathscr{S}(\mathbb{R}^n) \to \mathscr{O}$$

such that there exists a dense subspace $D \subset \mathbb{H}$ satisfying

1. For each $f \in \mathscr{S}$ the domain of definition $D_{\Phi(f)}$ contains D.
2. The induced map $\mathscr{S} \to \mathrm{End}(D), f \mapsto \Phi(f)|_D$, is linear.
3. For each $v \in D$ and $w \in \mathbb{H}$ the assignment $f \mapsto \langle w, \Phi(f)(v) \rangle$ is a tempered distribution.

The concept of a quantum field as an operator-valued distribution corresponds better to the actual physical situation than the more familiar notion of a field as a quantity defined at each point of spacetime. Indeed, in experiments the field strength is always measured not at a point x of spacetime but rather in some region of space and in a finite time interval. Therefore, such a measurement is naturally described by the expectation value of the field as a distribution applied to a test function with support in the given spacetime region. See also Proposition 8.15 below.

As a generalization of the Definition 8.8, it is necessary to consider *operator-valued tensor distributions* also. Here, the term *tensor* is used for a quantity which transforms according to a finite-dimensional representation of the Lorentz group L (resp. of its universal cover).

8.3 Wightman Axioms

In order to present the axiomatic quantum field theory according to Wightman we need the notion of a quantum field or field operator Φ as an operator-valued distribution which we have introduced in Definition 8.8 and some informations about properties on geometric invariance which we recall in the sequel.

Relativistic Invariance. As before, let $M = \mathbb{R}^{1,D-1}$ D-dimensional Minkowski space (in particular the usual four-dimensional Minkowski space $M = \mathbb{R}^{1,3}$ or the Minkowski plane $M = \mathbb{R}^{1,1}$) with the (Lorentz) metric

$$x^2 = \langle x,x \rangle = x^0 x^0 - \sum_{j=1}^{D-1} x^j x^j, x = (x^0, \ldots, x^{D-1}) \in M.$$

Two subsets $X, Y \subset M$ are called to be *space-like separated* if for any $x \in X$ and any $y \in Y$ the condition $(x-y)^2 < 0$ is satisfied, that is

$$(x^0 - y^0)^2 < \sum_{j=1}^{D-1} (x^j - y^j)^2.$$

The *forward cone* is $C_+ := \{x \in M : x^2 = <x,x> \geq 0, x^1 \geq 0\}$ and the *causal order* is given by $x \geq y \iff x - y \in C_+$.

Relativistic invariance of classical point particles in $M = \mathbb{R}^{1,D-1}$ or of classical field theory on M is described by the *Poincaré group* $P := P(1, D-1)$, the identity component of the group of all transformations of M preserving the metric. P is generated by the Lorentz group L, the identity component $L := SO_0(1, D-1) \subset GL(D, \mathbb{R})$ of the orthogonal group $O(1, D-1)$ of all *linear* transformations of M preserving the metric. (L is sometimes written $SO(1, D-1)$ by abuse of notation.) In fact, the Poincaré group P is the semidirect product (see Sect. 3.1) $L \ltimes \mathbb{R}^n \cong P$ of L and the translation group $M = \mathbb{R}^D$.

The Poincaré group P preserves the causal structure and the space-like separateness. Observe that the corresponding conformal group $SO(2, D)$ (cf. Theorem 2.9) which contains the Poincaré transformations also preserves the causal structure, but not the space-like separateness.

The Poincaré group P acts on $\mathscr{S} = \mathscr{S}(\mathbb{R}^D)$, the space of test functions, from the left by $h \cdot f(x) := f(h^{-1}x)$ with $g \cdot (h \cdot f) = (gh) \cdot f$ and this left action is continuous. It is mostly written in the form

$$(q, \Lambda) f(x) = f(\Lambda^{-1}(x - q)),$$

where the Poincaré transformations h are parameterized by $(q, \Lambda) \in L \ltimes M, q \in M$, $\Lambda \in L$.

The relativistic invariance of the quantum system with respect to Minkowski space $M = \mathbb{R}^{1,D-1}$ is in general given by a projective representation $P \to U(\mathbb{P}(\mathbb{H}))$ of the Poincaré group P, a representation in the space $\mathbb{P}(\mathbb{H})$ of states of the quantum system as we explain in Sect. 3.2. By Bargmann's Theorem 4.8 such a representation can be lifted to an essentially uniquely determined unitary representation of the 2-to-1 covering group of P, the simply connected universal cover of P. This group is isomorphic to the semidirect product $\text{Spin}(1, D-1) \ltimes \mathbb{R}^D$ for $D > 2$ where $\text{Spin}(1, D-1)$ is the corresponding spin group, the universal covering group of the Lorentz group $L = SO(1, D-1)$. In the sequel we often call these covering groups the Poincaré group and Lorentz group, respectively, and denote them simply again by P and L.

Note that in the two-dimensional case, the Lorentz group L is isomorphic to the abelian group $\mathbb{R}$ of real numbers (cf. Remark 1.15) and therefore agrees with its universal covering group.

We thus suppose to have a unitary representation of the Poincaré group P which will be denoted by

$$U : P \to U(\mathbb{H}), (q, \Lambda) \mapsto U(q, \Lambda),$$

$(q, \Lambda) \in M \times L = L \ltimes M$.

Since the transformation group $M \subset P$ is abelian one can apply Stone's Theorem 8.7 in order to obtain the restriction of the unitary representation U to M in the form

$$U(q, 1) = \exp iqP = \exp i(q^0 P_0 - q^1 P_1 - \ldots - q^{D-1} P_{D-1}), \qquad (8.2)$$

$q \in \mathbb{R}^{1,D-1}$, with self-adjoint commuting operators $P_0, \ldots, P_{D-1}$ on $\mathbb{H}$. P_0 is interpreted as the energy operator $P_0 = H$ and the $P_j, j > 0$, as the components of the momentum.

We are now in the position to formulate the axioms of quantum field theory.

Wightman Axioms. A *Wightman quantum field theory* (Wightman QFT) in dimension D consists of the following data:

- the space of states, which is the projective space $\mathbb{P}(\mathbb{H})$ of a separable complex Hilbert space $\mathbb{H}$,
- the vacuum vector $\Omega \in \mathbb{H}$ of norm 1,
- a unitary representation $U : P \to U(\mathbb{H})$ of P, the covering group of the Poincaré group,
- a collection of field operators $\Phi_a, a \in I$ (cf. Definition 8.8),

$$\Phi_a : \mathscr{S}(\mathbb{R}^D) \to \mathscr{O},$$

with a dense subspace $D \subset \mathbb{H}$ as their common domain (that is the domain $D_a(f)$ of Φ_a contains D for all $a \in A, f \in \mathscr{S}$) such that Ω is in the domain D.

These data satisfy the following three axioms:

Axiom W1 (Covariance)

1. Ω *is P-invariant, that is* $U(q, \Lambda)\Omega = \Omega$ *for all* $(q, \Lambda) \in P$, *and D is P-invariant, that is* $U(q, \Lambda)D \subset D$ *for all* $(q, \Lambda) \in P$,
2. *the common domain* $D \subset \mathbb{H}$ *is invariant in the sense that* $\Phi_a(f)D \subset D$ *for all* $f \in \mathscr{S}$ *and* $a \in I$,
3. *the actions on* $\mathbb{H}$ *and* $\mathscr{S}$ *are equivariant where P acts on* $\text{End}(D)$ *by conjugation. That is on D we have*

$$U(q, \Lambda)\Phi_a(f)U(q, \Lambda)^* = \Phi((q, \Lambda)f) \qquad (8.3)$$

for all $f \in \mathscr{S}$ *and for all* $(q, \Lambda) \in P$.

Axiom W2 (Locality) $\Phi_a(f)$ and $\Phi_b(g)$ commute on D if the supports of $f,g \in \mathscr{S}$ are space-like separated, that is on D

$$\Phi_a(f)\Phi_b(g) - \Phi_b(g)\Phi_a(f) = [\Phi_a(f),\Phi_b(g)] = 0. \qquad (8.4)$$

Axiom W3 (Spectrum Condition) The joint spectrum of the operators P_j is contained in the forward cone C_+.

Recall that the support of a function f is the closure of the points x with $f(x) \neq 0$.

If one represents the operator-valued distribution Φ_a symbolically by a function $\Phi_a = \Phi_a(x) \in \mathscr{O}$ the equivariance (8.3) can be written in the following form:

$$U(q,\Lambda)\Phi_a(x)U(q,\Lambda)^* = \Phi_a(\Lambda x + q).$$

This form is frequently used even if Φ_a cannot be represented as a function, and the equality is only valid in a purely formal way.

Remark 8.9. The relevant fields, that is the operators $\Phi_a(f)$ for real-valued test functions $f \in \mathscr{S}$, should be essentially self-adjoint. In the above axioms this has not been required from the beginning because often one considers a larger set of field operators so that only certain combinations are self-adjoint. In that situation it is reasonable to require Φ_a^* to be in the set of quantum fields, that is $\Phi_a^* = \Phi_{a'}$ for a suitable $a' \in A$ (where $a = a'$ if $\Phi_a(f)$ is essentially self-adjoint).

Remark 8.10. Axiom W1 is formulated for scalar fields only which transform under the trivial representation of L. In general, if fields have to be considered which transform according to a nontrivial (finite-dimensional) complex or real representation $R : \mathrm{L} \to \mathrm{GL}(W)$ of the (double cover of the) Lorentz group (like spinor fields, for example) the equivariance in (8.3) has to be replaced by

$$U(q,\Lambda)\Phi_j(f)U(q,\Lambda)^* = \sum_{k=1}^m R_{jk}(\Lambda^{-1})\Phi_k((q,\Lambda)f). \qquad (8.5)$$

Here, W is identified with $\mathbb{R}^m$ resp. $\mathbb{C}^m$, and the $R(\Lambda)$ are given by matrices $(R_{jk}(\Lambda))$. Moreover, the fields Φ_a are merely components and have to be grouped together to vectors $(\Phi_1,\ldots,\Phi_m)$.

Remark 8.11. In the case of $D = 2$ there exist nontrivial one-dimensional representations $R : \mathrm{L} \to \mathrm{GL}(1,\mathbb{C}) = \mathbb{C}^\times$ of the Lorentz group L, since the Lie algebra Lie L of L is $\mathbb{R}$ and therefore not semi-simple. In this situation the equivariance (8.3) has to be extended to

$$U(q,\Lambda)\Phi_a(f)U(q,\Lambda)^* = R(\Lambda^{-1})\Phi_a((q,\Lambda)f). \qquad (8.6)$$

Remark 8.12. Another generalization of the axioms of a completely different nature concerns the locality. In the above axioms only bosonic fields are considered. For the fermionic case one has to impose a grading into even and odd (see also Remark 10.19), and the commutator of odd fields in Axiom W2 has to be replaced with the anticommutator.

Remark 8.13. The spectrum condition (Axiom W3) implies that for eigenvalues p_μ of P_μ the vector $p = (p_0, \ldots, p_{D-1})$ satisfies $p \in C_+$. In particular, with the interpretation of $P_0 = H$ as the energy operator the system has no negative energy states: $p_0 \geq 0$. Moreover, $P^2 = P_0^2 - P_1^2 - \ldots - P_{D-1}^2$ has the interpretation of the mass-squared operator with the condition $p^2 \geq 0$ for each D–tuple of eigenvalues p_μ of P_μ in case Axiom W3 is satisfied.

Remark 8.14. In addition to the above axioms in many cases an irreducibility or completeness condition is required. For example, it is customary to require that the vacuum is cyclic in the sense that the subspace $D_0 \subset D$ spanned by all the vectors

$$\Phi_{a_1}(f_1)\Phi_{a_2}(f_2)\ldots\Phi_{a_m}(f_m)\Omega^1$$

is dense in D and thus dense in $\mathbb{H}$.

Moreover, as an additional axiom one can require the vacuum Ω to be unique:

Axiom W4 (Uniqueness of the Vacuum) *The only vectors in $\mathbb{H}$ left invariant by the translations $U(q, 1)$, $q \in M$, are the scalar multiples of the vacuum Ω.*

Although the above postulates appear to be quite evident and natural, it is by no means easy to give examples of Wightman quantum field theories even for the case of free theories. For the case of proper interaction no Wightman QFT is known so far in the relevant case of $D = 4$, and it is one of the millennium problems discussed in [JW06*] to construct such a theory. For $D = 2$, however, there are theories with interaction (cf. [Simo74*]), and many of the conformal field theories in two dimensions have nontrivial interaction.

Example: Free Bosonic QFT. In the following we sketch a Wightman QFT for a quantized boson of mass $m > 0$ in three-dimensional space (hence $D = 4$, the considerations work for arbitrary $D \geq 2$ without alterations). The basic differential operator, the Klein–Gordon operator $\Box + m^2$ with mass m, has already been studied in Sect. 8.1. We look for a field operator

$$\Phi : \mathscr{S} = \mathscr{S}(\mathbb{R}^4) \longrightarrow \mathscr{SO}(\mathbb{H})$$

on a Hilbert space $\mathbb{H}$ such that for all test function $f, g \in \mathscr{S}$:

1. Φ satisfies the Klein–Gordon equation in the following sense:

$$\Phi(\Box f + m^2 f) = 0 \text{ for all } f \in \mathscr{S}.$$

2. Φ obeys the commutation relation

$$[\Phi(f), \Phi(g)] = -i \int_{\mathbb{R}^4 \times \mathbb{R}^4} f(x) D_m(x - y) g(y) dx dy.$$

---

[1] As before, we write the composition $B \circ C$ of operators as multiplication BC and similarly the value $B(v)$ as multiplication Bv.

Here, D_m is the Pauli–Jordan function (cf. Definition 8.5)

$$D_m(x) := i(2\pi)^{-3} \int_{\mathbb{R}^D} \mathrm{sgn}(p_0)\delta(p^2 - m^2)e^{-ip \cdot x}dp.$$

The construction of such a field and the corresponding Hilbert space is a Fock space construction. Let $H_1 = \mathscr{S}(\Gamma_m) \cong \mathscr{S}(\mathbb{R}^3)$. The isomorphism is induced by the global chart

$$\xi : \mathbb{R}^3 \to \Gamma_m, \mathrm{p} \mapsto (\omega(\mathrm{p}), \mathrm{p}),$$

where $\omega(\mathrm{p}) = \sqrt{\mathrm{p}^2 + m^2}$. We denote the points in Γ_m by ξ or ξ_j in the following:

H_1 is dense in $\mathbb{H}_1 := \mathrm{L}^2(\Gamma_m, d\lambda_m)$, the complex Hilbert space of square-integrable functions on the upper hyperboloid Γ_m. Furthermore, let H_N denote the space of rapidly decreasing functions on the N-fold product of the upper hyperboloid Γ_m which are symmetric in the variables $(\mathrm{p}_1, \ldots, \mathrm{p}_N) \in \Gamma_m^N$. H_N has the inner product

$$\langle u, v \rangle := \int_{\Gamma_m^N} \bar{u}(\xi_1, \ldots, \xi_N)v(\xi_1, \ldots, \xi_N)d\lambda_m(\xi_1)\ldots d\lambda_m(\xi_N).$$

The Hilbert space completion of H_N will be denoted by $\mathbb{H}_N$. H_N contains the N-fold symmetric product of H_1 and this space is dense in H_N and thus also in $\mathbb{H}_N$. Now, the direct sum

$$D := \bigoplus_{N=0}^{\infty} H_N$$

($H_0 = \mathbb{C}$ with the vacuum $\Omega := 1 \in H_0$) has a natural inner product given by

$$\langle f, g \rangle := \overline{f_0}g_0 + \sum_{N \geq 1} \frac{1}{N!} \langle f_N, g_N \rangle,$$

where $f = (f_0, f_1, \ldots), g = (g_0, g_1, \ldots) \in D$. The completion of D with respect to this inner product is the Fock space $\mathbb{H}$. $\mathbb{H}$ can also be viewed as a suitable completion of the symmetric algebra

$$S(H_1) = \bigoplus H_1^{\odot N},$$

where $H_1^{\odot N}$ is the N-fold symmetric product

$$H_1^{\odot N} = H_1 \odot \ldots \odot H_1.$$

The operators $\Phi(f)$, $f \in \mathscr{S}$, will be defined on $g = (g_0, g_1, \ldots) \in D$ by

$$(\Phi(f)g)_N(\xi_1, \ldots, \xi_N) := \int_{\Gamma_m} \widehat{f}(\xi)g_{N+1}(\xi, \xi_1, \ldots, \xi_N)d\lambda_m(\xi)$$

$$+ \sum_{j=1}^{N} \widehat{f}(-\xi_j)g_{N-1}(\xi_1, \ldots \widehat{\xi}_j \ldots, \xi_N),$$

where $\hat{\xi}_j$ means that this variable has to be omitted. This completes the construction of the Wightman QFT for the free boson.

The various requirements and axioms are not too difficult to verify. For example, we obtain $\Phi(\Box f - m^2 f) = 0$ since

$$\mathscr{F}(\Box f - m^2 f) = (-p^2 + m^2)\hat{f}$$

vanishes on Γ_m, and similarly we obtain the second requirement on the commutators the formula

$$[\Phi(f), \Phi(g)] = -i \int_{\mathbb{R}^4 \times \mathbb{R}^4} f(x) D_m(x-y) g(y) dx dy.$$

Furthermore, we observe that the natural action of the Poincaré group on $\mathbb{R}^{1,3}$ and on $\mathscr{S}(\mathbb{R}^{1,3})$ induces a unitary representation U in the Fock space $\mathbb{H}$ leaving invariant the vacuum and the domain of definition D. Of course, Φ is a field operator in the sense of our Definition 8.8 with $\Phi(f)D \subset D$ and, moreover, it can be checked that Φ is covariant in the sense of Axiom W1 and that the joint spectrum of the operators P_j is supported in Γ_m hence in the forward light cone (Axiom W3). Finally, the construction yields locality (Axiom W2) according to the above formula for $[\Phi(f), \Phi(g)]$.

We conclude this section with the following result of Wightman which demonstrates that in QFT it is necessary to consider operator-valued distributions instead of operator-valued mappings:

Proposition 8.15. *Let Φ be a field in a Wightman QFT which can be realized as a map $\Phi : M \to \mathcal{O}$ and where Φ^* belongs to the fields. Moreover, assume that Ω is the only translation-invariant vector (up to scalars). Then $\Phi(x) = c\Omega$ is the constant operator for a suitable constant $c \in \mathbb{C}$.*

In fact, it is enough to require equivariance with respect to the transformation group only and the property that $\Phi(x)$ and $\Phi(y)^*$ commute if $x - y$ is spacelike.

8.4 Wightman Distributions and Reconstruction

Let $\Phi = \Phi_a$ be a field operator in a Wightman QFT acting on the space $\mathscr{S} = \mathscr{S}(\mathbb{R}^{1,D-1})$ of test functions

$$\Phi : \mathscr{S} \longrightarrow \mathcal{O}(\mathbb{H}).$$

We assume $\Phi(f)$ to be self-adjoint for real-valued $f \in \mathscr{S}$ (cf. 8.9), hence $\Phi(f)^* = \Phi(\bar{f})$ in general. Then for $f_1, \ldots, f_N \in \mathscr{S}$ one can define

$$W_N(f_1, \ldots, f_N) := \langle \Omega, \Phi(f_1) \ldots \Phi(f_N)\Omega \rangle$$

according to Axiom W1 part 2. Since Φ is a field operator the mapping

$$W_N : \mathscr{S} \times \mathscr{S} \ldots \times \mathscr{S} \longrightarrow \mathbb{C}$$

is multilinear and separately continuous. It is therefore jointly continuous and one can apply the nuclear theorem of Schwartz to obtain a uniquely defined distribution on the space in DN variables, that is a distribution in $\mathscr{S}'((\mathbb{R}^D)^N) = \mathscr{S}'(\mathbb{R}^{DN})$. This continuation of W_N will be denoted again by W_N.

The sequence (W_N) of distributions generated by Φ is called the sequence of *Wightman distributions*. The $W_N \in \mathscr{S}'(\mathbb{R}^{DN})$ are also called *vacuum expectation values* or *correlation functions*.

Theorem 8.16. *The Wightman distributions associated to a Wightman QFT satisfy the following conditions: Each $W_N, N \in \mathbb{N}$, is a tempered distribution*

$$W_N \in \mathscr{S}'(\mathbb{R}^{DN})$$

with

WD1 (Covariance) *W_N is Poincaré invariant in the following sense:*

$$W_N(f) = W_N((q,\Lambda)f) \ for \ all \ (q,\Lambda)) \in P.$$

WD2 (Locality) *For all $N \in \mathbb{N}$ and $j, 1 \le j < N$,*

$$W_N(x_1,\ldots,x_j,x_{j+1},\ldots,x_N) = W_N(x_1,\ldots,x_{j+1},x_j,\ldots,x_N),$$

if $(x_j - x_{j+1})^2 < 0.)$

WD3 (Spectrum Condition) *For each $N > 0$ there exists a distribution $M_N \in \mathscr{S}'(\mathbb{R}^{D(N-1)})$ supported in the product $(C_+)^{N-1} \subset \mathbb{R}^{D(N-1)}$ of forward cones such that*

$$W_N(x_1,\ldots,x_N) = \int_{\mathbb{R}^{D(N-1)}} M_N(p)e^{i\sum p_j \cdot (x_{j+1}-x_j)}dp,$$

where $p = (p_1,\ldots,p_{n-1}) \in (R^D)^{N-1}$ and $dp = dp_1 \ldots dp_{N-1}$.

WD4 (Positive Definiteness) *For any sequence $f_N \in \mathscr{S}(\mathbb{R}^{DN})N \in \mathbb{N}$ one has for all $m \in \mathbb{N}$:*

$$\sum_{M,N=0}^{k} W_{M+N}(\overline{f}_M \otimes f_N) \ge 0.$$

$f \otimes g$ for $f \in \mathscr{S}(\mathbb{R}^{DM}), g \in \mathscr{S}(\mathbb{R}^{DN})$ is defined by

$$f \otimes g(x_1,\ldots,x_{M+N}) = f(x_1,\ldots,x_M)g(x_{M+1},\ldots,x_{M+N}).$$

Proof. WD1 follows directly from W1. Observe that the unitary representation of the Poincaré group is no longer visible. And WD2 is a direct consequence of W2. WD4 is essentially the property that a vector of the form

$$\sum_{M=1}^{k} \Phi(f_M)\Omega \in \mathbb{H}$$

has a non-negative norm where $\Phi(f_M)\Omega$ is defined as follows: The map

$$(f_1,\ldots,f_M) \mapsto \Phi(f_1)\ldots\Phi(f_M)\Omega, (f_1,\ldots,f_M) \in \mathscr{S}(\mathbb{R}^D)^M,$$

is continuous and multilinear by the general assumptions on the field operator Φ and therefore induces by the nuclear theorem a vector-valued distribution Φ_M : $\mathscr{S}(\mathbb{R}^{DM}) \to \mathbb{H}$ which is symbolically written as $\Phi_M(x_1,\ldots,x_M)$. Now, $\Phi(f_M)\Omega := \Phi_M(f_M)\Omega$ and

$$0 \le \left\| \sum_{M=1}^{k} \Phi(f_M)\Omega \right\|^2 \le \left\langle \sum_{M=1}^{k} \Phi(f_M)\Omega, \sum_{N=1}^{k} \Phi(f_N)\Omega \right\rangle$$

$$\le \sum_{M,N} \langle \Omega, \Phi(f_M)^*\Phi(f_N)\Omega \rangle = \sum_{M,N} W_{M+N}(\overline{f}_M \otimes f_N).$$

WD3 will be proven in the next proposition. □

In the sequel we write the distributions Φ and W_N symbolically as functions $\Phi(x)$ and $W_N(x_1,\ldots,x_N)$ in order to simplify the notation and to work more easily with the supports of the distributions in consideration.

The covariance of the field operator Φ implies the covariance

$$W_N(x_1,\ldots,x_n) = W_N(\Lambda x_1 + q,\ldots,\Lambda x_N + q)$$

for every $(q,\Lambda) \in$ P. In particular, the Wightman distributions are translation-invariant:

$$W_N(x_1,\ldots,x_n) = W_N(x_1 + q,\ldots,x_N + q).$$

Consequently, W_N depends only on the differences

$$\xi_1 = x_1 - x_2,\ldots,\xi_{N-1} = x_{N-1} - x_N.$$

We define

$$w_N(\xi_1,\ldots,\xi_{N-1}) := W_N(x_1,\ldots,x_N).$$

Proposition 8.17. *The Fourier transform $\widehat{w}_N$ has its support in the product $(C_+)^{N-1}$ of the forward cone $C_+ \in \mathbb{R}^D$. Hence*

$$W_N(x) = (2\pi)^{-D(N-1)} \int_{\mathbb{R}^{D(N-1)}} \widehat{w}_N(p)e^{-i\sum p_j\cdot(x_j-x_{j+1})}dp.$$

Proof. Because of $U(x,1)^* = U(-x,1) = e^{-ix\cdot P}$ for $x \in \mathbb{R}^D$ (cf. 8.2) the spectrum condition W2 implies

$$\int_{\mathbb{R}^D} e^{ix\cdot P}U(x,1)^*vdx = 0$$

for every $v \in \mathbb{H}$ if $p \notin C_+$. Since $w_N(\xi_1, \ldots, \xi_j + x, \xi_{j+1}, \ldots, \xi_{N-1}) = W_N(x_1, \ldots, x_j, x_{j+1} - x, \ldots, x_N - x)$ the Fourier transform of w_N with respect to ξ_j gives

$$\int_{\mathbb{R}^D} w_N(\xi_1, \ldots, \xi_j + x, \xi_{j+1}, \ldots, \xi_{N-1}) e^{i p_j \cdot x} dx$$

$$= \left\langle \Omega, \Phi(x_1) \ldots \Phi(x_j) \int_{\mathbb{R}^D} \Phi(x_{j+1} - x) \ldots \Phi(x_N - x) e^{i p_j \cdot x} \Omega dx \right\rangle$$

$$= \left\langle \Omega, \Phi(x_1) \ldots \Phi(x_j) \int_{\mathbb{R}^D} e^{i x \cdot p_j} U^*(x, 1) \Phi(x_{j+1}) \ldots \Phi(x_N) \Omega dx \right\rangle = 0,$$

where the last identity is a result of applying the above formula to $v = \Phi(x_{j+1}) \ldots \Phi(x_N) \Omega$ whenever $p_j \notin C_+$. Hence,

$$\widehat{w}_N(p_1, \ldots, p_{N-1}) = 0$$

if $p_j \notin C_+$ for at least one index j.   □

Having established the basic properties of the Wightman functions we now explain how a sequence of distributions with the properties WD 1–4 induce a Wightman QFT by the following:

Theorem 8.18. (Wightman Reconstruction Theorem) *Given any sequence* (W_N), $W_N \in \mathscr{S}'(\mathbb{R}^{DN})$, *of tempered distributions obeying the conditions WD1–WD4, there exists a Wightman QFT for which the* W_N *are the Wightman distributions.*

Proof. We first construct the Hilbert space for the Wightman QFT. Let

$$\mathscr{L} := \bigoplus_{N=0}^{\infty} \mathscr{S}(\mathbb{R}^{DN})$$

denote the vector space of finite sequences $\underline{f} = (f_N)$ with $f_N \in \mathscr{S}(\mathbb{R}^{DN}) =: \mathscr{S}_N$. On $\mathscr{L}$ we define a multiplication

$$\underline{f} \times \underline{g} := (h_N), h_N := \sum_{k=0}^{N} f_k(x_1, \ldots, x_k) g_{N-k}(x_{k+1}, \ldots, x_N).$$

The multiplication is associative and distributive but not commutative. Therefore, $\mathscr{L}$ is an associative algebra with unit $1 = (1, 0, 0, \ldots)$ and with a convolution $\gamma(\underline{f}) := (\overline{f}_N) = \overline{\underline{f}}$. γ is complex antilinear and satisfies $\gamma^2 = \mathrm{id}$.

Our basic algebra $\mathscr{L}$ will be endowed with the direct limit topology and thus becomes a complete locally convex space which is separable. (The direct limit topology is the finest locally convex topology on $\mathscr{L}$ such that the natural inclusions $\mathscr{S}(\mathbb{R}^{DN}) \to \mathscr{L}$ are continuous.) The continuous linear functionals $\mu : \mathscr{L} \to \mathbb{C}$ are represented by sequences (μ_N) of tempered distributions $\mu_N \in \mathscr{S}'_N$: $\mu((f_N)) = \sum \mu_N(f_N)$.

Such a functional is called positive semi-definite if $\mu(\overline{\underline{f}} \times \underline{f}) \geq 0$ for all $\underline{f} \in \mathscr{L}$ because the associated bilinear form $\omega = \omega_\mu$ given by $\omega(\underline{f}, \underline{g}) := \mu(\overline{\underline{f}} \times \underline{g})$ is positive semi-definite. For a positive semi-definite continuous linear functional μ the subspace

$$J = \{\underline{f} \in \mathscr{L} : \mu(\overline{\underline{f}} \times \underline{f}) = 0\}$$

turns out to be an ideal in the algebra $\mathscr{L}$.

It is not difficult to show that in the case of a positive semi-definite $\mu \in \mathscr{L}'$ the form ω is hermitian and defines on the quotient $\mathscr{L}/J$ a positive definite hermitian scalar product. Therefore, $\mathscr{L}/J$ is a pre-Hilbert space and the completion of this space with respect to the scalar product is the Hilbert space $\mathbb{H}$ needed for the reconstruction. This construction is similar to the so-called GNS construction of Gelfand, Naimark, and Segal.

The vacuum $\Omega \in \mathbb{H}$ will be the class of the unit $1 \in \mathscr{L}$ and the field operator Φ is defined by fixing $\Phi(f)$ for any test function $f \in \mathscr{S}$ on the subspace $D = \mathscr{L}/J$ of classes $[\underline{g}]$ of elements of $\mathscr{L}$ by

$$\Phi(f)([\underline{g}]) := [\underline{g} \times f],$$

where f stands for the sequence $(0, f, 0, \ldots,)$. Evidently, $\Phi(f)$ is an operator defined on D depending linearly on f. Moreover, for $\underline{h}, \underline{g} \in \mathscr{L}$ the assignment

$$f \mapsto \langle [\underline{h}], \Phi(f)([\underline{g}]) \rangle = \mu(\underline{h} \times (\underline{g} \times f))$$

is a tempered distribution because μ is continuous. This means that Φ is a field operator in the sense of Definition 8.8. Obviously, $\Phi(f)D \subset D$ and $\Omega \in D$.

So far, the Wightman distributions W_N have not been used at all. We consider now the above construction for the continuous functional $\mu := (W_N)$. Because of property WD4 this functional is positive semi-definite and provides the Hilbert space $\mathbb{H}$ constructed above depending on (W_N) together with a vacuum Ω and a field operator Φ. The properties of the Wightman distributions which eventually ensure that the Wightman axioms for this construction are fulfilled are encoded in the ideal

$$J = \{\underline{f} = (f_N) \in \mathscr{L} : \sum W_N(\underline{f} \times \overline{\underline{f}}) = 0\}.$$

To show covariance, we first have to specify a unitary representation of the Poincaré group P in $\mathbb{H}$. This representation is induced by the natural action $\underline{f} \mapsto (q, \Lambda)\underline{f}$ of P on $\mathscr{L}$ given by

$$(q, \Lambda)f_n(x_1, \ldots, x_n) := f(\Lambda^{-1}(x_1 - q), \ldots, \Lambda^{-1}(x_n - q))$$

for $(q, \Lambda) \in L \ltimes M \cong P$. This action leads to a homomorphism $P \rightarrow GL(\mathscr{L})$ and the action respects the multiplication. Now, because of the covariance of the Wightman distributions (property WD1) the ideal J is invariant, that is for $\underline{f} \in J$ and $(q, \Lambda) \in P$ we have $(q, \Lambda)\underline{f} \in J$. As a consequence, $U(q, \Lambda)([\underline{f}]) := [(q, \Lambda)\underline{f}]$ is well-defined on $D \subset \mathbb{H}$ with

$$\langle U(q, \Lambda)([\underline{f}]), U(q, \Lambda)([\underline{f}]) \rangle = \langle [\underline{f}], [\underline{f}] \rangle.$$

Altogether, this defines a unitary representation of P in $\mathbb{H}$ leaving Ω invariant such that the field operator is equivariant. We have shown that the covariance axiom W1 is satisfied.

In a similar way, one can show that property WD2 implies W2 and property WD3 implies W3. Locality (property WD2) implies that J includes the ideal J_{lc} generated by the linear combinations of the form

$$f_N(x_1,\ldots,x_N) = g(x_1,\ldots,x_j,x_{j+1},\ldots,x_N) - g(x_1,\ldots,x_{j+1},x_j,\ldots,x_N)$$

with $g(x_1,\ldots,x_N) = 0$ for $(x_{j+1} - x_j)^2 \geq 0$. And property WD3 (spectrum condition) implies that the ideal

$$J_{sp} := \{(f_N) : f_0 = 0, \widehat{f}(p_1,\ldots,p_N) = 0 \text{ in a neighborhood of } C_N\},$$

where $C_N = \{p : p_1 + \ldots + p_j \in C_+, j = 1,\ldots,N\}$, is also contained in J. □

As a result of this section, in an axiomatic approach to quantum field theory the Wightman axioms W1–W3 on the field operators can be replaced by the equivalent properties or axioms WD1–WD4 on the corresponding correlation functions W_N, the Wightman distributions. This second approach is formulated without explicit reference to the Hilbert space.

In the next section we come to a different but again equivalent description of the axiomatics which is formulated completely in the framework of Euclidean geometry.

8.5 Analytic Continuation and Wick Rotation

In this section we explain how the Wightman axioms induce a Euclidean field theory through analytic continuation of the Wightman distributions.

We first collect some results and examples on analytic continuation of holomorphic functions. Recall that a complex-valued function $F : U \to \mathbb{C}$ on an open subset $U \subset \mathbb{C}^n$ is *holomorphic* or *analytic* if it has complex partial derivatives $\frac{\partial}{\partial z^j} F = \partial_j F$ on U with respect to each of its variables z^j or, equivalently, if F can be expanded in each point $a \in U$ into a convergent power series $\sum c_\alpha z^\alpha$ such that

$$F(a+z) = \sum_{\alpha \in \mathbb{N}^n} c_\alpha z^\alpha$$

for z in a suitable open neighborhood of 0. The partial derivatives of F in a of any order exist and appear in the power series expansions in the form $\partial^\alpha F(a) = \alpha! c_\alpha$.

A holomorphic function F on a connected domain $U \subset \mathbb{C}^n$ is completely determined by the restriction $F|_W$ to any nonempty open subset $W \subset U$ or by any of its germs (that is power series expansion) at a point $a \in U$. This property leads to the phenomenon of analytic continuation, namely that a holomorphic function g on an open subset $W \subset \mathbb{C}^n$ may have a so-called *analytic continuation* to a holomorphic $F : U \to \mathbb{C}$, that is $F|_W = g$, which is uniquely determined by g.

A different type of analytic continuation occurs if a real analytic function $g : W \to \mathbb{C}$ on an open subset $W \subset \mathbb{R}^n$ is regarded as the restriction of a holomorphic $F : U \to \mathbb{C}$ where U is an open subset in $\mathbb{C}^n$ with $U \cap \mathbb{R}^n = W$. Such a holomorphic function F is obtained by simply exploiting the power series expansions of the real analytic function g: For each $a \in W$ there are $c_\alpha \in \mathbb{C}$ and $r^j(a) > 0, j = 1, \ldots, n$, such that $g(a+x) = \sum_\alpha c_\alpha x^\alpha$ for all x with $|x^j| < r^j(a)$. By inserting $z \in \mathbb{C}, |z^j| < r^j(a)$, instead of x into the power series we get such an analytic continuation defined on the open neighborhood $U = \{a+z \in \mathbb{C}^n : a \in W, |z^j| < r^j(a)\} \subset \mathbb{C}^n$ of W.

Another kind of analytic continuation is given by the Laplace transform. As an example in one dimension let $u : \mathbb{R}_+ \to \mathbb{C}$ be a polynomially bounded continuous function on $\mathbb{R}_+ = \{t \in \mathbb{R} : t > 0\}$.

Then the integral ("Laplace transform")

$$\mathscr{L}(u)(z) = F(z) := \int_0^\infty u(t)e^{itz}dt, \text{ Im } z \in \mathbb{R}_+,$$

defines a holomorphic function F on the "tube" domain $U = \mathbb{R} \times \mathbb{R}_+ \subset \mathbb{C}$ such that,

$$\lim_{y \searrow 0} F(x+iy) = g(x) \text{ where } g(x) := \int_0^\infty u(t)e^{itx}dt.$$

In this situation the $g(x)$ are sometimes called the boundary values of $F(z)$. The analytic continuation is given by the Laplace transform.

Of course, the integral exists because of $|u(t)e^{itz}| = |u(t)e^{-ty}| \leq |u(x)|$ for $z = x+iy \in U$ and $t \in \mathbb{R}_+$. F is holomorphic since we can interchange integration and derivation to obtain

$$\frac{d}{dz}F(z) = F'(z) = i\int_0^\infty tu(t)e^{itz}dt.$$

We now present a result which shows how in a similar way even a distribution $T \in \mathscr{S}(\mathbb{R}^n)'$ can, in principle, be continued analytically from $\mathbb{R}^n$ into an open neighborhood $U \subset \mathbb{C}^n$ of $\mathbb{R}^n$ and in which sense T is a boundary value of this analytic continuation.

Let $C \subset \mathbb{R}^n$ be a convex cone with its dual $C' := \{p \in \mathbb{R}^n : p \cdot x \geq 0 \forall x \in C\}$ and assume that C' has a nonempty interior C°. Let $\mathscr{T} := \mathbb{R}^n \times (-C^\circ)$ be the induced open tube in $\mathbb{C}^n$. Here, the dot "$\cdot$" represents any scalar product on $\mathbb{R}^n$, that is any symmetric and nondegenerate bilinear form.

The particular case in which we are mainly interested is the case of the forward cone $C = C_+$ in $\mathbb{R}^D = \mathbb{R}^{1,D-1}$ with respect to the Minkowki scalar product. Here, the cone C is self-dual $C' = C$ and C° is the open forward cone

$$C^\circ = \{x \in \mathbb{R}^{1,D-1} : x^2 = <x,x> > 0, x^0 > 0\}$$

and $\mathscr{T} = \mathbb{R}^n \times (-C^\circ)$ is the backward tube.

Theorem 8.19. *For every distribution $T \in \mathscr{S}(\mathbb{R}^n)'$ whose Fourier transform has its support in the cone C there exists an analytic function F on the tube $\mathscr{T} \subset \mathbb{C}^n$ with*

- $|F(z)| \le c(1+|z|)^k(1+d^e(z,\partial\mathcal{T}))^{-m}$ for suitable $c \in \mathbb{R}$, $k,m \in \mathbb{N}$. (Here, d^e is the Euclidean distance in $\mathbb{C}^n = \mathbb{R}^{2n}$.)
- T is the boundary value of the holomorphic function F in the following sense. For any $f \in \mathscr{S}$ and $y \in -C^\circ \subset \mathbb{R}^n$:

$$\lim_{t \searrow 0} \int_{\mathbb{R}^n} f(x)F(x+ity)dx = T(f),$$

where the convergence is the convergence in $\mathscr{S}'$.

Proof. Let us first assume that $\widehat{T}$ is a polynomially bounded continuous function $g = g(p)$ with support in C. In that case the (Laplace transform) formula

$$F(z) := (2\pi)^{-n} \int_{\mathbb{R}^n} g(p)e^{-ip\cdot z}dp, \; z \in \mathcal{T},$$

defines a holomorphic function fulfilling the assertions of the theorem. Indeed, since the exponent $-ip\cdot z = -ip\cdot x + p\cdot y$ has a negative real part $p\cdot y < 0$ for all $z = x+iy \in \mathcal{T} = \mathbb{R}^n \times (-C^\circ)$ the integral is well-defined. F is holomorphic in z since one can take derivatives under the integral. To show the bounds is straightforward. Finally, for $y \in -C^\circ$ and $f \in \mathscr{S}(\mathbb{R}^n)$ the limit of

$$\int f(x)F(x+ity)dx = \int f(x)\left((2\pi)^{-n}\int g(p)e^{-ip\cdot x}e^{tp\cdot y}dp\right)dx$$

for $t \searrow 0$ is $\int f(x)\mathscr{F}^{-1}g(x)dx = T(f)$.

Suppose now that $\widehat{T}$ is of the form $P(-i\partial)g$ for a polynomial $P \in \mathbb{C}[X^1,\ldots,X^n]$ and g a polynomially bounded continuous function with support in C. Then

$$F(z) = P(z)(2\pi)^{-n}\int_{\mathbb{R}^n} g(p)e^{-ip\cdot z}dp, z \in \mathcal{T},$$

satisfies all conditions since $\mathscr{F}(P(x)\mathscr{F}^{-1}g) = P(-i\partial)g = \widehat{T}$.

Now the theorem follows from a result of [BEG67*] which asserts that for any distribution $S \in \mathscr{S}'$ with support in a convex cone C there exists a polynomial P and a polynomially bounded continuous function g with support in C and with $S = P(-i\partial)g$. $\qquad\Box$

We now draw our attention to the Wightman distributions.

Analytic Continuation of Wightman Functions. Given a Wightman QFT with field operator $\Phi: \mathscr{S}(\mathbb{R}^{1,D-1}) \longrightarrow \mathscr{O}$ (cf. Sect. 8.3) we explain in which sense and to which extent the corresponding Wightman distributions (cf. Sect. 8.4)

$$W_N \in \mathscr{S}'(\mathbb{R}^{DN})$$

can be continued analytically to an open connected domain $U_N \subset \mathbb{C}^{DN}$ of the complexification

$$\mathbb{C}^{DN} \cong \mathbb{R}^{DN} \otimes \mathbb{C}$$

of $\mathbb{R}^{DN}$.

The Minkowski inner product will be continued to a complex-bilinear form on $\mathbb{C}^D$ by $\langle z, w \rangle = z \cdot w = z^0 w^0 - \sum_{j=1}^{D-1} z^j w^j$.

An important and basic observation in this context is the possibility of identifying the Euclidean $\mathbb{R}^D$ with the real subspace

$$E := \{(it, x^1, \ldots, x^{D-1}) \in \mathbb{C}^D : (t, x^1, \ldots, x^{D-1}) \in \mathbb{R}^D\}$$

the "Euclidean points" of $\mathbb{C}^D$, since

$$\langle (it, x^1, \ldots, x^{D-1}), (it, x^1, \ldots, x^{D-1}) \rangle = -t^2 - \sum_{j=1}^{D-1} x^j x^j.$$

The Wightman distributions W_N will be analytically continued in three steps into open subsets U_N containing a great portion of the Euclidean points E^N, so that the restrictions of the analytically continued Wightman functions W_N to $U_N \cap E^N$ define a Euclidean field theory.

We have already used the fact that W_N is translation-invariant and therefore depends only on the differences $\xi_j := x_j - x_{j+1}$, $j = 1, \ldots, N-1$:

$$w_N(\xi_1, \ldots, \xi_{N-1}) := W_N(x_1, \ldots, x_N).$$

Each w_N is the inverse Fourier transform of its Fourier transform $\widehat{w}_N$, that is

$$w_N(\xi_1, \ldots, \xi_{N-1}) =$$
$$(2\pi)^{-D(N-1)} \int_{\mathbb{R}^{D(N-1)}} \widehat{w}_N(\omega_1, \ldots, \omega_{N-1}) e^{-i \sum_k \omega_k \cdot \xi_k} d\omega_1 \ldots d\omega_{N-1} \quad (8.7)$$

with

$$\widehat{w}_N(\omega_1, \ldots, \omega_{N-1}) = \int_{\mathbb{R}^{D(N-1)}} w_N(\xi_1, \ldots, \xi_{N-1}) e^{i \sum_k \omega_k \cdot \xi_k} d\xi_1 \ldots d\xi_{N-1}.$$

By the spectrum condition the Fourier transform $\widehat{w}_N(\omega_1, \ldots, \omega_{N-1})$ vanishes if one of the $\omega_1, \ldots, \omega_{N-1}$ lies outside the forward cone C_+ (cf. 8.17).

If we now take complex vectors $\zeta_k = \xi_k + i\eta_k \in \mathbb{C}^D$ instead of the ξ_k in the above formula for w_N, then the integrand in (8.7) has the form

$$\widehat{w}_N(\omega) e^{-i \sum_k \omega_k \cdot \xi_k} e^{\sum_k \omega_k \cdot \eta_k},$$

and the corresponding integral will converge if η_k fulfills $\sum_k \omega_k \cdot \eta_k < 0$ for all ω_k in the forward cone. With the N-fold backward tube $\mathscr{T}_N = (\mathbb{R}^D \times (-C^\circ))^N \subset (\mathbb{C}^D)^N$ this approach leads to the following result whose proof is similar to the proof of Theorem 8.19.

Proposition 8.20. *The formula*

$$w_N(\zeta) = (2\pi)^{-D(N-1)} \int \widehat{w}_N(\omega) e^{-i \sum_k \omega_k \cdot \zeta_k} d\omega, \zeta \in \mathscr{T}_{N-1},$$

provides a holomorphic function in $\mathscr{T}_{N-1}$ with the property

$$\lim_{t \searrow 0} w_N(\xi + it\eta) = w_N(\xi)$$

if $\xi + i\eta \in \mathscr{T}_{N-1}$ and where the convergence is the convergence in $\mathscr{S}'(\mathbb{R}^{D(N-1)})$.

As a consequence, the Wightman distributions have analytic continuations to $\{z \in (\mathbb{C}^D)^N : \mathrm{Im}(z_{j+1} - z_j) \in C^\circ\}$.

This first step of analytic continuation is based on the spectrum condition. In a second step the covariance is exploited.

The covariance implies that the identity

$$w_N(\zeta_1, \ldots, \zeta_{N-1}) = w_N(\Lambda \zeta_1, \ldots, \Lambda \zeta_{N-1}) \tag{8.8}$$

holds for $(\zeta_1, \ldots, \zeta_{N-1}) \in (R^D)^{N-1}$ and $\Lambda \in L$. Since analytic continuation is unique the identity also holds for $(\zeta_1, \ldots, \zeta_{N-1}) \in \mathscr{T}_{N-1}$ for those $(\zeta_1, \ldots, \zeta_{N-1})$ satisfying $(\Lambda \zeta_1, \ldots, \Lambda \zeta_{N-1}) \in \mathscr{T}_{N-1}$.

Moreover, the identity (8.8) extends to transformations Λ in the (proper) complex Lorentz group $L(\mathbb{C})$. This group $L(\mathbb{C})$ is the component of the identity of the group of complex $D \times D$-matrices Λ satisfying $\Lambda z \cdot \Lambda w = z \cdot w$ with respect to the complex Minkowski scalar product. This follows from the covariance and the fact that

$$\Lambda \mapsto w_N(\Lambda \zeta_1, \ldots, \Lambda \zeta_{N-1})$$

is holomorphic in a neighborhood of $\mathrm{id}_{\mathbb{C}^D}$ in $L(\mathbb{C})$. By the identity (8.8) one obtains an analytic continuation of w_N to $(\Lambda^{-1}(\mathscr{T}_{N-1}))^{N-1}$.
Let

$$\mathscr{T}_N^e = \bigcup \{\Lambda(\mathscr{T}_N) : \Lambda \in L(\mathbb{C})\}$$

be the *extended tube* where $\Lambda(\mathscr{T}_N) = \{(\Lambda \zeta_1, \ldots, \Lambda \zeta_N) : (\zeta_1, \ldots, \zeta_N) \in \mathscr{T}_N\}$. We have shown

Proposition 8.21. *w_N has an analytic continuation to the extended tube $\mathscr{T}_{N-1}^e$.*

While the tube $\mathscr{T}_N$ has no real points (that is points with only real coordinates $z_j \in \mathbb{R}^D$) as is clear from the definition of the tube, the extended tube contains real points.

For example, in the case $N = 1$ let $x \in \mathbb{R}^D$ be a real point with $x \cdot x < 0$. We can assume $x_2 = x_3 = \ldots = x_{D-1} = 0$ with $|x^1| > |x^0|$ by rotating the coordinate system. The complex Lorentz transformation $w = \Lambda z$, $w^0 = iz^1$, $w^1 = iz^0$ produces $w = \Lambda x$ with $\mathrm{Im}\, w^0 = x^1$, $\mathrm{Im}\, w^1 = x^0$, thus $\mathrm{Im}\, w \cdot \mathrm{Im}\, w = (x^1)^2 - (x^0)^2 > 0$ and $\Lambda x \in C^\circ$ if $x^1 < 0$. In the case $x^1 > 0$ one takes the transformation $w = \Lambda' z$, $w^0 = -iz^1$, $w^1 = -iz^0$. These two transformations are indeed in $L(\mathbb{C})$ since they can be connected with the identity by $\Lambda(\theta)$ acting on the first two variables by

$$\Lambda(\theta) = \begin{pmatrix} \cosh i\theta & \sinh i\theta \\ \sinh i\theta & \cosh i\theta \end{pmatrix} = \begin{pmatrix} \cos\theta & i\sin\theta \\ i\sin\theta & \cos\theta \end{pmatrix}$$

and leaving the remaining coordinates invariant.

We have proven that any real x with $x \cdot x < 0$ is contained in the extended tube $\mathscr{T}_1^e$. Similarly, one can show the converse, namely that a real x point of $\mathscr{T}_1^e$ satisfies $x \cdot x < 0$. In particular, the subset $\mathbb{R}^D \cap \mathscr{T}_1^e$ is open and not empty.

For general N, we have the following theorem due to Jost:

Theorem 8.22. *A real point* $(\zeta_1, \ldots, \zeta_N)$ *lies in the extended tube* $\mathscr{T}_N^e$ *if and only if all convex combinations*

$$\sum_{j=1}^{N} t_j \zeta_j, \ \sum_{j=1}^{N} t_j = 1, t_j \geq 0,$$

are space-like, that is $(\sum_{j=1}^{N} t_j \zeta_j)^2 < 0$.

In the third step of analytic continuation we exploit the locality. For a permutation $\sigma \in S_N$, that is a permutation of $\{1, \ldots, N\}$, let W_N^σ denote the Wightman distribution where the coordinates are interchanged by σ:

$$W_N^\sigma(x_1, \ldots, x_N) := W_N(x_{\sigma(1)}, \ldots, x_{\sigma(N)}),$$

and denote $w_N^\sigma(\xi_1, \ldots, \xi_{N-1}) = W_N(x_{\sigma(1)}, \ldots, x_{\sigma(N)}), \xi_j = x_j - x_{j+1}$.

Proposition 8.23. *Let* w_N *and* w_N^σ *be the holomorphic functions defined on the extended tube* $\mathscr{T}_{N-1}^e$ *by analytic continuation of the Wightman distributions* w_N *and* w_N^σ *according to Proposition 8.21. Then these holomorphic functions* w_N *and* w_N^σ *agree on their common domain of definition, which is not empty, and therefore define a holomorphic continuation on the union of their domains of definition.*

This result will be obtained by regarding the permuted tube $^\sigma\mathscr{T}_{N-1}'$ which is defined in analogy to $\Lambda\mathscr{T}_{N-1}$. The two domains $\mathscr{T}_{N-1}^e$ and $^\sigma\mathscr{T}_{N-1}^e$ have a nonempty open subset V of real points ξ with $\xi^2 < 0$ in common according to Theorem 8.22. Since all $\xi_j = x_j - x_{j+1}$ are space-like, this implies that w_N and w_N^σ agree on this open subset V and therefore w_N and w_N^σ agree in the intersection of the domains of definition.

We eventually have the following result:

Theorem 8.24. w_N *has an analytic continuation to the permuted extended tube* $\mathscr{T}_{N-1}^{pe} = \bigcup\{^\sigma\mathscr{T}_{N-1}^e : \sigma \in S_N\}$ *and similarly* W_N *has a corresponding analytic continuation to the permuted extended tube* $\mathscr{T}_N^{pe}$. *Moreover this tube contains all non-coincident points of* E^N.

Here E is the space of Euclidean points, $E := \{(it, x^1, \ldots, x^{D-1}) \in \mathbb{C}^D : (t, x^1, \ldots, x^{D-1}) \in \mathbb{R}^D\}$, and the last statement asserts that $E^N \setminus \Delta$ is contained in $\mathscr{T}_N^{pe}$ where $\Delta = \{(x_1, \ldots, x_N) \in E^N : x_j = x_k \text{ for some } j \neq k\}$.

As a consequence the W_N have an analytic continuation to $E^N \setminus \Delta$ and define the so-called Schwinger functions

$$S_N := W_N|_{E^N \setminus \Delta}.$$

8.6 Euclidean Formulation

In order to state the essential properties of the Schwinger functions S_N we use the Euclidean *time reflection*

$$\theta : E \to E, (it, x^1, \ldots, x^{D-1}) \mapsto (-it, x^1, \ldots, x^{D-1})$$

and its action Θ on

$$\mathscr{S}_+(\mathbb{R}^{DN}) = \{f : E^N \to \mathbb{C} : f \in \mathscr{S}(E^N) \text{ with support in } Q_+^N\},$$

where

$$Q_+^N = \{(x_1, \ldots, x_N) : x_j = (it_j, x_j^1, \ldots, x_j^{D-1}), 0 < t_1 < \ldots < t_N\} :$$
$$\Theta : \mathscr{S}_+(\mathbb{R}^{DN}) \to \mathscr{S}(\mathbb{R}^{DN}), \Theta f(x_1, \ldots, x_N) := \overline{f}(\theta x_1, \ldots, \theta x_N).$$

Theorem 8.25. *The Schwinger functions S_N are analytic functions $S_N : E^N \setminus \Delta \to \mathbb{C}$ satisfying the following axioms:*

S1 (Covariance) $S_N(gx_1, \ldots, gx_N) = S_N(x_1, \ldots, x_N)$ *for Euclidean motions* $g = (q, R), q \in \mathbb{R}^D, R \in \mathrm{SO}(D)$ *(or $R \in \mathrm{Spin}(D)$).*

S2 (Locality) $S_N(x_1, \ldots, x_N) = S_N(x_{\sigma(1)}, \ldots, x_{\sigma(N)})$ *for any permutation σ.*

S3 (Reflection Positivity)

$$\sum_{M,N} S_{M+N}(\Theta f_M \otimes f_N) \geq 0$$

for finite sequences $(f_N), f_N \in \mathscr{S}_+(\mathbb{R}^{DN})$, where, as before,

$$g_M \otimes f_N(x_1, \ldots, x_{M+N}) = g_M(x_1, \ldots, x_M) f_N(x_{M+1}, \ldots, x_{M+N}).$$

These properties of correlation functions are called the Osterwalder–Schrader axioms.

Reconstruction. Several slightly different concepts are called *reconstruction* in the context of axiomatic quantum field theory when Wightman's axioms are involved and also the Euclidean formulation (Osterwalder–Schrader axioms) is considered.

For example from the axioms S1–S3 one can deduce the Wightman distributions satisfying WD1–WD4 and this procedure can be called reconstruction. Moreover, after this step one can reconstruct the Hilbert space (cf. Theorem 8.18) with the relativistic fields Φ as in W1–W3. Altogether, on the basis of Schwinger functions and its properties we thus have reconstructed the relativistic fields and the corresponding Hilbert space of states. This procedure is also called reconstruction.

But starting with S1–S3 one could, as well, build a Euclidean field theory by constructing a Hilbert space directly with the aid of S3 and then define the Euclidean fields as operator-valued distributions similar to the reconstruction of the relativistic fields as described in Sect. 8.4, in particular in the proof of the Wightman Reconstruction Theorem 8.18. Of course, this procedure is also called reconstruction. In the next chapter this kind of reconstruction is described with some additional details in Sects. 9.2 and 9.3 in the two-dimensional case.

8.7 Conformal Covariance

The theories described in this chapter do not incorporate conformal symmetry, so far. Let us describe how the covariance with respect to conformal mappings can be formulated within the framework of the axioms. Recall (cf. Theorem 1.9) that the conformal mappings not already included in the Poincaré group resp. the Euclidean group of motions are the special conformal transformations

$$q \mapsto q^b = \frac{q - \langle q, q \rangle b}{1 - 2\langle q, b \rangle + \langle q, q \rangle \langle b, b \rangle}, \, q \in \mathbb{R}^n,$$

where $b \in \mathbb{R}^n$, and the dilatations

$$q \mapsto q^\lambda = e^\lambda q, q \in \mathbb{R}^n,$$

where $\lambda \in \mathbb{R}$.

The Wightman Axioms 8.3 are now extended in such a way that one requires U to be a unitary representation $U = U(q, \Lambda, b)$ of the conformal group $SO(n, 2)$ or $SO(n, 2)/\{\pm 1\}$ (cf. Sect. 2.2), resp. of its universal covering, such that in addition to the Poincaré covariance

$$U(q, \Lambda)\Phi_a(x)U(q, \Lambda)^* = \Phi_a(\Lambda x),$$

the following has to be satisfied:

$$U(0, E, b)\Phi_a(x)U(0, E, b)^* = N(q, b)^{-h_a}\Phi_a(x^b),$$

where $N(x, b) = 1 - 2\langle q, b \rangle + \langle q, q \rangle \langle b, b \rangle$ is the corresponding denominator and where $h_a \in \mathbb{R}$ is a so-called *conformal weight* of the field Φ_a. Moreover, the conformal covariance for the dilatations is

$$U(\lambda)\Phi_a(x)U(\lambda)^* = e^{\lambda d_a}\Phi_a(x^\lambda),$$

with a similar weight d_a. Observe that $N(x,b)^{-n}$ resp. $e^{n\lambda}$ is the Jacobian of the transformation x^b resp. x^λ.

We now turn our attention to the two-dimensional case. Since the Lorentz group of the Minkowski plane is isomorphic to the abelian group $\mathbb{R}$ (cf. Remark 1.15) and the rotation group of the Euclidean plane is isomorphic to $\mathbb{S}$, the one-dimensional representations of the isometry groups are no longer trivial (as in the higher-dimensional case). Consequently, in the covariance condition, in principle, these one-dimensional representations could occur, see also Remark 8.11. As an example, one can expect that the Lorentz boosts

$$\Lambda = \begin{pmatrix} \cosh\chi & \sinh\chi \\ \sinh\chi & \cosh\chi \end{pmatrix}, \quad \chi \in \mathbb{R},$$

in the two-dimensional case satisfy the following covariance condition:

$$U(\Lambda)\Phi_a(x)U(\Lambda)^* = e^{\chi s_a}\Phi_a(\Lambda x),$$

where s_a would represent a spin of the field. Similarly, in the Euclidean case

$$U(\Lambda)\Phi_a(x)U(\Lambda)^* = e^{i\alpha s_a}\Phi_a(\Lambda x),$$

if α is the angle of the rotation Λ.

It turns out that in two-dimensional conformal field theory this picture is even refined further when formulating the covariance condition for the other conformal transformations. The light cone coordinates are regarded separately in the Minkowski case and similarly in the Euclidean case the coordinates are split into the complex coordinate and its conjugate.

With respect to the Minkowski plane one first considers the restricted conformal group (cf. Remark 2.16) only which is isomorphic to $SO(2,2)/\{\pm 1\}$ and not the full infinite dimensional group of conformal transformations. With respect to the light cone coordinates the restricted conformal group $SO(2,2)/\{\pm 1\}$ acts in the form of two copies of $SL(2,\mathbb{R})/\{\pm 1\}$ (cf. Proposition 2.17). For a conformal transformation $g = (A_+, A_-), A_\pm \in SL(2,\mathbb{R})$,

$$A_+ = \begin{pmatrix} a_+ & b_+ \\ c_+ & d_+ \end{pmatrix}, \quad A_- = \begin{pmatrix} a_- & b_- \\ c_- & d_- \end{pmatrix},$$

with the action

$$(A_+, A_-)(x_+, x_-) = \left(\frac{a_+ x_+ + b_+}{c_+ x_+ + d_+}, \frac{a_- x_- + b_-}{c_- x_- + d_-} \right),$$

the covariance condition now reads

$$U(g)\Phi_a(x)U(g)^* = \left(\frac{1}{(c_+x_+ + d_+)^2}\right)^{h_a^+} \left(\frac{1}{(c_-x_- + d_-)^2}\right)^{h_a^-} \Phi_a(gx),$$

where the conformal weights h_a^+, h_a^- are in general independent of each other. Note that the factor

$$\frac{1}{(c_+x_+ + d_+)^2}$$

is the derivative of

$$x_+ \mapsto A_+(x_+) = \frac{a_+x_+ + b_+}{c_+x_+ + d_+},$$

and therefore essentially the conformal factor.

The boost described above is given by $g = (A_+, A_-)$ with $a_+ = \exp\frac{1}{2}\chi = d_-, d_+ = \exp-\frac{1}{2}\chi = a_-$, the bs and cs being zero. By comparison we obtain

$$s_a = h_a^+ - h_a^-,$$

for the spins s_a and, similarly, for the weights d_a related to the dilatations:

$$d_a = h_a^+ + h_a^-.$$

In the Euclidean case one writes the general point in the Euclidean plane as $z = x + iy$ or $t + iy$ and $\bar{z} = x - iy$. The conformal covariance for the rotation $w(z) = e^{i\alpha}z$ will correspondingly be formulated by

$$U(\Lambda)\Phi_a(z)U(\Lambda)^* = \left(\frac{dw}{dz}\right)^{h_a} \left(\overline{\frac{dw}{dz}}\right)^{\bar{h}_a} \Phi_a(w),$$

where again $h_a, \bar{h}_a$ are independent. Equivalently, one writes

$$U(\Lambda)\Phi_a(z,\bar{z})U(\Lambda)^* = \left(\frac{dw}{dz}\right)^{h_a} \left(\overline{\frac{dw}{dz}}\right)^{\bar{h}_a} \Phi_a(w,\bar{w}),$$

emphasizing the two components of z resp. w (cf. the Axiom 2 in the following chapter). This is the formulation of covariance for other conformal transformations as well.

References

[BLT75*]   N.N. Bogolubov, A.A. Logunov, and I.T. Todorov. *Introduction to Axiomatic Quantum Field Theory*. Benjamin, Reading, MA, 1975.

[BEG67*]   J. Bros, H. Epstein, and V. Glaser. On the connection between analyticity and Lorentz covariance of Wightman functions. *Comm. Math. Phys.* **6** (1967), 77–100.

[Haa93*]   R. Haag. *Local Quantum Physics*. Springer-Verlag, Berlin, 2nd ed., 1993.

[JW06*]   A. Jaffe and E. Witten. Quantum Yang-Mills theory. In: *The Millennium Prize Problems*, 129–152. Clay Mathematics Institute, Cambridge, MA, 2006.
[OS73]   K. Osterwalder and R. Schrader. Axioms for Euclidean Green's functions I. *Comm. Math. Phys.* **31** (1973), 83–112.
[OS75]   K. Osterwalder and R. Schrader. Axioms for Euclidean Green's functions II. *Comm. Math. Phys.* **42** (1975), 281–305.
[RS80*]   M. Reed and B. Simon. *Methods of modern Mathematical Physics, Vol. 1: Functional Analysis*. Academic Press, New York, 1980.
[Rud73*]   W. Rudin. *Functional Analysis*. McGraw-Hill, New York, 1973.
[Simo74*] B. Simon. *The $P(\phi)_2$ Euclidian (Quantum) Field Theory*. Princeton Series in Physics, Princeton University Press, Princeton, NJ, 1974.
[SW64*]   R. F. Streater and A. S. Wightman. *PCT, Spin and Statistics, and All That*. Princeton University Press, Princeton, NJ, 1964 (Corr. Third printing 2000).

Chapter 9
Foundations of Two-Dimensional Conformal Quantum Field Theory

In this chapter we study *two-dimensional conformally invariant quantum field theory* (*conformal field theory* for short) by some basic concepts and postulates – that is using a system of axioms as presented in [FFK89] and based on the work of Osterwalder and Schrader [OS73], [OS75]. We will assume the Euclidean signature $(+, +)$ on $\mathbb{R}^2$ (or on surfaces), as it is customary because of the close connection of conformal field theory to statistical mechanics (cf. [BPZ84] and [Gin89]) and its relation to complex analysis.

We do not use the results of Chap. 8 where the axioms of quantum field theory are investigated in detail and for arbitrary spacetime dimensions nor do we assume the notations to be known in order to keep this chapter self-contained. However, the preceding chapter may serve as a motivation for several concepts and constructions. In particular, the presentation of the axioms explains why locality for the correlation functions in Axiom 1 below is expressed as the independence of the order of the indices, and why the covariance in Axiom 2 does not refer to the unitary representation of the Poincaré group. Moreover, in the light of the results of the preceding chapter the reconstruction used below on p. 158 is a general principle in quantum field theory relating the formulation based on field operators with an equivalent formulation based on correlation functions.

9.1 Axioms for Two-Dimensional Euclidean Quantum Field Theory

The basic objects of a two-dimensional quantum field theory (cf. [BPZ84], [IZ80], [Gaw89], [Gin89], [FFK89], [Kak91], [DMS96*]) are the fields Φ_i, $i \in B_0$, subject to a number of properties. These fields are also called field operators or operators. They are defined as maps on open subsets M of the complex plane $\mathbb{C} \cong \mathbb{R}^{2,0}$ (or on Riemann surfaces M). They take their values in the set $\mathcal{O} = \mathcal{O}(\mathbb{H})$ of (possibly unbounded and mostly self-adjoint) operators on a fixed Hilbert space $\mathbb{H}$. To be precise, these field operators are usually defined only on spaces of test functions on M, e.g. on the Schwartz space $\mathscr{S}(M)$ of rapidly decreasing functions or on other

Schottenloher, M.: *Foundations of Two-Dimensional Conformal Quantum Field Theory*. Lect. Notes Phys. **759**, 153–170 (2008)
DOI 10.1007/978-3-540-68628-6_10   © Springer-Verlag Berlin Heidelberg 2008

suitable spaces of test functions. Hence, they can be regarded as operator-valued distributions (cf. Definition 8.8).

The *matrix coefficients* $\langle v|\Phi_i(z)|w\rangle$ of the field operators are supposed to be well-defined for $v, w \in D$ in a dense subspace $D \subset \mathbb{H}$. Here, $\langle v, w\rangle$, $v, w \in \mathbb{H}$, denotes the inner product of $\mathbb{H}$ and $\langle v|\Phi_i(z)|w\rangle$ is the same as $\langle v, \Phi_i(z)w\rangle$.

The essential parameters of the theory, which connect the theory with experimental data, are the *correlation functions*

$$G_{i_1\ldots i_n}(z_1,\ldots,z_n) := \langle \Omega|\Phi_{i_1}(z_1)\ldots\Phi_{i_n}(z_n)|\Omega\rangle.$$

These functions are also called *n-point functions* or *Green's functions*. Here, $\Omega \in \mathbb{H}$ is the vacuum vector. These correlation functions have to be interpreted as *vacuum expectation values* of time-ordered products $\Phi_{i_1}(z_1)\ldots\Phi_{i_n}(z_n)$ of the field operators (*time ordered* means $\mathrm{Re}\, z_n > \ldots > \mathrm{Re}\, z_1$, or $|z_n| > \ldots > |z_1|$ for the radial quantization). They usually can be analytically continued to

$$M_n := \{(z_1,\ldots,z_n) \in \mathbb{C}^n : z_i \neq z_j \quad \text{for } i \neq j\},$$

the space of configurations of n points. (To be precise, they have a continuation to the universal covering $\tilde{M}_n$ of M_n and thus they are no longer single valued on M_n, in general. In this manner, the pure braid group P_n appears, which is the fundamental group $\pi_1(M_n)$ of M_n.) For simplification we will assume in the formulation of the axioms that the $G_{i_1\ldots i_n}$ are defined on M_n.

The positivity of the hermitian form, that is the inner product of $\mathbb{H}$, can be expressed by the so-called reflection positivity of the correlation functions. This property is defined by fixing a reflection axis – which typically is the imaginary axis in the simplest case – and requiring the correlation of operator products of fields on one side of the axis with their reflection on the other side to be non-negative (cf. Axiom 3 below).

Now, the two-dimensional quantum field theory can be described completely by the properties of the correlation functions using a system of axioms (Axiom 1–6 in these notes, see below). The field operators and the Hilbert space do not have to be specified a priori, they are determined by the correlation functions (cf. Lemma 9.2 and Theorem 9.3).

To state the axioms we need a few notations:

$$M_n^+ := \{(z_1,\ldots,z_n) \in M_n : \mathrm{Re}\, z_j > 0 \quad \text{for } j = 1,\ldots,n\},$$
$$\mathscr{S}_0^+ := \mathbb{C},$$
$$\mathscr{S}_n^+ := \{f \in \mathscr{S}(\mathbb{C}^n) : \mathrm{Supp}(f) \subset M_n^+\}.$$

Here, $\mathscr{S}(\mathbb{C}^n)$ is the Schwartz space of rapidly decreasing smooth functions, that is the complex vector space of all functions $f \in C^\infty(\mathbb{C}^n)$ for which

$$\sup_{|\alpha|\leq p} \quad \sup_{x\in\mathbb{R}^{2n}} |\partial^\alpha f(x)|(1+|x|^2)^k < \infty,$$

for all $p, k \in \mathbb{N}$. We have identified the spaces $\mathbb{C}^n$ and $\mathbb{R}^{2n}$ and have used the real coordinates $x = (x_1, \ldots, x_{2n})$ as variables. ∂^α is the partial derivative for the multi-index $\alpha \in \mathbb{N}^{2n}$ with respect to x. Supp(f) denotes the support of f, that is the closure of the set $\{ x \in \mathbb{R}^{2n} : f(x) \neq 0 \}$.

It makes sense to write $z \in \mathbb{C}$ as $z = t + iy$ with $t, y \in \mathbb{R}$, and to interpret $\bar{z} = t - iy$ as a quantity not depending on z. In this sense one sometimes writes $G(z, \bar{z})$ instead of $G(z)$, to emphasize that $G(z)$ is not necessarily holomorphic. In the notation $z = t + iy$, y is the "space coordinate" and t is the (imaginary) "time coordinate".

The group $E = E_2$ of Euclidean motions, that is the Euclidean group (which corresponds to the Poincaré group in this context), is generated by the rotations

$$r_\alpha : \mathbb{C} \to \mathbb{C}, \quad z \mapsto e^{i\alpha} z, \quad \alpha \in \mathbb{R},$$

and the translations

$$t_a : \mathbb{C} \to \mathbb{C}, \quad z \mapsto z + a, \quad a \in \mathbb{C}.$$

Further Möbius transformations are the dilatations

$$d_\tau : \mathbb{C} \to \mathbb{C}, \quad z \mapsto e^\tau z, \quad \tau \in \mathbb{R},$$

and the inversion

$$i : \mathbb{C} \to \mathbb{C}, \quad z \mapsto z^{-1}, \quad z \in \mathbb{C} \setminus \{0\}.$$

These conformal transformations generate the Möbius group Mb (cf. Sect. 2.3). All other global conformal transformations (cf. Definition 2.10) of the Euclidean plane (with possibly one singularity) are generated by Mb and the *time reflection*

$$\theta : \mathbb{C} \to \mathbb{C}, \quad z = t + iy \mapsto -t + iy = -\bar{z}.$$

(cf. Theorems 1.11 and 2.11 and the discussion after Definition 2.12)

Osterwalder–Schrader Axioms ([OS73], [OS75], [FFK89])
Let B_0 be a countable index set. For multi-indices $(i_1, \ldots, i_n) \in B_0^n$ we also use the notation $i = i_1 \ldots i_n = (i_1, \ldots, i_n)$. Let $B = \bigcup_{n \in \mathbb{N}_0} B_0^n$. The quantum field theory is described by a family $(G_i)_{i \in B}$ of continuous and polynomially bounded correlation functions

$$G_{i_1 \ldots i_n} : M_n \to \mathbb{C}, \quad G_\emptyset = 1,$$

subject to the following axioms:

Axiom 1 (Locality) *For all* $(i_1, \ldots, i_n) \in B_0^n, (z_1, \ldots, z_n) \in M_n$, *and every permutation* $\pi : \{1, \ldots, n\} \to \{1, \ldots, n\}$ *one has*

$$G_{i_1, \ldots, i_n}(z_1, \ldots, z_n) = G_{i_{\pi(1)} \ldots i_{\pi(n)}}(z_{\pi(1)}, \ldots, z_{\pi(n)}).$$

Axiom 2 (Covariance) *For every* $i \in B_0$ *there are* conformal weights $h_i, \bar{h}_i \in \mathbb{R}$ ($\bar{h}_i$ *is not the complex conjugate of* h_i, *but completely independent of* h_i), *such that for all* $w \in E$ *and* $n \geq 1$ *one has*

$$G_{i_1 \ldots i_n}(z_1, \overline{z}_1, \ldots, z_n, \overline{z}_n)$$

$$= \prod_{j=1}^{n} \left(\frac{dw}{dz}(z_j) \right)^{h_j} \left(\overline{\frac{dw}{dz}(z_j)} \right)^{\overline{h}_j} G_{i_1 \ldots i_n}(w_1, \overline{w}_1, \ldots, w_n, \overline{w}_n), \qquad (9.1)$$

with $w_j := w(z_j)$, $\overline{w}_j := \overline{w(z_j)}$, $h_j := h_{i_j}$.

Here, $s_i := h_i - \overline{h}_i$ is called the *conformal spin* for the index i and $d_i := h_i + \overline{h}_i$ is called the *scaling dimension*.

Furthermore, we assume

$$h_i - \overline{h}_i, h_i + \overline{h}_i \in \mathbb{Z}, \quad i \in B_0.$$

As a consequence, there do not occur any ambiguities concerning the exponents. In particular, this is satisfied whenever

$$h_i, \overline{h}_i \in \frac{1}{2}\mathbb{Z}.$$

See Hawley/Schiffer [HS66] for a discussion of this condition.

The covariance of the correlation functions formulated in Axiom 2 corresponds to the transformation behavior of tensors or generalized differential forms under change of coordinates when extended to more general conformal transformations (see also p. 164).

The covariance conditions severely restricts the form of 2-point functions and 3-point functions. Because of the covariance with respect to translations, all correlation functions $G_{i_1 \ldots i_n}$ for $n \geq 2$ depend only on the differences $z_{ij} := z_i - z_j$, $i \neq j$, $i, j \in \{1, \ldots, n\}$. Typical 2-point functions $G_{i_1 i_1} = G$, which satisfy Axiom 2, are

$$G = \text{const.} \qquad \text{with} \quad h = \overline{h} = 0,$$
$$G(z_1, \overline{z}_1, z_2, \overline{z}_2) = C z_{12}^{-2} \overline{z}_{12}^{-2} \quad \text{with} \quad h = \overline{h} = 1,$$
$$G(z_1, z_2) = C z_{12}^{-4} \qquad \text{with} \quad h = 2, \overline{h} = 0.$$

A general example is

$$G(z_1, z_2) = C z_{12}^{-2h} \overline{z}_{12}^{-2\overline{h}} \quad \text{with} \quad h, \overline{h} \in \frac{1}{2}\mathbb{Z}.$$

Hence, for the case $h = \overline{h}$,

$$G(z_1, \overline{z}_1, z_2, \overline{z}_2) = C|z_{12}|^{-4h} = C|z_{12}|^{-2d}.$$

Typical 2-point functions $G = G_{i_1 i_2}$ with $i_1 \neq i_2$, for which Axiom 2 is valid, are

$$G(z_1, \overline{z}_1, z_2, \overline{z}_2) = C z_{12}^{-h_1} z_{12}^{-h_2} \overline{z}_{12}^{-\overline{h}_1} \overline{z}_{12}^{-\overline{h}_2}.$$

All these examples are also Möbius covariant.

For the function $F = G_{i_1 i_1}$ with

$$F(z_1, \bar{z}_1, z_2, \bar{z}_2) = -\log |z_{12}|^2$$

Axioms 1 and 2 hold as well (with arbitrary $h, \bar{h}, h = \bar{h}$). However, this function is not Möbius covariant because one has e.g., for $w(z) = e^\tau z$, $\tau \neq 0$, and in the case $h = \bar{h} \neq 0$,

$$\prod_{j=1}^{2} \left(\frac{dw}{dz}(z_j) \right)^h \left(\frac{\overline{dw}}{dz}(z_j) \right)^{\bar{h}} F(w_1, w_2)$$

$$= (e^\tau)^{2h + 2\bar{h}} (-\log e^{2\tau} |z_{12}|^2) \neq -\log |z_{12}|^2.$$

In particular, F is not scaling covariant in the sense of Axiom 4 (see below). A typical 3-point function is

$$G(z_1, \bar{z}_1, z_2, \bar{z}_2, z_3, \bar{z}_3)$$
$$= z_{12}^{-h_1 - h_2 + h_3} z_{23}^{-h_2 - h_3 + h_1} z_{13}^{-h_3 - h_1 + h_2}$$
$$\bar{z}_{12}^{-\bar{h}_1 - \bar{h}_2 + \bar{h}_3} \bar{z}_{23}^{-\bar{h}_2 - \bar{h}_3 + \bar{h}_1} \bar{z}_{13}^{-\bar{h}_3 - \bar{h}_1 + \bar{h}_2}, \qquad (9.2)$$

as can be checked easily. It is not difficult to see that this 3-point function is also Möbius covariant, hence conformally covariant.

We describe a rather simple example involving all correlation functions.

Example 9.1. Let $B_0 = \{1\}$ and $n := (1, \ldots 1) \in B_0^n = \{n\}$. The functions G_n are supposed to be zero if n is odd and

$$G_{2n}(z_1, \ldots, z_{2n}) = \frac{k^n}{2^n n!} \sum_{\sigma \in S_{2n}} \prod_{j=1}^{n} \frac{1}{(z_{\sigma(j)} - z_{\sigma(n+j)})^2},$$

where S_N is the group of permutations of N elements and where $k \in \mathbb{C}$ is a constant. The weights are $h_1 = 1$, $\bar{h}_1 = 0$.

If the exponent "2" in the denominator is replaced with $2m$ we get another example with conformal weight $h = m$ instead of 1 and $\bar{h} = 0$.

The dependence in z and $\bar{z}$ can be treated independently, as in the example. The example can be extended by defining $F_{2n}(z, \bar{z}) = G_{2n}(z) G_{2n}(\bar{z})$, and the resulting theory has the weights $h_1 = 1 = \bar{h}_1$.

Note that the correlation functions in Example 9.1 are covariant with respect to general Möbius transformations, even if the $\bar{z}$-dependence is included. Möbius covariance (and hence conformal covariance) holds as well if the exponent 2 is replaced by $2m$.

In the following, we mostly treat only the dependence in z in order to simplify the formulas. The general case can easily be derived from the formulas respecting only the dependence on z (see p. 88 for an explanation).

Next, we formulate reflection positivity (cf. Sect. 8.6). Let $\mathscr{L}^+$ be the space of all sequences $\underline{f} = (f_i)_{i \in B}$ with $f_i \in \mathscr{S}_n^+$ for $i \in B_0^n$ and $f_i \neq 0$ for at most finitely many $i \in B$.

Axiom 3 (Reflection Positivity) *There is a map* $* : B_0 \to B_0$ *with* $*^2 = \mathrm{id}_{B_0}$ *and a continuation* $* : B \to B, i \mapsto i^*$, *so that*

1. *$G_i(z) = G_{i^*}(\theta(z)) = G_{i^*}(-z^*)$ for $i \in B$, where z^* is the complex conjugate of z.*
2. *$\langle \underline{f}, \underline{f} \rangle \geq 0$ for all $\underline{f} \in \mathscr{L}^+$.*

Here, $\langle \underline{f}, \underline{f} \rangle$ *is defined by*

$$\sum_{i,j \in B} \sum_{n,m} \int_{M_{n+m}} G_{i^* j}(\theta(z_1), \ldots, \theta(z_n), w_1, \ldots, w_m) f_i(z)^* f_j(w) d^n z d^m w.$$

In the Example 9.1 for $*1 = 1$ the two conditions of Axiom 3 are satisfied.

Lemma 9.2 (Reconstruction of the Hilbert Space). *Axiom 3 yields a positive semi-definite hermitian form H on $\mathscr{L}^+$ and hence the Hilbert space $\mathbb{H}$ as the completion of $\mathscr{L}^+ / \ker H$ with the inner product $\langle \, , \, \rangle$.*

We now obtain the field operators by using a multiplication in $\mathscr{L}^+$ in the same way as in the proof of the Wightman Reconstruction Theorem 8.18. Indeed, Φ_j for $j \in B_0$ shall be defined on the space $\mathscr{S}^+ = \mathscr{S}_1^+$ of distributions with values in a space of operators on $\mathbb{H}$. Given $f \in \mathscr{S}_1^+$ and $\underline{g} \in \mathscr{L}^+$, $\underline{g} = (g_i)_{i \in B}$, we define $\Phi_j(f)([\underline{g}])$ to be the equivalence class (with respect to $\ker H$) of $\underline{g} \times f$ (the expected value of Φ_j at f), with

$$\underline{g} \times f = ((\underline{g} \times f)_{i_1 \ldots i_{n+1}})_{i_1 \ldots i_{n+1} \in B},$$

where

$$(\underline{g} \times f)_{i_1 \ldots i_{n+1}}(z_1, \ldots, z_{n+1}) := g_{i_1 \ldots i_n}(z_1, \ldots, z_n) f(z_{n+1}) \delta_{j i_{n+1}}.$$

It can be shown (cf. [OS73], [OS75]) that this construction yields a unitary representation U of the group E of Euclidean motions of the plane in $\mathbb{H}$. Moreover, there exists a dense subspace $D \subset \mathbb{H}$ left invariant by the unitary representation such that the maps $\Phi_j(f) : [\underline{g}] \mapsto [\underline{g} \times f]$ are defined on D for all $j \in B_0$ and $\Phi_j(f)(D) \subset D$. In addition, with the vacuum $\Omega \in \mathbb{H}$ (namely $\Omega = [\underline{f}]$, with $f_{\emptyset} = 1$ and $f_i = 0$ for $i \neq \emptyset$) the following properties are satisfied:

Theorem 9.3. (Reconstruction of the Field Operators)

1. *For all $j \in B_0$ the mapping $\Phi_j : \mathscr{S}^+ \to \mathrm{End}(D)$ is linear, and Φ_j is a field operator. Moreover, $\Phi_j(D) \subset D$, $\Omega \in D$, and the unitary representation U leaves Ω invariant.*

2. *The fields Φ_j transform covariantly with respect to the representation U:*

$$U(w) \Phi_j(z) U(w)^* = \left(\frac{\partial w}{\partial z} \right)^{h_i} \Phi_j(w(z)).$$

3. *The matrix coefficients $\langle\Omega|\Phi_i(f)|\Omega\rangle$ can be represented by analytic functions and for* $\operatorname{Re} z_n > \ldots > \operatorname{Re} z_1 > 0$ *the correlation functions agree with the given functions*

$$\langle\Omega|\Phi_{i_1}(z_1)\ldots\Phi_{i_n}(z_n)|\Omega\rangle = G_{i_1\ldots i_n}(z_1,\ldots,z_n).$$

Furthermore, if the dependence on z and $\bar{z}$ is taken into account the corresponding correlation functions $G_{i_1\ldots i_n}(z_1,\bar{z}_1,\ldots,z_n,\bar{z}_n)$ are holomorphic in $M_n^> \times M_n^>$, where

$$M_n^> := \{z \in M_n^+ : \operatorname{Re} z_n > \ldots > \operatorname{Re} z_1 > 0\}.$$

They can be analytically continued into a larger domain $N \subset \mathbb{C}^n \times \mathbb{C}^n$. A general description of the largest domain (the domain of holomorphy for the $G_{i_1\ldots i_n}$) is not known.

Similar results are true for other regions in $\mathbb{C}$ instead of the right half plane

$$\{w \in \mathbb{C} : \operatorname{Re} w > 0\},$$

e.g., for the disc (radial quantization). In this case the points $z \in \mathbb{C}$ are parameterized as $z = e^{\tau+i\alpha}$ with the time variable τ and the space variable α, which is cyclic. The time order becomes $|z_n| > \ldots > |z_1|$.

The Axioms 1–3 describe essentially a general two-dimensional Euclidian field theory as in Sect. 8.6 where no conformal invariance is required.

9.2 Conformal Fields and the Energy–Momentum Tensor

A two-dimensional quantum field theory with field operators

$$(\Phi_i)_{i\in B_0},$$

satisfying Axioms 1–3, is a *conformal field theory* if the following conditions hold:

- the theory is covariant with respect to dilatations (Axiom 4),
- it has a divergence-free energy–momentum tensor (Axiom 5), and
- it has an associative operator product expansion for the primary fields (Axiom 6).

Axiom 4 (Scaling Covariance) *The correlation functions*

$$G_i, i \in B,$$

satisfy (34) also for the dilatations $w(z) = e^\tau z$, $\tau \in \mathbb{R}$. *Hence*

$$G_i(z_1,\ldots,z_n) = (e^\tau)^{h_1+\ldots+h_n+\bar{h}_1+\ldots+\bar{h}_n} G_i(e^\tau z_1,\ldots,e^\tau z_n)$$

for $(z_1,\ldots,z_n) \in M$, $i = (i_1,\ldots,i_n)$ *and* $h_j = h_{i_j}$.

The correlation functions in the Example 9.1 are scaling covariant.

Lemma 9.4. *In a quantum field theory satisfying Axioms 1–4, any 2-point function* G_{ij} *has the form*

$$G_{ij}(z_1, z_2) = C_{ij} z_{12}^{-(h_i + h_j)} \bar{z}_{12}^{-(\bar{h}_i + \bar{h}_j)} \quad (z_{12} = z_1 - z_2)$$

with a suitable constant $C_{ij} \in \mathbb{C}$. *Hence, for* $i = j$,

$$G_{ii}(z_1, z_2) = C_{ii} z_{12}^{-2h} \bar{z}_{12}^{-2\bar{h}}.$$

Similarly, any 3-point function G_{ijk} *is a constant multiple of the function G in (9.1):*

$$G_{ijk} = C_{ijk} G, \text{ with } C_{ijk} \in \mathbb{C}.$$

In particular, the 2- and 3-point functions are completely determined by the constants C_{ij}, C_{ijk}.

Proof. As a consequence of the covariance with respect to translations, $G := G_{ij}$ depends only on $z_{12} = z_1 - z_2$, that is $G(z_1, z_2) = G_{ij}(z_1 - z_2, 0)$. For $z = re^{i\alpha} = e^\tau e^{i\alpha}$ one has $G(z, 0) = G(e^{\tau + i\alpha} 1, 0)$. From Axioms 2 and 4 it follows

$$G(1, 0) = (e^{\tau + i\alpha})^{h_i} (e^{\tau - i\alpha})^{\bar{h}_i} (e^{\tau + i\alpha})^{h_j} (e^{\tau - i\alpha})^{\bar{h}_j} G(e^{\tau + i\alpha} 1, 0).$$

This implies $G(z, 0) = z^{-(h_i + h_j)} \bar{z}^{-(\bar{h}_i + \bar{h}_j)} G(1, 0)$, $C := G(1, 0)$.
A similar consideration leads to the assertion on 3-point functions. □

The 4-point functions are less restricted, but they have a specific form for all theories satisfying Axioms 1–3 where the correlation functions are Möbius covariant. To show this, one can use the following differential equations:

Proposition 9.5 (Conformal Ward Identities). *Under the assumption that the correlation function* $G = G_{i_1 \ldots i_n}(z_1, \ldots, z_n)$ *satisfies the covariance condition (9.1) for all Möbius transformations the following Ward identities hold:*

$$0 = \sum_{j=1}^{n} \partial_{z_j} G(z_1, \ldots, z_n),$$

$$0 = \sum_{j=1}^{n} (z_j \partial_{z_j} + h_j) G(z_1, \ldots, z_n),$$

$$0 = \sum_{j=1}^{n} (z_j^2 \partial_{z_j} + 2 h_j z_j) G(z_1, \ldots, z_n)$$

Proof. These identities are shown in the same way as Lemma 9.4. We focus on the third identity. The Möbius covariance applied to the conformal transformation

$$w = w(z) = \frac{z}{1 - \zeta z}$$

with a complex parameter ζ yields

$$G(z_1,\ldots,z_n) = \prod_{i=1}^{n}\left(\frac{1}{1-\zeta z_i}\right)^{2h_i} G(w_1,\ldots,w_n)$$

because of

$$\frac{\partial w}{\partial z} = \frac{1}{(1-\zeta z)^2},$$

where $w_j = w(z_j)$. The derivative of this equality with respect to ζ is

$$0 = \prod_{i=1}^{n}\left(\frac{1}{1-\zeta z_i}\right)^{2h_i}\sum_{j=1}^{n} 2h_j \frac{1}{1-\zeta z_j} z_j G(w_1,\ldots,w_n)$$

$$+ \prod_{i=1}^{n}\left(\frac{1}{1-\zeta z_i}\right)^{2h_i}\sum_{j=1}^{n}\frac{z_j^2}{(1-\zeta z_j)^2}\partial_{z_j}G(w_1,\ldots,w_n),$$

from which the identity follows by setting $\zeta = 0$. $\qquad\square$

It can be seen that the solutions of these differential equations in the case of $n = 4$ are of the following form:

$$G(z_1,z_2,z_3,z_4) = F(r(z),\overline{r(z)})\prod_{i<j}z_{ij}^{-(h_i+h_j)+\frac{1}{3}h}\prod_{i<j}\overline{z}_{ij}^{-(\overline{h}_i+\overline{h}_j)+\frac{1}{3}\overline{h}},$$

where $h = h_1 + h_2 + h_3 + h_4$ and correspondingly for $\overline{h}$, and where F is a holomorphic function in the cross-ratio

$$r(z) := (z_{12}z_{34})/(z_{13}z_{24})$$

of the $z_{12}, z_{34}, z_{13}, z_{24}$ and in $\overline{r(z)}$.

Analogous statements hold for the n-point functions, $n \geq 5$. As an essential feature of conformal field theory we observe that the form of the n-point functions can be determined by using the global conformal symmetry. They turn out to be Laurent monomials in the $z_{ij}, \overline{z}_{ij}$ up to a factor similar to F.

Axiom 5 (Existence of the Energy–Momentum Tensor)
Among the fields $(\Phi_i)_{i\in B_0}$ there are four fields $T_{\mu\nu}$, $\mu,\nu \in \{0,1\}$, with the following properties:

- $T_{\mu\nu} = T_{\nu\mu}$, $T_{\mu\nu}(z)^* = T_{\nu\mu}(\theta(z))$,
- $\partial_0 T_{\mu 0} + \partial_1 T_{\mu 1} = 0$ with $\partial_0 := \frac{\partial}{\partial t}$, $\partial_1 := \frac{\partial}{\partial y}$,
- $d(T_{\mu\nu}) = h_{\mu\nu} + \overline{h}_{\mu\nu} = 2$, $s(T_{00} - T_{11} \pm 2iT_{01}) = \pm 2$.

Theorem 9.6 (Lüscher–Mack). [LM76] *The Axioms 1–5 imply*

- $\operatorname{tr}(T_{\mu\nu}) = T_\mu^\mu = T_{00} + T_{11} = 0$.
  Therefore, $T := T_{00} - iT_{01} = \frac{1}{2}(T_{00} - T_{11} - 2iT_{01})$ *is independent of* $\bar{z}$, *that is* $\bar{\partial} T = 0$. *Hence, T is holomorphic. In the same way* $\overline{T} := T_{00} + iT_{01}$ *is independent of z, and therefore antiholomorphic. For the corresponding conformal weights we have* $h(T) = \bar{h}(\overline{T}) = 2$ *and* $\bar{h}(T) = h(\overline{T}) = 0$.
- *By*

$$L_{-n} := \frac{1}{2\pi i} \oint_{|\zeta|=1} \frac{T(\zeta)}{\zeta^{n+1}} d\zeta, \quad \overline{L}_{-n} := \frac{1}{2\pi i} \oint_{|\zeta|=1} \frac{\overline{T}(\zeta)}{\zeta^{n+1}} d\zeta \qquad (9.3)$$

the operators $L_n, \overline{L}_n$ *on* $D \subset \mathbb{H}$ *are defined, which satisfy the commutation relations of two commuting Virasoro algebras with the same central charge* $c \in \mathbb{C}$:

$$[L_n, L_m] = (n - m)L_{n+m} + \frac{c}{12} n(n^2 - 1)\delta_{n+m},$$

$$[\overline{L}_n, \overline{L}_m] = (n - m)\overline{L}_{n+m} + \frac{c}{12} n(n^2 - 1)\delta_{n+m},$$

$$[L_n, \overline{L}_m] = 0.$$

- *The representations of the Virasoro algebra defined by* L_n *and* $\overline{L}_n$, *respectively, are unitary:* $L_n^* = L_{-n}$ *and* $\overline{L}_n^* = \overline{L}_{-n}$.

Incidentally, the proof given in [LM76] is based on the Minkowski signature.

The $L_n, \overline{L}_n$ can be interpreted as Fourier coefficients of $T, \overline{T}$, since

$$T(z) = \sum_{n\in\mathbb{Z}} L_n z^{-(n+2)}, \quad \overline{T}(z) = \sum_{n\in\mathbb{Z}} \overline{L}_n \bar{z}^{-(n+2)}. \qquad (9.4)$$

This is how conformal symmetry in the sense of the representation theory of the Virasoro algebra (cf. Sect. 6) appears in the axiomatic presentation of conformal field theory. The operators $L_n, \overline{L}_n$ define a unitary representation of Vir × $\overline{\text{Vir}}$. In general, this representation decomposes into unitary highest-weight representations as follows:

$$\bigoplus W(c,h) \otimes W(c,\bar{h}),$$

where one has to sum over a suitable collection of central charges c and conformal weights $h, \bar{h}$. The theory is called *minimal*, if this sum is finite.

An important tool in conformal field theory is the operator product expansion of two operators A and B of the form $A = \Phi(z_1)$ and $B = \Psi(z_2)$, where Φ, Ψ are field operators. Before we treat operator product expansions in the next section (and also in the next chapter on vertex algebras) let us briefly note that in the case of $\Phi = \Psi = T$ the product $T(z_1)T(z_2)$ has the operator product expansion

$$T(z_1)T(z_2) \sim \frac{c}{2} \frac{1}{(z_1 - z_2)^4} + \frac{2T(z_2)}{(z_1 - z_2)^2} + \frac{dT}{dz_2}(z_2) \frac{1}{(z_1 - z_2)}. \qquad (9.5)$$

The symbol "$\sim$" signifies asymptotic expansion, that is "$=$" modulo a regular function $R(z_1, z_2)$.

The validity of (9.5) turns out to be equivalent to the commutation relations of the $L_n, \bar{L}_n$ (see also Theorem 9.6 and the formula (10.2) in Sect. 10.2).

9.3 Primary Fields, Operator Product Expansion, and Fusion

The primary fields are distinguished by the property that their correlation functions have the covariance property as in Axiom 2 for arbitrary local (that is defined on open subsets of $\mathbb{C}$) holomorphic transformations $w = w(z)$ as well. This covariance expresses the full conformal symmetry. However, the covariance property (9.1) for general w only holds "infinitesimally". This infinitesimal version of (9.1) leads to the following concept of a primary field.

Definition 9.7 (Primary Field). A conformal field Φ_i, $i \in B_0$, is called a *primary field* if

$$[L_n, \Phi_i(z)] = z^{n+1} \partial \Phi_i(z) + h_i(n+1)z^n \Phi_i(z) \qquad (9.6)$$

for all $n \in \mathbb{Z}$, where $\partial = \frac{\partial}{\partial z}$ (and correspondingly for the $\bar{z}$-dependence, which we shall not consider in the following).

The primary field property can be characterized in the following way: the primary fields are precisely those field operators $\Phi_i, i \in B_0$, which have the following *operator product expansion* (OPE) with the energy–momentum tensor T (cf. Corollary 10.43):

$$T(z_1)\Phi_i(z_2) \sim \frac{h_i}{(z_1 - z_2)^2}\Phi_i(z_2) + \frac{1}{z_1 - z_2}\frac{\partial}{\partial z_2}\Phi_i(z_2). \qquad (9.7)$$

(Note that this condition and other formulas used in physics as well as several calculations and formal manipulations become clearer within the formalism of vertex algebras which we introduce in the next chapter.)

The invariance required by (9.6) can also be interpreted as a formal infinitesimal version of (9.1) in Axiom 2 for the transformation $w = w(z) = z + z^{n+1}$. Assume that there would exist a Virasoro group, that is Lie group for Vir with a reasonable exponential map (which is not the case, cf. Sect. 5.4), and assume that we would have a corresponding unitary representation of this symmetry group (or of a central extension of $\mathrm{Diff}_+(\mathbb{S})$ according to Chap. 3) denoted by U. This would imply the formal identity

$$U(e^{tL_n})\Phi_i(z)U(e^{-tL_n}) = \left(\frac{dw_t}{dz}\right)^{h_i}\Phi_i(w_t(z)) \qquad (9.8)$$

for $w_t(z) = z + tz^{n+1}$ (here we take $L_n = -(z^{n+1})\frac{d}{dz}$, cf. Sect. 5.2). Since U is unitary, the globalized formal analogue of (9.8) for holomorphic transformations leads to (9.1) for w_t:

$$G_i(z) = \left(\frac{dw_t}{dz}\right)^{h_i} G_i(w_t(z)).$$

Applying $\frac{d}{dt}\big|_{t=0}$ to the equation (9.8) we obtain

$$[L_n, \Phi_i(z)]$$

on the left-hand side and

$$\frac{d}{dt}(1 + t(n+1)z^n)^{h_i}\Phi_i(z)\Big|_{t=0} + \frac{d}{dt}\Phi_i(w_t(z))\Big|_{t=0}$$

$$= h_i(n+1)z^n\Phi_i(z) + z^{n+1}\frac{\partial}{\partial z}\Phi_i(z)$$

on the right-hand side. This discussion motivates the notion of a primary field, and in particular (9.6).

The correlation functions of primary fields satisfy more than the three identities in Proposition 9.5.

Proposition 9.8 (Conformal Ward Identities). *For every correlation function* $G = G_{i_1\dots i_n}(z_1,\dots,z_n)$ *where all the fields* Φ_{i_j} *are primary the Ward identities*

$$0 = \sum_{j=1}^{n}(z_j^{m+1}\partial_{z_j} + (m+1)h_j z_j^m)G(z_1,\dots,z_n)$$

are satisfied for all $m \in \mathbb{Z}$.

To show these identities one proceeds as in the proof of Proposition 9.5, but with the conformal transformation $w(z) = z + \zeta z^{m+1}$.

The energy–momentum tensor T is not a primary field, as one can see by comparing the expansions (9.5) and (9.7), except for the special case of $c = 0$ and $h = 2$. The deviation from T being primary can be described by the Schwarzian derivative.

From a more geometrical point of view, a primary field with $h = 1, \overline{h} = 0$ or better its matrix coefficient $G_i = \langle \Omega, \Phi_i \Omega \rangle$ corresponds to a meromorphic differential form. In general, it has the transformation property of a quantity like

$$G_i(z,\overline{z})(dz)^h(d\overline{z})^{\overline{h}} = G_i(w,\overline{w})(dw)^h(d\overline{w})^{\overline{h}},$$

where $w = w(z)$ is a local conformal transformation. In geometric terms such a G_i could be understood as a meromorphic section in the vector bundle $K^h \otimes \overline{K}^{\overline{h}}$ where K is the canonical bundle of the respective Riemann surface.

Let $\Phi_i = \Phi$ be a primary field of conformal weight $h_i = h$ and assume that the asymptotic state $v = \lim_{z\to 0}\Phi(z)\Omega$ exists as a vector in the Hilbert space $\mathbb{H}$ of states (v is often denoted by $|h\rangle$).

We have $[L_0, \Phi(z)]\Omega = L_0\Phi(z)\Omega$ and $[L_0, \Phi(z)]\Omega = z\partial\Phi(z)\Omega + h\Phi(z)\Omega$. Therefore v is an eigenvector of L_0 with eigenvalue h. Moreover, for $n > 0$ we deduce in the same way $L_n v = 0$ by using $L_n\Phi(z)\Omega = [L_n, \Phi(z)]\Omega = z^{n+1}\partial\Phi(z)\Omega + h(n+1)z^n\Phi\Omega$. Consequently,

$$L_0 v = hv, L_n v = 0, n > 0.$$

According to our exposition on Virasoro modules in Chapt. 6 we come to the following conclusion:

Remark 9.9. The asymptotic state $v = \lim_{z \to 0} \Phi(z)\Omega$ of a primary field defines a Virasoro module

$$\{L_{-n_1} \ldots L_{-n_k} v : n \geq 0, k \in \mathbb{N}\} \subset \mathbb{H}$$

with highest-weight vector v.

The states $L_{-n_1} \ldots L_{-n_k} v$ can be viewed as excited states of the ground state and they are called *descendants* of v.

It is in general required that the collection of all descendants of the asymptotic states belonging to the primary fields has a dense span in the Hilbert space $\mathbb{H}$ of states. In this case, we obtain a decomposition of $\mathbb{H}$ into Virasoro modules as described above but more concretely given by the primary fields.

Definition 9.10. In a quantum field theory satisfying Axioms 1–5 let

$$B_1 := \{i \in B_0 : \Phi_i \text{ is a primary field}\}.$$

The associated *conformal family* $[\Phi_i]$ for $i \in B_1$ is the complex vector space generated by

$$\Phi_i^\alpha(z) := L_{-\alpha_1}(z) \ldots L_{-\alpha_N}(z)\Phi_i(z) \tag{9.9}$$

for $\alpha = (\alpha_1, \ldots, \alpha_N) \in \mathbb{N}^N$, $\alpha_1 \geq \ldots \geq \alpha_N > 0$, where

$$L_{-n}(z) := \frac{1}{2\pi i} \oint \frac{T(\zeta)}{(\zeta - z)^{n+1}} d\zeta$$

for $z \in \mathbb{C}$. The operators $\Phi_i^\alpha(z)$ are called *secondary fields* or *descendants*.

The operators $L_{-n}(z)$ are in close connection with the Virasoro generators L_n because of

$$L_{-n} = \frac{1}{2\pi i} \oint \frac{T(\zeta)}{\zeta^{n+1}} d\zeta = L_{-n}(0)$$

(cf. Theorem 9.6). The secondary fields Φ_i^α can be expressed as integrals as well. For instance, for Φ_i^k, $k \in \mathbb{N}$,

$$\Phi_i^k(z) = L_{-k}(z)\Phi_i(z) = \frac{1}{2\pi i} \oint \frac{T(\zeta)}{(\zeta - z)^{k+1}} \Phi_i(z) d\zeta.$$

Moreover, the correlation functions of the secondary fields can be determined in terms of correlation functions of primary fields by means of certain specific linear differential equations. It therefore suffices for many purposes to know the correlation functions of the primary fields and in particular the constants C_{ijk} for $i, j, k \in B_1$.

For any fixed $z \in \mathbb{C}$ the conformal family $[\Phi_i]$ of a given primary field Φ_i defines a highest-weight representation with weight (c_i, h_i) (cf. Sect. 6) in a natural manner. $v := \Phi_i(z)$ is the highest-weight vector, $L_0(v) = h_i v$, $L_n(v) := 0$ for $n \in \mathbb{N}$, and $L_{-n}(v) := \Phi_i^n(z)$ for $n \in \mathbb{N}$.

Remark 9.11 (State Field Correspondence). Assume that the asymptotic states of the primary fields together with their descendants generate a dense subspace V of $\mathbb{H}$. Then to each state $a \in V$ there corresponds a field Φ such that $\lim_{z\to 0} \Phi(z)\Omega = a$.

To show this property we only have to observe that for a descendant state of the form $w = L_{-\alpha_1} \ldots L_{-\alpha_N} \Phi_i(0)\Omega$ with respect to a primary field Φ_i one has

$$w = \lim_{z\to 0} \Phi_i^\alpha(z)\Omega = \lim_{z\to 0} L_{-\alpha_1}(z) \ldots L_{-\alpha_N}(z)\Phi_i(z)\Omega.$$

Of course, the remark does not assert that a field corresponding to a state is already of the form Φ_i with $i \in B_0$. It rather means that there is always a suitable field among the descendants of the primary fields.

Note that the state field correspondence is one of the basic requirements in the definition of vertex algebras (see Sect. 10.4). If we denote the field $\Phi(z)$ in the last remark by $Y(a, z)$ we are close to a vertex algebra, where $Y(a, z)$ is supposed to be a formal series with coefficients in $\mathrm{End}\, V$.

Operator Product Expansion. For the primary fields of a conformal field theory it is postulated (according to the fundamental article of Belavin, Polyakov, and Zamolodchikov [BPZ84]) that they obey the following operator product expansion (OPE)

$$\Phi_i(z_1)\Phi_j(z_2) \sim \sum_{k \in B_0} C_{ijk}(z_1 - z_2)^{h_k - h_i - h_j}\Phi_k(z_2) \tag{9.10}$$

with the constants C_{ijk} that occur already in the expression (9.2) of the 3-point functions (cf. Lemma 9.4). Similar expansions hold for the descendants.

The **central object of conformal field theory** is the determination of

- the scaling dimensions $d_i = h_i + \overline{h}_i$,
- the *central charge* c_i for the family $[\Phi_i]$, and
- the coefficients C_{ijk} (structure constants)

from the operator product expansion (9.10) using the conformal symmetry. When all these constants are calculated one has a complete solution.

Proposition 9.12 (Bootstrap Hypothesis). *This can be achieved if the OPE (9.10) is required in addition to be associative. (See also Axiom 6 below.)*

Some comments are due concerning the use of terms like "operator product" and its "associativity". First of all, the expansion (9.10) can only be valid for the corresponding matrix coefficients or better for the vacuum expectation values. In particular, we do not have an algebra of operators with a nice expansion of the product. Therefore the associativity constraint does not refer to the associativity of a true multiplication in a ring as the term suggests from the mathematical viewpoint, but simply means that the respective behavior of the expansions of the product of three or more primary fields is independent of the order the expansions are executed. And this equality concerns again only the vacuum expectation values and it is restricted to the singular terms in the expansions.

Note that in the language of vertex algebras the "associativity" constraint has a nice and clear formulation, cf. Theorem 10.36. Furthermore, the associativity is a consequence of the basic properties of a vertex algebra and not an additional postulate.

In any case, the associativity of the OPE (9.10) in this sense is strong enough to determine all generic 4-point functions

$$G_{i_1i_2i_3i_4}(z_1,z_2,z_3,z_4,\bar{z}_1,\bar{z}_2,\bar{z}_3,\bar{z}_4), (i_1,i_2,i_3,i_4) \in B_1^4.$$

This can be done by using the associativity of the OPE to obtain several expansions of $G_{i_1i_2i_3i_4}$ differing by the order in which we expand. For instance, one can first expand with respect to the indices i_1,i_2 and i_3,i_4 and then expand the resulting two expansions to obtain a series $\sum_m \alpha_m G_m$ or one expands first with respect to the indices i_1,i_4 and i_2,i_3 (here we need locality) and then expand the resulting expansions to obtain another series $\sum_m \beta_m G_m$. Associativity means that the resulting two expansions are the same. This gives infinitely many equations for the structure constants C_{ijk} of the 3-pointfunctions and allows in turn to determine $G_{i_1i_2i_3i_4}$.

We know already that such a function depends only on the cross-ratios $r(z) := (z_{12}z_{34})/(z_{13}z_{24})$ and $\bar{r}(z)$ (see p. 161). Since these ratios are invariant under global conformal transformations on the extended plane we can set $z_1 = \infty, z_2 = 1, z_3 = z$, and $z_4 = 0$. The above correlation function reduces under this change of coordinates to

$$G(z,\bar{z}) = \lim_{z_1,\bar{z}_1 \to \infty} G_{i_1i_2i_3i_4}(z_1,1,z,0,\bar{z}_1,1,\bar{z},0).$$

The associativity of the OPE (9.10) allows to represent G with the aid of so-called (holomorphic and antiholomorphic, respectively) "conformal blocks" $\mathscr{F}^r, \overline{\mathscr{F}}^s$:

$$G(z,\bar{z}) = \sum_{k \in B_1} C_{i_1i_2k}C_{i_3i_4k}\mathscr{F}^k(z)\overline{\mathscr{F}}^k(\bar{z}),$$

where the $C_{i_1i_2k}, C_{i_3i_4k} \in \mathbb{C}$ are the coefficients of the 3-point functions in Lemma 9.4.

The associativity can be indicated schematically in diagrammatic language:

The diagram has a physical interpretation as *crossing symmetry.*

Note that there is an additional way applying the associativity of the OPE in case of the 4-point function leading to another diagram and two further equalities.

A conformal field theory can also be defined on arbitrary Riemann surfaces instead of $\mathbb{C}$. Then the $\mathscr{F}^r, \overline{\mathscr{F}}^s$ depend only on the complex structure of the surface. Finally, they can be considered as holomorphic sections on the appropriate moduli spaces with values in suitable line bundles (cf. [FS87], [TUY89], [KNR94], [Uen95], [Sor95], [Bea95], [Tyu03*] and Chap. 11).

In any case a conformal field theory has to satisfy – in addition to the Axioms 1–5 – the following axiom:

Axiom 6 (Operator Product Expansion) *The primary fields have the OPE (9.10). This OPE is associative.*

Concluding Remarks:

1. All n-point functions of the primary fields can be derived from the G_i for $i \in B_1^4$.
2. The expansions (9.10) are the fusion rules, which can be written formally as

$$[\Phi_i] \times [\Phi_j] = \sum_{l \in B_1} [\Phi_l],$$

   or, carrying more information, as

$$\Phi_i \times \Phi_j = \sum_l N_{ij}^l \Phi_l,$$

   where $N_{ij}^l \in \mathbb{N}_0$ is the number of occurrences of elements of the family $[\Phi_l]$ in the OPE of $\Phi_i(z)\Phi_j(0)$. The coefficients N_{ij}^k define the structure of a fusion ring, cf. Sect. 11.4.
3. We have sometimes passed over to radial quantization, e.g., by using Cauchy integrals in Sect. 9.2, for instance

$$L_{-n}(z) = \frac{1}{2\pi i} \oint \frac{T(\zeta)}{(\zeta - z)^{n+1}} d\zeta.$$

4. To construct interesting examples of conformal field theories satisfying Axioms 1–6 it is reasonable to begin with string theory. On a more algebraic level this amounts to study Kac–Moody algebras (cf. pp. 65 and 196). This subject is surveyed, e.g., in [Uen95] where an interesting connection with the presentation of conformal blocks as sections in certain holomorphic vector bundles is described (cf. also [TUY89] or [BF01*]). For other examples, see [FFK89].

9.4 Other Approaches to Axiomatization

In order to lay down the foundations of conformal field theory introduced in [BPZ84], Moore and Seiberg proposed the following axioms for a conformal field theory in [MS89]:

A conformal field theory is a Virasoro module

$$V = \bigoplus_{i \in B_1} W(c_i, h_i) \otimes W(\overline{c}_i, \overline{h}_i)$$

with unitary highest-weight modules $W(c_i, h_i)$, $W(\overline{c}_i, \overline{h}_i)$ (cf. Sect. 6), subject to the following axioms:

P 1. There is a uniquely determined *vacuum vector* $\Omega = |0\rangle \in V$ with $\Omega \in W(c_{i_0}, h_{i_0}) \otimes W(\overline{c}_{i_0}, \overline{h}_{i_0})$, $h_{i_0} = \overline{h}_{i_0} = 0$. Ω is $SL(2, \mathbb{C}) \times SL(2, \mathbb{C})$-invariant.

P 2. To each vector $\alpha \in V$ there corresponds a field Φ_α, i.e. an operator $\Phi_\alpha(z)$ on V, $z \in \mathbb{C}$. Moreover, there exists a conjugate $\Phi_{\alpha'}$ such that the OPE of $\Phi_\alpha \Phi_{\alpha'}$ contains a descendant of the unit operator.

P 3. The highest-weight vectors $\alpha = i = v_i$ of $W(c_i, h_i)$ determine primary fields Φ_i. Similarly for the highest-weight vectors of $W(\overline{c}_i, \overline{h}_i)$.

P 4. $G_i(z) = \langle \Omega | \Phi_{i_1}(z_1) \ldots \Phi_{i_n}(z_n) | \Omega \rangle$, $|z_1| > \ldots > |z_n|$, always has an analytical continuation to M_n.

P 5. The correlation functions and the one-loop partition functions are modular invariant (cf. [MS89]).

Another axiomatic description of conformal field theory was proposed by Segal in [Seg91], [Seg88b], [Seg88a]. The basic object in this ansatz is the set of equivalence classes of Riemann surfaces with boundaries, which becomes a semi-group by defining the product of two such Riemann surfaces by a suitable fusion or sewing (cf. Sect. 6.5).

Friedan and Shenker introduced in [FS87] a different, interesting system of axioms, which also uses the collection of all Riemann surfaces as a starting point.

All these approaches can be formulated in the language of vertex algebras which seems to be the right theory to describe conformal field theory. In the next chapter we present a short introduction to vertex algebras and their relation to conformal field theory.

Along these lines, the course of V. Kac [Kac98*] describes the structure of conformal field theories as well as the book of E. Frenkel and D. Ben-Zvi [BF01*]. A more general point of view is taken by Beilinson and Drinfeld in their work on chiral algebras [BD04*] where the theory of vertex algebras turns out to be a special case of a much wider theory of chiral algebras.

A comprehensive account of different developments in conformal field theory is collected in the Princeton notes on strings and quantum field theory of Deligne and others [Del99*].

References

[Bea95]   A. Beauville. Vector bundles on curves and generalized theta functions: Recent results and open problems. In: *Current Topics in Complex Algebraic Geometry*. Math. Sci. Res. Inst. Publ. **28**, 17–33, Cambridge University Press, Cambridge, 1995.

[BD04*]   A. Beilinson and V. Drinfeld. *Chiral Algebras*. AMS Colloquium Publications **51**, AMS, Providence, RI, 2004.

[BPZ84]   A.A. Belavin, A.M. Polyakov, and A.B. Zamolodchikov. In- finite conformal symme-
            try in two-dimensional quantum field theory. *Nucl. Phys.* **B 241** (1984), 333–380.
[BF01*]   D. Ben-Zvi and E. Frenkel. *Vertex Algebras and Algebraic Curves.* AMS, Providence,
            RI, 2001.
[Del99*]   P. Deligne et al. *Quantum Fields and Strings: A Course for Mathematicians I, II.*
            AMS, Providence, RI, 1999.
[DMS96*]   P. Di Francesco, P. Mathieu, and D. Sénéchal. *Conformal Field Theory.* Springer-
            Verlag, Berlin, 1996.
[FFK89]   G. Felder, J. Fröhlich, and J. Keller. On the structure of unitary conformal field theory,
            I. Existence of conformal blocks. *Comm. Math. Phys.* **124** (1989), 417–463.
[FS87]   D. Friedan and S. Shenker. The analytic geometry of two-dimensional conformal field
            theory. *Nucl. Phys.* B **281** (1987), 509–545.
[Gaw89]   K. Gawedski. Conformal field theory. *Sém. Bourbaki 1988–89*, Astérisque **177–178**
            (no 704) (1989) 95–126.
[Gin89]   P. Ginsparg. *Introduction to Conformal Field Theory. Fields, Strings and Critical Phe-
            nomena*, Les Houches 1988, Elsevier, Amsterdam, 1989.
[HS66]   N.S. Hawley and M. Schiffer. Half-order differentials on Riemann surfaces. *Acta
            Math.* **115** (1966), 175–236.
[IZ80]   C. Itzykson and J.-B. Zuber. *Quantum Field Theory.* McGraw-Hill, New York, 1980.
[Kac98*]   V. Kac. *Vertex Algebras for Beginners.* University Lecture Series 10, AMS, Provi-
            dencs, RI, 2nd ed., 1998.
[Kak91]   M. Kaku. *Strings, Conformal Fields and Topology.* Springer Verlag, Berlin, 1991.
[KNR94]   S. Kumar, M. S. Narasimhan, and A. Ramanathan. Infinite Grassmannians and moduli
            spaces of G-bundles. *Math. Ann.* **300** (1994), 41–75.
[LM76]   M. Lüscher and G. Mack. The energy-momentum tensor of critical quantum field
            theory in $1 + 1$ dimensions. Unpublished Manuscript, 1976.
[MS89]   G. Moore and N. Seiberg. Classical and conformal field theory. *Comm. Math. Phys.*
            123 (1989), 177–254.
[OS73]   K. Osterwalder and R. Schrader. Axioms for Euclidean Green's functions I. *Comm.
            Math. Phys.* **31** (1973), 83–112.
[OS75]   K. Osterwalder and R. Schrader. Axioms for Euclidean Green's functions II. *Comm.
            Math. Phys.* **42** (1975), 281–305.
[Seg88a]   G. Segal. The definition of conformal field theory. Unpublished Manuscript, 1988.
            Reprinted in *Topology, Geometry and Quantum Field Theory*, U. Tillmann (Ed.), 432–
            574, Cambridge University Press, Cambridge, 2004.
[Seg88b]   G. Segal. Two dimensional conformal field theories and modular functors. In: *Proc.
            IXth Intern. Congress Math. Phys.* Swansea, 22–37, 1988.
[Seg91]   G. Segal. Geometric aspects of quantum field theory. *Proc. Intern. Congress Kyoto
            1990, Math. Soc.* Japan, 1387–1396, 1991.
[Sor95]   C. Sorger. La formule de Verlinde. Preprint, 1995. (to appear in *Sem. Bourbaki*, année
            95 (1994), no 793)
[TUY89]   A. Tsuchiya, K. Ueno, and Y. Yamada. Conformal field theory on the universal family
            of stable curves with gauge symmetry. In: Conformal field theory and solvable lattice
            models. *Adv. Stud. Pure Math.* **16** (1989), 297–372.
[Tyu03*]   A. Tyurin. *Quantization, Classical and Quantum Field Theory and Theta Functions*,
            CRM Monograph Series 21 AMS, Providence, RI, 2003.
[Uen95]   K. Ueno. On conformal field theory. In: *Vector Bundles in Algebraic Geometry*, N.J.
            Hitchin et al. (Eds.), 283–345. Cambridge University Press, Cambridge, 1995.

Chapter 10
Vertex Algebras

In this chapter we give a brief introduction to the basic concepts of vertex algebras. Vertex operators have been introduced long ago in string theory in order to describe propagation of string states. The mathematical concept of a vertex (operator) algebra has been introduced later by Borcherds [Bor86*], and it has turned out to be extremely useful in various areas of mathematics. Conformal field theory can be formulated and analyzed efficiently in terms of the theory of vertex algebras because of the fact that the associativity of the operator product expansion of conformal field theory is already encoded in the associativity of a vertex algebra and also because many formal manipulations in conformal field theory which are not always easy to justify become more accessible and true assertions for vertex algebras. As a result, vertex algebra theory has become a standard way to formulate conformal field theory, and therefore cannot be neglected in an introductory course on conformal field theory.

In a certain way, vertex operators are the algebraic counterparts of field operators investigated in Chap. 8 and the defining properties for a vertex algebra have much in common with the axioms for a quantum field theory in the sense of Wightman and Osterwalder–Schrader. This has been indicated by Kac in [Kac98*] in some detail.

The introduction to vertex algebras in this chapter intends to be self-contained including essentially all proofs. Therefore, we cannot present much more than the basic notions and results together with few examples.

We start with the notion of a formal distribution and familiarize the reader with basic properties of formal series which are fundamental in understanding vertex algebras. Next we study locality and normal ordering as well as fields in the setting of formal distributions and we see how well these concepts from physics are described even before the concept of a vertex algebra has been introduced. In particular, an elementary way of operator expansion can be studied directly after knowing the concept of normal ordering. After the definition of a vertex algebra we are interested in describing some examples in detail which have in parts appeared already at several places in the notes (like the Heisenberg algebra or the Virasoro algebra) but, of course, in a different formulation. In this context conformal vertex algebras are introduced which appear to be the right objects in conformal field theory. Finally, the associativity of the operator product expansion is proven in detail. We conclude this chapter with a section on induced representation of Lie algebras because they

Schottenloher, M.: *Vertex Algebras*. Lect. Notes Phys. **759**, 171–212 (2008)
DOI 10.1007/978-3-540-68628-6_11   © Springer-Verlag Berlin Heidelberg 2008

have been used implicitly throughout the notes and show a common feature in many of our constructions.

The presentation in these notes is based mainly on the course [Kac98*] and to some extent also on the beginning sections of the book [BF01*]. Furthermore, we have consulted other texts like, e.g., [Bor86*], [FLM88*], [FKRW95*], [Hua97*], [Bor00*], and [BD04*].

10.1 Formal Distributions

Let $Z = \{z_1, \ldots, z_n\}$ be a set of indeterminates and let R be a vector space over $\mathbb{C}$. A *formal distribution* is a formal series

$$A(z_1, \ldots, z_n) = \sum_{j \in \mathbb{Z}^n} A_j z^j = \sum_{j \in \mathbb{Z}^n} A_{j_1, \ldots, j_n} z_1^{j_1} \cdots z_n^{j_n}$$

with coefficients $A_j \in R$. The vector space of formal distributions will be denoted by $R[[z_1^{\pm}, \ldots, z_n^{\pm}]] = R\left[[z_1, \ldots, z_n, z_1^{-1} \ldots, z_n^{-1}]\right]$ or $R[[Z^{\pm}]]$ for short. It contains the subspace of *Laurent polynomials*

$$R[z_1^{\pm}, \ldots, z_n^{\pm}] = \{A \in R[[z_1^{\pm}, \ldots, z_n^{\pm}]] \mid$$
$$\exists k, l : A_j = 0 \text{ except for } k \le j \le l\}.$$

Here, the partial order on $\mathbb{Z}^n$ is defined by $i \le j :\Longleftrightarrow i_v \le j_v$ for all $v = 1, \ldots, n$. $R[[z_1^{\pm}, \ldots, z_n^{\pm}]]$ also contains the subspace

$$R[[z_1, \ldots, z_n]] := \{A : A = \sum_{j \in \mathbb{N}^n} A_{j_1, \ldots, j_n} z_1^{j_1} \cdots z_n^{j_n}\}$$

of *formal power series* (here $\mathbb{N} = \{0, 1, 2, \ldots\}$). The space of *formal Laurent series* will be defined only in one variable

$$R((z)) = \{A \in R\left[[z^{\pm}]\right] \mid \exists k \in \mathbb{Z} \ \forall j \in \mathbb{Z} : j < k \Rightarrow A_j = 0\}.$$

When R is an algebra over $\mathbb{C}$, the usual Cauchy product for power series

$$AB(z) = A(z)B(z) := \sum_{j \in \mathbb{Z}^n} \left(\sum_{i+k=j} A_i B_k\right) z^j$$

is not defined for all formal distributions. However, given $A, B \in R[[Z^{\pm}]]$, the product is well-defined whenever A and B are formal Laurent series or when B is a Laurent polynomial. Moreover, the product $A(z)B(w) \in R[[Z^{\pm}, W^{\pm}]]$ is well-defined.

In case of $R = \mathbb{C}$, the ring of formal Laurent series $\mathbb{C}((z))$ is a field and this field can be identified with the field of fractions of the ring $\mathbb{C}[[z]]$ of formal power series in z. In several variables we define $\mathbb{C}((z_1, \ldots, z_n))$ to be the field of fractions of the ring $\mathbb{C}[[z_1,\dots,z_n]]$. This field cannot be identified directly with a field of suitable series. For example, $\mathbb{C}((z,w))$ contains $f = (z-w)^{-1}$, but the following two possible expansions of f,

$$\frac{1}{z}\sum_{n\geq 0} z^{-n}w^n = \sum_{n\geq 0} z^{-n-1}w^n \ , \quad -w\sum_{n\geq 0} z^n w^{-n} = -\sum_{n\geq 0} z^n w^{-n-1},$$

give no sense as elements of $\mathbb{C}((z,w))$. Furthermore, these two series represent two different elements in $\mathbb{C}[[z^{\pm},w^{\pm}]]$. This fact and its precise description are an essential ingredient of vertex operator theory. We come back to these two expansions in Remark 10.16.

Definition 10.1. In the case of one variable $z = z_1$ the *residue* of a formal distribution $A \in R[[z^{\pm}]]$, $A(z) = \sum_{j\in\mathbb{Z}} A_j z^j$, is defined to be

$$\mathrm{Res}_z A(z) = A_{-1} \in R.$$

The *formal derivative* $\partial = \partial_z : R[[z^{\pm}]] \to R[[z^{\pm}]]$ is given by

$$\partial\left(\sum_{j\in\mathbb{Z}} A_j z^j\right) = \sum_{j\in\mathbb{Z}} (j+1)A_{j+1} z^j.$$

One gets immediately the formulas

$$\mathrm{Res}_z A(z)B(z) = \sum_{k\in\mathbb{Z}} A_k B_{-k-1},$$

$$\mathrm{Res}_z \partial A(z)B(z) = -\mathrm{Res}_z A(z)\partial B(z) = \sum_{k\in\mathbb{Z}} kA_k R_{\ k}$$

provided the product AB is defined. The following observation explains the name "formal distribution":

Lemma 10.2. *Every $A \in R[[z^{\pm}]]$ acts on $\mathbb{C}[z^{\pm}]$ as a linear map*

$$\widehat{A} : \mathbb{C}[z^{\pm}] \to R,$$

given by $\widehat{A}(f(z)) := \mathrm{Res}_z A(z)f(z), \phi \in \mathbb{C}[z^{\pm}]$, thereby providing an isomorphism $R[[z^{\pm}]] \to \mathrm{Hom}(\mathbb{C}[z^{\pm}],R)$.

Proof. Of course, $\widehat{A} \in \mathrm{Hom}(\mathbb{C}[z^{\pm}],R)$, and the map $A \mapsto \widehat{A}$ is well-defined and linear. Due to $\widehat{A}(f) = \sum_{j\in\mathbb{Z}} A_j f_{-(j+1)}$ for $f = \sum f_j z^j$ it is injective. Moreover, any $\mu \in \mathrm{Hom}(\mathbb{C}[z^{\pm}],R)$ defines coefficients $A_j := \mu(z^{-j-1}) \in R$, and the distribution $A := \sum A_j z^j$ satisfies $\widehat{A}(z^{-j-1}) = A_j = \mu(z^{-j-1})$. Hence, $\widehat{A} = \mu$ and the map $A \mapsto \widehat{A}$ is surjective. $\qquad\square$

This lemma shows that Laurent polynomials $f \in \mathbb{C}[z^{\pm}]$ can be viewed as to be test functions on which the distributions $A \in R[[z^{\pm}]]$ act.

Definition 10.3. The *formal delta function* is the formal distribution $\delta \in \mathbb{C}[[z^{\pm}, w^{\pm}]]$ in the two variables z, w with coefficients in $\mathbb{C}$ given by

$$\delta(z-w) = \sum_{n \in \mathbb{Z}} z^{n-1} w^{-n} = \sum_{n \in \mathbb{Z}} z^n w^{-n-1} = \sum_{n \in \mathbb{Z}} z^{-n-1} w^n.$$

Note that δ is the difference of the two above-mentioned expansions of $(z-w)^{-1}$:

$$\delta(z-w) = \sum_{n \geq 0} z^{-n-1} w^n - \left(-\sum_{n \geq 0} z^n w^{-n-1} \right).$$

We have

$$\delta(z-w) = \sum_{k+n+1=0} z^k w^n = \delta(w-z)$$

and

$$\delta(z-w) = \sum D_{kn} z^k w^n \in \mathbb{C}[[z^{\pm}, w^{\pm}]]$$

with coefficients $D_{kn} = \delta_{k,-n-1}$. Hence, for all $f \in R[[z^{\pm}]]$, the product $f(z)\delta(z-w)$ is well-defined and can be regarded as a distribution in $R[[w^{\pm}]])[[z^{\pm}]]$. From the formula

$$f(z)\delta(z-w) = \sum_{n,k \in \mathbb{Z}} f_k z^{k-n-1} w^n = \sum_{k \in \mathbb{Z}} \left(\sum_{n \in \mathbb{Z}} f_{k+n+1} w^n \right) z^k$$

for $f = \sum f_k z^k$ one can directly read off

Lemma 10.4. *For every* $f \in R[[z^{\pm}]]$

$$\mathrm{Res}_z f(z)\delta(z-w) = f(w)$$

and

$$f(z)\delta(z-w) = f(w)\delta(z-w).$$

The last formula implies the first of the following related identities. We use the following convenient abbreviation

$$D_w^j := \frac{1}{j!} \partial_w^j$$

during the rest of this chapter.

Lemma 10.5.

1. $(z-w)\delta(z-w) = 0$,
2. $(z-w)D^{k+1}\delta(z-w) = D^k \delta(z-w)$ *for* $k \in \mathbb{N}$,
3. $(z-w)^n D^j \delta(z-w) = D^{j-n} \delta(z-w)$ *for* $j, n \in \mathbb{N}, n \leq j$,
4. $(z-w)^n D^n \delta(z-w) = \delta(z-w)$ *for* $n \in \mathbb{N}$,
5. $(z-w)^{n+1} D^n \delta(z-w) = 0$ *for* $n \in \mathbb{N}$, *and therefore* $(z-w)^{n+m+1} D^n \delta(z-w) = 0$ *for* $n, m \in \mathbb{N}$.

Proof. 3 and 4 follow from 2, and 5 is a direct consequence of 4 and 1. Hence, it only remains to show 2. One uses $\delta(z-w) = \sum_{m\in\mathbb{Z}} z^{-m-1}w^m$ to obtain the expansion

$$\partial_w^{k+1}\delta(z-w) = \sum_{m\in\mathbb{Z}} m\ldots(m-k)z^{-m-1}w^{m-k-1}, \text{ and one gets}$$

$$(z-w)\partial_w^{k+1}\delta(z-w) = \sum_{m\in\mathbb{Z}} m\ldots(m-k)(z^{-m}w^{m-k-1} - z^{-m-1}w^{m-k})$$

$$= \sum_{m\in\mathbb{Z}} ((m+1)m\ldots(m-k+1)) - (m\ldots(m-k))z^{-m-1}w^{m-k}$$

$$= (k+1)\sum_{m\in\mathbb{Z}} m\ldots(m-k+1)z^{-m-1}w^{m-k} = (k+1)\partial_w^k\delta(z-w),$$

which is property 2 of the Lemma. $\qquad\square$

As a consequence, for every $N \in \mathbb{N}, N > 0$, the distribution $(z-w)^N$ annihilates all linear combinations of $\partial_w^k\delta(z-w)$, $k = 0,\ldots,N-1$, with coefficients in $R[[w^{\pm}]]$. The next result (due to Kac [Kac98*]) states that these linear combinations already exhaust the subspace of $R[[z^{\pm}, w^{\pm}]]$ annihilated by $(z-w)^N$.

Proposition 10.6. *For a fixed $N \in \mathbb{N}$, $N > 0$, each*

$$f \in R[[z^{\pm}, w^{\pm}]] \text{ with } (z-w)^N f = 0$$

can be written uniquely as a sum

$$f(z,w) = \sum_{j=0}^{N-1} c^j(w)D_w^j\delta(z-w), \ c^j \in R[[w^{\pm}]].$$

Moreover, for such f the formula

$$c^n(w) = \mathrm{Res}_z(z-w)^n f(z,w)$$

holds for $0 \leq n < N$.

Proof. We have stated already that each such sum is annihilated by $(z-w)^N$ according to the last identity of Lemma (10.5).

The converse will be proven by induction. In the case $N = 1$ the condition $(z-w)f(z,w) = 0$ for $f(z,w) = \sum f_{nm}z^n w^m \in R[[z^{\pm}, w^{\pm}]]$ implies

$$0 = \sum f_{nm}z^{n+1}w^m - f_{nm}z^n w^{m+1} = \sum(f_{n,m+1} - f_{n+1,m})z^{n+1}w^{m+1},$$

and therefore $f_{n,m+1} = f_{n+1,m}$ for all $n, m \in \mathbb{Z}$. As a consequence, $f_{0,m+1} = f_{1,m} = f_{k,m-k-1}$ for all $m, k \in \mathbb{Z}$ which implies

$$f = \sum_{m,k\in\mathbb{Z}} f_{k,m-k-1}z^k w^{m-k-1} = \sum_{m\in\mathbb{Z}} f_{1,m}w^m \sum_{k\in\mathbb{Z}} z^k w^{-k-1} = c^0(w)\delta(z-w)$$

with $c^0(w) = \sum f_{1,m}w^m$. This concludes the proof for $N = 1$.

For a general $N \in \mathbb{N}, N > 0$, let f satisfy

$$0 = (z-w)^{N+1} f(z,w) = (z-w)^N (z-w) f(z,w).$$

The induction hypothesis gives

$$(z-w) f(z,w) = \sum_{j=0}^{N-1} d^j(w) D^j \delta(z-w),$$

hence, by differentiating with respect to z

$$f + (z-w)\partial_z f = \sum_{j=0}^{N-1} d^j(w)\partial_z D^j \delta(z-w) = - \sum_{j=0}^{N-1} d^j(w)(j+1) D^{j+1} \delta(z-w).$$

Here, we use $\partial_z \delta(z-w) = -\partial_w \delta(z-w)$. Now, applying the induction hypothesis once more to

$$\partial_z((z-w)^{N+1} f) = (z-w)^N((N+1)f + (z-w)\partial_z f) = 0$$

we obtain

$$(N+1)f + (z-w)\partial_z f = \sum_{j=0}^{N-1} e^j(w) D^j \delta(z-w).$$

By subtracting the two relevant equations we arrive at

$$Nf = \sum_{j=0}^{N-1} e^j(w) D_w^j \delta(z-w) + \sum_{j=1}^{N} j d^{j-1}(w) D^j \delta(z-w),$$

and get

$$f(z,w) = \sum_{j=0}^{N} c^j(w) D^j \delta(z-w)$$

for suitable $c^j(w) \in R[[w^{\pm}]]$.

The uniqueness of this representation of f follows from the formula $c^n(w) = \mathrm{Res}_z (z-w)^n f(z,w)$ which in turn follows from

$$(z-w)^n f(z,w) = c^n(w) f(z,w), 0 \le n \le N-1, \text{if} f(z,w) = \sum_{j=0}^{N-1} c^j(w) D^j \delta(z,w)$$

by applying Lemma 10.4. Finally, the identities $(z-w)^n f(z,w) = c^n(w) f(z,w)$ are immediate consequences of

$$(z-w)^n D_w^j \delta(z-w) = 0 \text{ for } n > j$$

and

$$(z-w)^n D_w^j \delta(z-w) = D^{j-n} \delta(z-w)$$

for $n \le j$ (cf. Lemma 10.5).   □

Analytic Aspects. For a rational function $F(z, w)$ in two complex variables z, w with poles only at $z = 0, w = 0$, or $|z| = |w|$ one denotes the power series expansion of F in the domain $\{|z| > |w|\}$ by $\iota_{z,w}F$ and correspondingly the power series expansion of F in the domain $\{|z| < |w|\}$ by $\iota_{w,z}F$. For example,

$$\iota_{z,w}\frac{1}{(z-w)^{j+1}} = \sum_{m=0}^{\infty} \binom{m}{j} z^{-m-1} w^{m-j},$$

$$\iota_{w,z}\frac{1}{(z-w)^{j+1}} = -\sum_{m=1}^{\infty} \binom{-m}{j} z^{m-1} w^{-m-j}.$$

In particular, as formal distributions

$$\iota_{z,w}\frac{1}{(z-w)} - \iota_{w,z}\frac{1}{(z-w)} = \sum_{m\geq 0} z^{-m-1} w^m + \sum_{m>0} z^{m-1} w^{-m}$$

$$= \sum_{m\in\mathbb{Z}} z^{-m-1} w^m = \delta(z-w) \qquad (10.1)$$

and similarly for the derivatives of δ,

$$D^j \delta(z-w) = \iota_{z,w}\frac{1}{(z-w)^{j+1}} - \iota_{w,z}\frac{1}{(z-w)^{j+1}} = \sum \binom{m}{j} z^{-m-1} w^{m-j}.$$

10.2 Locality and Normal Ordering

Let R be an associative $\mathbb{C}$-algebra. On R one has automatically the *commutator* $[S, T] = ST - TS$, for $S, T \in R$.

Definition 10.7 (Locality). Two formal distributions $A, B \in R[[z^{\pm}]]$ are *local* with respect to each other if there exists $N \in \mathbb{N}$ such that

$$(z - w)^N [A(z), B(w)] = 0$$

in $R[[z^{\pm}, w^{\pm}]]$.

Remark 10.8. Differentiating $(z - w)^N [A(z), B(w)] = 0$ and multiplying by $(z - w)$ yields $(z - w)^{N+1} [\partial A(z), B(w)] = 0$. Hence, if A and B are mutually local, ∂A and B are mutually local as well.

In order to formulate equivalent conditions of locality we introduce some notations. For $A = \sum A_m z^m$ we mostly write $A = \sum A_{(n)} z^{-n-1}$ such that we have the following convenient formula:

$$A_{(n)} = A_{-n-1} = \operatorname{Res}_z A(z) z^n.$$

We break A into

$$A(z)_- := \sum_{n\geq 0} A_{(n)} z^{-n-1}, \quad A(z)_+ := \sum_{n<0} A_{(n)} z^{-n-1}.$$

This decomposition has the property

$$(\partial A(z))_{\pm} = \partial (A(z)_{\pm}),$$

and conversely, this property determines this decomposition.

Definition 10.9. The *normally ordered product* for distributions $A, B \in R[[z^{\pm}]]$ is the distribution

$$:A(z)B(w): \; := A(z)_+B(w) + B(w)A(z)_- \in R[[z^{\pm}, w^{\pm}]].$$

Equivalently,

$$:A(z)B(w): \; = \sum_{n \in \mathbb{Z}} \left(\sum_{m < 0} A_{(m)}B_{(n)}z^{-m-1} + \sum_{m \geq 0} B_{(n)}A_{(m)}z^{-m-1} \right) w^{-n-1},$$

and the definition leads to the formulas

$$A(z)B(w) = +[A(z)_-, B(w)] + :A(z)B(w):,$$
$$B(w)A(z) = -[A(z)_+, B(w)] + :A(z)B(w):.$$

With this new notation the result of Proposition 10.6 can be restated as follows.

Theorem 10.10. *The following properties are equivalent for* $A, B \in R[[z^{\pm}]]$ *and* $N \in \mathbb{N}$:

1. A, B *are mutually local with* $(z-w)^N[A(z), B(w)] = 0$.

2. $[A(z), B(w)] = \sum_{j=0}^{N-1} C^j(w)D^j\delta(z-w)$ *for suitable* $C^j \in R[[w^{\pm}]]$.

3. $[A(z)_-, B(w)] = \sum_{j=0}^{N-1} \iota_{z,w} \frac{1}{(z-w)^{j+1}} C^j(w),$

   $-[A(z)_+, B(w)] = \sum_{j=0}^{N-1} \iota_{w,z} \frac{1}{(z-w)^{j+1}} C^j(w)$

   for suitable $C^j \in R[[w^{\pm}]]$.

4. $A(z)B(w) = \sum_{j=0}^{N-1} \iota_{z,w} \frac{1}{(z-w)^{j+1}} C^j(w) + :A(z)B(w):,$

   $B(w)A(z) = \sum_{j=0}^{N-1} \iota_{w,z} \frac{1}{(z-w)^{j+1}} C^j(w) + :A(z)B(w):$

   for suitable $C^j \in R[[w^{\pm}]]$.

5. $[A_{(m)}, B_{(n)}] = \sum_{j=1}^{N-1} \binom{m}{j} C^j_{(m+n-j)}, \quad m, n \in \mathbb{Z},$ *for suitable* $C^j = \sum_{k \in \mathbb{Z}} C^j_{(k)} w^{-k-1}$

   $\in R[[w^{\pm}]].$

The notation of physicists for the first equation in 4 is

$$A(z)B(w) = \sum_{j=0}^{N-1} \frac{C^j(w)}{(z-w)^{j+1}} + :A(z)B(w):$$

with the implicit assumption of $|z| > |w|$ in order to justify

$$\frac{1}{(z-w)^{j+1}} = \iota_{z,w} \frac{1}{(z-w)^{j+1}}.$$

Another frequently used notation for this circumstance by restricting to the singular part is

$$A(z)B(w) \sim \sum_{j=0}^{N-1} \frac{C^j(w)}{(z-w)^{j+1}}.$$

Here, $\sim$ denotes as before (Sect. 9.2, in particular (9.5)) the asymptotic expansion neglecting the regular part of the series. This is a kind of operator product expansion as in Sect. 9.3, in particular (9.13).

As an example for the operator product expansion in the context of formal distributions and vertex operators, let us consider the Heisenberg algebra H and its generators $a_n, Z \in$ H, with the relations (cf. (4.1) in Sect. 4.1)

$$[a_m, a_n] = m\delta_{m+n}Z , \quad [a_m, Z] = 0$$

for $m, n \in \mathbb{Z}$. Let $U(\mathsf{H})$ denote the universal enveloping algebra (cf. Definition 10.45) of H. Then $A(z) = \sum_{n \in \mathbb{Z}} a_n z^{-n-1}$ defines a formal distribution $a \in U(\mathsf{H})[[z^{\pm}]]$. It is easy to see that

$$[A(z), A(w)] = \partial\delta(z-w)Z,$$

since

$$\sum_{m,n \in \mathbb{Z}} [a_m, a_n] z^{-m-1} w^{-n-1} = \sum_{m \in \mathbb{Z}} mz^{-m-1} w^{m-1} Z.$$

As a result, the distribution A is local with respect to itself. Because of $C^1(w) = Z$ and $C^j(w) = 0$ for $j \neq 1$ in the expansion of $A(z)A(w)$ according to 4 in Lemma 10.5 the operator product expansion has the form

$$A(z)A(w) \sim \frac{Z}{(z-w)^2}.$$

Another example of a typical operator product expansion which is of particular importance in the context of conformal field theory can be derived by replacing the Heisenberg algebra H in the above consideration with the Virasoro algebra Vir. As we know, Vir is generated by $L_n, n \in \mathbb{Z}$, and the central element Z with the relations

$$[L_m, L_n] = (m-n)L_{m+n} + \frac{m}{12}(m^2 - 1)\delta_{m+n}Z , \quad [a_m, Z] = 0,$$

for $m, n \in \mathbb{Z}$. We consider any representation of Vir in a vector space V with $L_n \in$ End V and $Z = c \mathrm{id}_V$. Then

$$T(z) = \sum_{n \in \mathbb{Z}} L_n z^{-n-2}$$

defines a formal distribution (with coefficients in End V). A direct calculation (see below) shows

$$[T(z), T(w)] = \frac{Z}{12} \partial^3 \delta(z - w) + 2T(w) \partial_w \delta(z - w) + \partial_w T(w) \delta(z - w)$$

and, therefore, according to our Theorem 10.5 with $N = 4$ the following OPE holds (observe the factor $3! = 6$ in the first equation of property 4 of the theorem):

$$T(z)T(w) \sim \frac{c}{2} \frac{1}{(z-w)^4} + \frac{2T(w)}{(z-w)^2} + \frac{\partial_w T(w)}{(z-w)}, \tag{10.2}$$

which we have encountered already in (9.5).

In order to complete the derivation of this result let us check the identity for $[T(z), T(w)]$ stated above:

$$[T(z), T(w)] = \sum_{m,n} [L_m, L_n] z^{-m-2} w^{-n-2}$$

$$= \sum_{m,n} (m - n) L_{m+n} z^{-m-2} w^{-n-2} + \sum_m \frac{m}{12} (m^2 - 1) z^{-m-2} w^{m-2} Z.$$

Substituting $k = m + n$ in the first term and then $l = m + 1$ we obtain

$$\sum_{m,n} (m - n) L_{m+n} z^{-m-2} w^{-n-2}$$

$$= \sum_{k,m} (2m - k) L_k z^{-m-2} w^{-k+m-2}$$

$$= \sum_{k,l} (2l - k - 2) L_k z^{-l-1} w^{-k+l-3}$$

$$= 2 \sum_{k,l} L_k w^{-k-2} l z^{-l-1} w^{l-1} + \sum_{k,l} (-k - 2) L_k w^{-k-3} z^{-l-1} w^l$$

$$= 2T(w) \partial_w \delta(z - w) + \partial_w T(w) \delta(z - w).$$

The second term is (substituting $m + 1 = n$)

$$\frac{Z}{12} \sum_n n(n - 1)(n - 2) z^{-n-1} w^{n-3} = \frac{Z}{12} \partial_w^3 \delta(z - w).$$

Note that the expansion (10.2) can also be derived by using property 5 in Lemma 10.5 by explicitly determining the related $C_{(n)}^j$ to obtain $C^j(w)$.

Without proof we state the following result:

Lemma 10.11 (Dong's Lemma). *Assume $A(z), B(z), C(z)$ are distributions which are pairwise local to each other, than the normally ordered product $:A(z)B(z):$ is local with respect to $C(z)$ as well.*

10.3 Fields and Locality

From now on we restrict our consideration to the case of the endomorphism algebra $R = \operatorname{End} V$ of a complex vector space consisting of the linear operators $b : V \to V$ defined on all of V. The value $b(v)$ of b at $v \in V$ is written $b(v) = b.v$ or simply bv.

Definition 10.12. A formal distribution

$$a \in \operatorname{End} V\left[\left[z^{\pm}\right]\right], a = \sum a_{(n)} z^{-n-1},$$

is called a *field* if for every $v \in V$ there exists $n_0 \in \mathbb{N}$ such that for all $n \geq n_0$ the condition

$$a_{(n)}(v) = a_{(n)}.v = a_{(n)}v = 0$$

is satisfied.

Equivalently, $a(z).v = \sum (a_{(n)}.v)z^{-n-1}$ is a formal Laurent series with coefficients in V, that is $a(z).v \in V((z))$. We denote the vector space of fields by $\mathscr{F}(V)$. As a general rule, fields will be written in small letters $a, b, \ldots$ in these notes whereas $A, B, \ldots$ are general formal distributions.

We come back to the example given by the Heisenberg algebra and replace the universal enveloping algebra by the Fock space $S = \mathbb{C}[T_1, T_2, \ldots]$ (cf. (7.12) in Sect. 7.2) in order to have the coefficients in the endomorphism algebra End S and also to relate the example with our previous considerations concerning quantized fields in Sect. 7.2. Hence, we define

$$\Phi(z) := \sum_{n \in \mathbb{Z}} a_n z^{-n-1},$$

where now the $a_n : S \to S$ are given by the representing endomorphisms $a_n = \rho(a_n) \in \operatorname{End} S$: For a polynomial $P \in S$ and $n \in \mathbb{N}, n > 0$, we have

$$a_n(P) = \frac{\partial}{\partial T_n} P,$$

$$a_0(P) = 0,$$

$$a_{-n}(P) = nT_n P,$$

$$Z(P) = P.$$

The calculation above shows that Φ is local with respect to itself, and it satisfies the operator product expansion

$$\Phi(z)\Phi(w) \sim \frac{1}{(z-w)^2}$$

with the understanding that a scalar $\lambda \in \mathbb{C}$ (here $\lambda = 1$) as an operator is the operator $\lambda \mathrm{id}_S$. Moreover, Φ is a field: Each polynomial $P \in S$ depends on finitely many variables T_n, for example on $T_1, \ldots, T_k$ and, hence, $a_n P = 0$ for $n > k$. Consequently,

$$\Phi(z)P = \sum_{n \in \mathbb{Z}} a_n(P) z^{-n-1} = \sum_{n \leq k} a_n(P) z^{-n-1} = \sum_{m \geq -k-1} a_{-m-1}(P) z^m$$

is a Laurent series. The field Φ is the quantized field of the infinite set of harmonic oscillators (cf. Sect. 7.2) and thus represents the quantized field of a free boson.

In many important cases the vector space V has a natural $\mathbb{Z}$-grading

$$V = \bigoplus_{n \in \mathbb{Z}} V_n$$

with $V_n = \{0\}$ for $n < 0$ and $\dim V_n < \infty$. An endomorphism $T \in \mathrm{End}\, V$ is called *homogeneous of degree* g if $T(V_n) \subset V_{n+g}$. A formal distribution $a = \sum a_{(n)} z^{-n-1} \in \mathrm{End}\, V\,[[z^\pm]]$ is called *homogeneous of (conformal) weight* $h \in \mathbb{Z}$ if each $a_{(k)} : V \to V$ is homogeneous of degree $h - k - 1$. In this case, for a given $v \in V_m$ it follows that $a_{(k)} v \in V_{m+h-k-1}$, and this implies $a_{(k)} v = 0$ for $m + h - k - 1 < 0$, that is $k \geq m + h$. Therefore, $\sum_{k \geq m+h} (a_{(k)} v) z^{-k-1}$ is a Laurent series and we have shown the following assertion:

Lemma 10.13. *Any homogeneous distribution* $a \in \mathrm{End}\, V\,[[z^\pm]]$ *is a field.*

In our example of the free bosonic field $\Phi \in \mathrm{End}\, S\,[[z^\pm]]$ there is a natural grading on the Fock space S given by the degree

$$\deg(\lambda T_{n_1} \ldots T_{n_m}) := \sum_{j=1}^{m} n_j$$

of the homogeneous polynomials $P = \lambda T_{n_1} \ldots T_{n_m}$:

$$S_n := \mathrm{span}\{P : P \text{ homogeneous with } \deg(P) = n\}$$

with $S = \bigoplus S_n$, $S_n = \{0\}$ for $n < 0$ and $\dim S_n < \infty$. Because of $\deg(a_{(k)} P) = \deg(P) - k$ if $a_{(k)} P \neq 0$ ($a_{(k)} = a_k$ in this special example) we see that $a_{(k)}$ is homogeneous of degree $-k$ and the field Φ is homogeneous of weight $h = 1$.

Remark 10.14. The derivative ∂a of a field $a \in \mathscr{F}(V)$ is a field and the normally ordered product $:a(z)b(z):$ of two fields $a(z), b(z)$ is a field as well. Because of $\partial(a(z)_\pm) = (\partial a(z))_\pm$, the derivative $\partial : \mathscr{F}(V) \to \mathscr{F}(V)$ acts as a derivation with respect to the normally ordered product:

$$\partial(:a(z)b(z):) = :(\partial a(z))b(z): + :a(z)(\partial b(z)):.$$

Moreover, using Dong's Lemma 10.11 we conclude that in the case of three pairwise mutually local fields $a(z), b(z), c(z)$ the normally ordered product $:a(z)b(z):$ is a field which is local with respect to $c(z)$. The corresponding assertion holds for the normally ordered product of more than two fields $a^1(z), a^2(z), \ldots, a^n(z)$ which is defined inductively by

$$:a^1(z)\ldots a^n(z)a^{n+1}(z): := :a^1(z)\ldots :a^n(z)a^{n+1}(z): \ldots :.$$

It is easy to check the following behavior of the weights of homogeneous fields.

Lemma 10.15. *For a homogeneous field a of weight h the derivative ∂a has weight $h+1$, and for another homogeneous field b of weight h' the weight of the normally ordered product $:a(z)b(z):$ is $h+h'$.*

We want to formulate the locality of two fields $a, b \in \mathscr{F}(V)$ by matrix coefficients. For any $v \in V$ and any linear functional $\mu \in V^* = \mathrm{Hom}(V, \mathbb{C})$ the evaluation

$$\langle \mu, a(z).v \rangle = \mu(a(z).v) = \sum \mu(a_{(n)}.v)z^{-n-1}$$

yields a formal Laurent series with coefficients in $\mathbb{C}$, i.e., $\langle \mu, a(z).v \rangle \in \mathbb{C}((z))$. The matrix coefficients satisfy $\langle \mu, a(z)b(w).v \rangle \in \mathbb{C}[[z^{\pm}, w^{\pm}]]$ in any case, since they are formal distributions. A closer inspection regarding the field condition for a and b shows

$$\langle \mu, a(z)b(w).v \rangle = \sum_{n<n_0} \mu(a(z)b_{(n)}.v)w^{-n-1} \in \mathbb{C}((z))((w)).$$

Similarly,

$$\langle \mu, b(w)a(z).v \rangle \in \mathbb{C}((w))((z)).$$

In which sense can such matrix coefficients commute? Commutativity in this context can only mean that the equality

$$\langle \mu, a(z)b(w).v \rangle = \langle \mu, b(w)a(z).v \rangle$$

holds in the intersection of $\mathbb{C}((z))((w))$ and $\mathbb{C}((w))((z))$. Consequently, these matrix coefficients of the fields a, b to μ, v commute if and only if the two series are expansions of one and the same element in $\mathbb{C}[[z^{\pm}, w^{\pm}]][z^{-1}, w^{-1}]$. Fields $a, b \in \mathscr{F}(V)$ whose matrix coefficients commute in this sense for all μ, v are local to each other, but locality for fields in general as given in Definition 10.7 is a weaker condition as stated in the following proposition.

Before formulating the proposition we want to emphasize that it is particularly important to be careful with equalities of series regarding the various identifications or embeddings of spaces of series. This is already apparent with our main example, the delta function. Observe that we have two embeddings

$$\mathbb{C}((z, w)) \hookrightarrow \mathbb{C}((z))((w)), \mathbb{C}((z, w)) \hookrightarrow \mathbb{C}((w))((z))$$

of the field of fractions $\mathbb{C}((z, w))$ of $\mathbb{C}[[z, w]]$ induced by the natural embeddings

$$\mathbb{C}[[z,w]] \hookrightarrow \mathbb{C}((z))((w)), \mathbb{C}[[z,w]] \hookrightarrow \mathbb{C}((w))((z))$$

and the universal property of the field of fractions $\mathbb{C}((z,w))$. Moreover, the two spaces $\mathbb{C}((z))((w))$ and $\mathbb{C}((w))((z))$ both have a natural embedding into $\mathbb{C}[[z^\pm, w^\pm]]$ the full space of formal distributions in the two variables z, w. Now, for a Laurent polynomial $P(z,w) \in \mathbb{C}[z^\pm, w^\pm]$ considered as an element in $\mathbb{C}((z,w))$ the two embeddings of P agree in $\mathbb{C}[[z^\pm, w^\pm]]$. However, this is no longer true for general elements $f \in \mathbb{C}((z,w))$.

Remark 10.16. For example, the element $f = (z-w)^{-1} \in \mathbb{C}((z,w))$ induces the element

$$\delta_-(z-w) = \sum_{n \geq 0} z^{-n-1} w^n = \iota_{z,w} \frac{1}{(z-w)}$$

in $\mathbb{C}((z))((w))$ and the element $-\delta_+(z-w)$ in $\mathbb{C}((w))((z))$ where

$$\delta_+(z-w) = \sum_{n > 0} w^{-n} z^{n-1} = \iota_{w,z} \frac{1}{(z-w)}.$$

Hence their embeddings in $\mathbb{C}[[z^\pm, w^\pm]]$ do not agree; the difference $\delta_- - \delta_+$ is, in fact, the delta distribution $\delta(z-w) = \sum_{n \in \mathbb{Z}} z^{-n-1} w^n$, cf. (10.1).

If we now multiply f by $z - w$ we obtain 1 which remains 1 after the embedding into $\mathbb{C}[[z^\pm, w^\pm]]$. Therefore, if we multiply δ_- and $-\delta_+$ by $z - w$ we obtain the same element 1 in $\mathbb{C}[[z^\pm, w^\pm]]$. We are now ready for the content of the proposition.

Proposition 10.17. *Two fields $a, b \in \mathscr{F}(V)$ are local with respect to each other if and only if for all $\mu \in V^*$ and $v \in V$ the matrix coefficients $\langle \mu, a(z)b(w).v \rangle$ and $\langle \mu, b(w)a(z).v \rangle$ are expansions of one and the same element $f_{\mu,v} \in \mathbb{C}[[z,w]][z^{-1}, w^{-1}, (z-w)^{-1})]$ and if the order of pole in $z - w$ is uniformly bounded for the $\mu \in V^*, v \in V$.*

Proof. When $N \in \mathbb{N}$ is a uniform bound of the order of pole in $z - w$ of the $f_{\mu,v}$ one has $(z-w)^N f_{\mu,v} \in \mathbb{C}[[z^\pm, w^\pm]][z^{-1}, w^{-1}]$ uniformly for all $\mu \in V^*, v \in V$. The expansion condition implies

$$(z-w)^N \langle \mu, a(z)b(w).v \rangle = (z-w)^N f_{\mu,v} = (z-w)^N \langle \mu, b(w)a(z).v \rangle.$$

Consequently, $(z-w)^N \langle \mu, [a(z), b(w)].v \rangle = 0$, and therefore

$$(z-w)^N [a(z), b(w)].v = 0,$$

and finally $(z-w)^N [a(z), b(w)] = 0$.

Conversely, if the fields a, b are local with respect to each other, that is if they satisfy $(z-w)^N [a(z), b(w)] = 0$ for a suitable $N \in \mathbb{N}$, we know already by property 4 of Theorem 10.10 that

$$a(z)b(w) = \sum_{j=0}^{N-1} \iota_{z,w} \frac{1}{(z-w)^{j+1}} c^j(w) + :a(z)b(w):,$$

$$b(w)a(z) = \sum_{j=0}^{N-1} \iota_{w,z} \frac{1}{(z-w)^{j+1}} c^j(w) + :a(z)b(w):$$

for suitable fields $c^j \in R[[w^\pm]]$ given by $\mathrm{Res}_z(z-w)^j[a(z),b(w)]$. This shows that $\langle \mu, a(z)b(w).v \rangle$ and $\langle \mu, b(w)a(z).v \rangle$ are expansions of

$$\sum_{j=0}^{N-1} \frac{1}{(z-w)^{j+1}} \mu(c^j(w).v) + \mu(:a(z)b(w):v).$$

$\square$

10.4 The Concept of a Vertex Algebra

Definition 10.18. A *vertex algebra* is a vector space V with a distinguished vector Ω (the *vacuum vector*)[1], an endomorphism $T \in \mathrm{End}\ V$ (the *infinitesimal transla-tion operator*)[2], and a linear map $Y : V \to \mathscr{F}(V)$ to the space of fields (the *vertex operator* providing the *state field correspondence*)

$$a \mapsto Y(a,z) = \sum_{n \in \mathbb{Z}} a_{(n)} z^{-n-1}, a_{(n)} \in \mathrm{End}\ V,$$

such that the following properties are satisfied: For all $u, b \in V$

Axiom V1 (Translation Covariance)

$$[T, Y(a,z)] = \partial Y(a,z),$$

Axiom V2 (Locality)

$$(z-w)^N[Y(a,z), Y(b,w)] = 0$$

for a suitable $N \in \mathbb{N}$ (depending on a, b),

Axiom V3 (Vacuum)

$$T\Omega = 0, Y(\Omega, z) = \mathrm{id}_V, Y(a,z)\Omega|_{z=0} = a.$$

The last condition $Y(a,z)\Omega|_{z=0} = a$ is an abbreviation for $a_{(n)}\Omega = 0, n \geq 0$ and $a_{(-1)}\Omega = a$ when $Y(a,z) = \sum a_{(n)} z^{-n-1}$. In particular,

$$Y(a,z)\Omega = a + \sum_{n < -1} (a_{(n)}\Omega)z^{-n-1} = a + \sum_{k > 0} (a_{(-k-1)}\Omega)z^k \in V[[z]].$$

---

[1] We keep the notation Ω for the vacuum in accordance with the earlier chapters although it is common in vertex algebra theory to denote the vacuum by $|0\rangle$.

[2] Not to be mixed up with the energy–momentum tensor $T(z)$.

Several variants of this definition are of interest.

Remark 10.19. For example, as in the case of Wightman's axioms (cf. Remark 8.12) one can adopt the definition to the supercase in order to include anticommuting fields and therefore the fermionic case. One has to assume that the vector space V is $\mathbb{Z}/2\mathbb{Z}$-graded (i.e., a superspace) and the Locality Axiom V2 is generalized accordingly by replacing the commutator with the anticommutator for fields of different parity. Then we obtain the definition of a *vertex superalgebra*.

Remark 10.20. A different variant concerns additional properties of V since in many important examples V has a natural direct sum decomposition $V = \bigoplus_{n=0}^{\infty} V_n$ into finite-dimensional subspaces V_n. In addition to the above axioms one requires Ω to be an element of V_0 or even $V_0 = \mathbb{C}\Omega$, T to be homogeneous of degree 1 and $Y(a,z)$ to be homogeneous of weight m for $a \in V_m$. We call such a vertex algebra a *graded vertex algebra*.

Remark 10.21. The notation in the axioms could be reduced, for example, the infinitesimal translation operator T can equivalently be described by $Ta = a_{(-2)}\Omega$ for all $a \in V$.

Proof. In fact, the Axiom V1 reads for $Y(a,z) = \sum a_{(n)} z^{-n-1}$:

$$\sum [T, a_{(n)}] z^{-n-1} = \sum (-n-1) a_{(n)} z^{-n-2} = \sum -n a_{(n-1)} z^{-n-1}.$$

Hence, $[T, a_{(n)}] = -n a_{(n-1)}$. Because of $T\Omega = 0$, this implies $Ta_{(n)}\Omega = -n a_{(n-1)}\Omega$. For $n = -1$ we conclude $a_{(-2)}\Omega = Ta_{(-1)}\Omega = Ta$, where $a_{(-1)}\Omega = a$ is part of the Vacuum Axiom V3. $\square$

Vertex Algebras and Quantum Field Theory. To bring the new concept of a vertex algebra into contact to the axioms of a quantum field theory as presented in the last two chapters we observe that the postulates for a vertex algebra determine a structure which is similar to axiomatic quantum field theory.

In fact, on the one hand a field in Chap. 8 is an operator-valued distribution

$$\Phi_a : \mathscr{S} \to \text{End } V$$

indexed by $a \in I$ with $V = D$ a suitable common domain of definition for all the operators $\Phi(f), f \in \mathscr{S}$. On the other hand, a field in the sense of vertex algebra theory is a formal series $Y(a,z) \in \text{End } V[[z^{\pm}]]$, $a \in V$, which acts as a map

$$\widehat{Y}(a, \ ) : \mathbb{C}[[z^{\pm}]] \to \text{End } V$$

as has been shown in Lemma 10.2. This map resembles an operator-valued distribution with $\mathbb{C}[[z^{\pm}]]$ as the space of test functions.

Locality in the sense of Chap. 9 is transferred into the locality condition in Axiom V2. The OPE and its associativity is automatically fulfilled in vertex algebras (see Theorem 10.36 below). However, the reflection positivity or the spectrum condition has no place in vertex algebra theory since we are not dealing with a Hilbert space. Moreover, the covariance property is not easy to detect due to the absence of an inner product except for the translation covariance in Axiom V2. Finally, the existence of the energy–momentum tensor as a field and its properties according to the presentation in Chap. 9 is in direct correspondence to the existence of a conformal vector in the vertex algebra as described below in Definition 10.30.

Under suitable assumptions a two-dimensional conformally invariant field theory in the sense of Chap. 9 determines a vertex algebra as is shown below (p. 190).

We begin now the study of vertex algebras with a number of consequences of the Translation Covariance Axiom V1. Observe that it splits into the following two conditions:

$$[T, Y(a,z)_\pm] = \partial Y(a,z)_\pm.$$

The significance of Axiom V1 is explained by the following:

Proposition 10.22. *Any element $a \in V$ of a vertex algebra V satisfies*

$$Y(a,z)\Omega = e^{zT}a,$$
$$e^{wT}Y(a,z)e^{-wT} = Y(a,z+w),$$
$$e^{wT}Y(a,z)_\pm e^{-wT} = Y(a,z+w)_\pm,$$

where the last equalities are in End $V[[z^\pm]][[w]]$ *which means that $(z+w)^n$ is replaced by its expansion* $\iota_{z,w}(z+w)^n = \sum_{k\geq 0} \binom{n}{k} z^{n-k} w^k \in \mathbb{C}[[z^\pm]][[w]]$.

For the proof we state the following technical lemma which is of great importance in the establishment of equalities.

Lemma 10.23. *Let W be a vector space with an endomorphism $S \in$ End W. To each element $f_0 \in W$ there corresponds a uniquely determined solution*

$$f = \sum_{n \geq 0} f_n z^n \in W[[z]]$$

of the initial value problem

$$\frac{d}{dz}f(z) = Sf(z), f(0) = f_0.$$

In fact, $f(z) = e^{Sz}f_0 = \sum \frac{1}{n!}S^n f_0 z^n$.

Proof. The differential equation means $\sum(n+1)f_{n+1}z^n = \sum Sf_n z^n$, and therefore $(n+1)f_{n+1} = Sf_n$ for all $n \geq 0$, which is equivalent to $f_n = \frac{1}{n!}S^n f_0$. □

Proof. (Proposition 10.22) By the translation covariance we obtain for $f(z) = Y(a,z)\Omega$ ($\in V[[z]]$ by the Vacuum Axiom) the differential equation $\partial f(z) = Tf(z)$. Applying Lemma 10.23 to $W = V$ and $S = T$ yields $f(z) = e^{Tz}a = e^{zT}a$. This proves the first equality. To show the second, we apply Lemma 10.23 to $W =$

End $V[[z^{\pm}]]$ and $S = \mathrm{ad}T$. We have $\partial_w(e^{wT}Y(a,z)e^{-wT}) = [T, e^{wT}Y(a,z)e^{-wT}] = \mathrm{ad}T(e^{wT}Y(a,z)e^{-wT})$ by simply differentiating, and $\partial_w Y(a,z+w) = [T, Y(a,z+w)]$ by translation covariance. Because of $Y(a,z) = Y(a,z+w)|_{w=0}$ the two solutions of the differential equation $\partial_w f = (\mathrm{ad}T)(f)$ have the same initial value $f_0 = Y(a,z) \in$ End $V[[z^{\pm}]]$ and thus agree. The last equalities follow by observing the splitting $[T, Y(a,z)_{\pm}] = \partial Y(a,z)_{\pm}$. $\square$

In order to describe examples the following existence result is helpful:

Theorem 10.24 (Existence). *Let V be a vector space with an endomorphism T and a distinguished vector $\Omega \in V$. Let $(\Phi_a)_{a \in I}$ be a collection of fields*

$$\Phi_a(z) = \sum a_{(k)} z^{-k-1} = a(z) \in \text{End } V\left[\left[z^{\pm}\right]\right]$$

indexed by a linear independent subset $I \subset V$ such that the following conditions are satisfied for all $a, b \in I$:

1. *$[T, \Phi_a(z)] = \partial \Phi_a(z)$.*
2. *$T\Omega = 0$ and $\Phi_a(z)\Omega|_{z=0} = a$.*
3. *Φ_a and Φ_b are local with respect to each other.*
4. *The set $\{a^1_{(-k_1)} a^2_{(-k_2)} \cdots a^n_{(-k_n)}\Omega : a^j \in I, k_j \in \mathbb{Z}, k_j > 0\}$ of vectors along with Ω forms a basis of V.*

Then the formula

$$Y(a^1_{(-k_1)} \cdots a^n_{(-k_n)}\Omega, z) := {:}D^{k_1-1}\Phi_{a^1}(z) \ldots D^{k_n-1}\Phi_{a^n}(z){:} \qquad (10.3)$$

together with $Y(\Omega, z) = \mathrm{id}_V$ defines the structure of a unique vertex algebra with translation operator T, vacuum vector Ω, and

$$Y(a,z) = \Phi_a(z) \text{ for all } a \in I.$$

Proof. First of all, we note that the requirement $\Phi_a(z)\Omega|_{z=0} = a$ in condition 2, that is $\sum a_{(n)}(\Omega)z^{-n-1}|_{z=0} = a$, implies that $a = a_{(-1)}\Omega$ for each $a \in I$. Therefore, $Y(a,z) = Y(a_{(-1)}\Omega, z) = {:}D^0\Phi_a(z){:} = \Phi_a(z)$ for $a \in I$ if everything is well-defined. According to condition 4 the fields $Y(a,z)$ will be well-defined by formula (10.3).

To show the Translation Axiom V1 one observes that for any endomorphism $T \in$ End V the adjoint $\mathrm{ad}T : \mathscr{F}(V) \to \mathscr{F}(V)$ acts as a derivation with respect to the normal ordering:

$$[T, {:}a(z)b(z){:}] = {:}[T, a(z)]b(z){:} + {:}a(z)[T, b(z)]{:}.$$

Moreover, $\mathrm{ad}T$ commutes with $D^k, k \in \mathbb{N}$. Since the derivative ∂ is a derivation with respect to the normal ordering as well (cf. Remark 10.14) commuting with D^k, and since $\mathrm{ad}T$ and ∂ agree on all $\Phi_a, a \in I$, by condition 1, they agree on all repeated normally ordered products of the fields $D^k\Phi_a(z)$ for all $a \in I, k \in \mathbb{N}$, and hence on all $Y(b,z), b \in V$, because of condition 4 and the formula (10.3).

To check the Locality Axiom V2 one observes that all the fields

$$D^k \Phi_a(z), \ a \in I, k \in \mathbb{N},$$

are pairwise local to each other by condition 3 and Remark 10.8. As a consequence, this property also holds for arbitrary repeated normally ordered products of the $D^k \phi_a(z)$ by and Dong's Lemma 10.11 and Remark 10.14.

Finally, the requirements of the Vacuum Axiom V3 are directly satisfied by assumption 2 and the definition of Y.   □

The condition of being a basis in Theorem 10.24 can be relaxed to the requirement that $\{a^1_{(-k_1)} a^2_{(-k_2)} \cdots a^n_{(-k_n)} \Omega : a^j \in I, k_j \in \mathbb{Z}, k_j > 0\} \cup \{\Omega\}$ spans V (cf. [FKRW95*]). With this result one can deduce that in a vertex algebra the field formula (10.3) holds in general.

Heisenberg Vertex Algebra. Let us apply the Existence Theorem 10.24 to determine the vertex algebra of the free boson. In Sect. 10.3 right after the Definition 10.12 we have already defined the generating field

$$\Phi(z) = \sum a_n z^{-n-1}$$

with $a_n \in \mathrm{End}\ S$. We use the representation $\mathsf{H} \to \mathrm{End}\ \mathsf{S} = \mathbb{C}[T_1, T_2, \ldots]$ of the Heisenberg Lie algebra H in the Fock space S which describes the canonical quantization of the infinite dimensional harmonic oscillator (cf. p. 114). The vacuum vector is $\Omega = 1$, as before, and the definition of the action of the a_n on S yields immediately $a_n \Omega = 0$ for $n \in \mathbb{Z}, n \geq 0$. It follows

$$\Phi(z)\Omega = \sum_{n<0} (a_n \Omega) z^{-n-1} = \sum_{k \geq 0} (a_{-k-1}\Omega) z^k.$$

Consequently, $\Phi(z)\Omega|_{z=0} = a_{-1}\Omega$. Hence, to apply Theorem 10.24 we set $\Phi_a = \Phi$ with $u := u_{-1}\Omega = T_1 \in \mathsf{S}$ and $I = \{a\}$. We know that the properties 3 and 4 of the theorem are satisfied.

In order to determine the infinitesimal translation operator T we observe that T has to satisfy

$$[T, a_n] = -n a_{n-1}, \quad T\Omega = 0,$$

by property 1 and the first condition of property 2. This is a recursion for T determining T uniquely. We can show that

$$T = \sum_{m>0} a_{-m-1} a_m. \tag{10.4}$$

In fact, the endomorphism

$$T' = \sum_{m>0} a_{-m-1} a_m \in \mathrm{End}\ \mathsf{H}$$

is well-defined and has to agree with T since $T'\Omega = 0$ and T' satisfies the same recursion $[T', a_n] = -n a_{n-1}$: If $n > 0$ then $a_m a_n = a_n a_m$ and $[a_{-m-1}, a_n] = (-m - 1)\delta_{n-m-1}$ for $m > 0$, hence $[a_{-m-1}a_m, a_n] = [a_{-m-1}, a_n]a_m = -(m+1)\delta_{n-m-1}a_m$, and therefore

$$[T', a_n] = \sum_{m>0} -(m+1)\delta_{n-m-1}a_m = -na_{n-1}.$$

Similarly, if $n < 0$ we have $[a_m, a_n] = m\delta_{m+n}$ and $a_{-m-1}a_n = a_n a_{-m-1}$ for $m > 0$, hence $[a_{-m-1}a_m, a_n] = m\delta_{m+n}a_{-m-1}$, and therefore again $[T', a_n] = -na_{n-1}$.

Now, the theorem guarantees that with the definition of the vertex operation as

$$Y(a, z) := \Phi(z) \text{ for } a = T_1 \text{ and}$$

$$Y(T_{k_1} \dots T_{k_n}, z): = :D^{k_1-1}\Phi(z) \dots D^{k_n-1}\Phi(z):$$

for $k_j > 0$ we have defined a vertex algebra structure on S, the vertex algebra associated to the Heisenberg algebra H. This vertex algebra will be called the *Heisenberg vertex algebra* S.

In the preceding section we have introduced the natural grading of the Fock space $S = \bigoplus S_n$ and we have seen that $\Phi(z)$ is homogeneous of degree 1. Using Lemma 10.15 on the weight of the derivative of a homogeneous field it follows that $D^{k-1}\Phi(z)$ is homogeneous of weight k for $k > 0$ and therefore, again using Lemma 10.15 on the weight of a normally ordered product of homogeneous fields, that $Y(T_{k_1} \dots T_{k_n}, z)$ has weight $k_1 + \dots + k_n = \deg(T_{n_1} \dots T_{k_n})$. As a consequence, for $b \in S_m$ the vertex operator $Y(b, z)$ is homogeneous of weight m and thus the requirements of Remark 10.20 are satisfied. The Heisenberg vertex algebra is a graded vertex algebra.

Vertex Algebras and Osterwalder–Schrader Axioms. Most of the models satisfying the six axioms presented in Chap. 9 can be transformed into a vertex algebra thereby yielding a whole class of examples of vertex algebras. To sketch how this can be done we start with a conformal field theory given by a collection of correlation functions satisfying the six axioms in Chap. 9. According to the reconstruction in Theorem 9.3 there is a collection of fields Φ_a defined as endomorphisms on a common dense subspace $D \subset \mathbb{H}$ of a Hilbert space $\mathbb{H}$ with $\Omega \in D$.

Among the fields Φ_a in the sense of Definition 9.3 we select the primary fields $(\Phi_a)_{a \in B_1}$. We assume that the asymptotic states $a := \Phi_a(z)\Omega|_{z=0} \in D$ exist. Without loss of generality we can assume, furthermore that $\{a : a \in B_1\}$ is linearly independent. Otherwise, we delete some of the fields.

The operator product expansion (Axiom 6 on p. 168) of the primary fields allows to understand the fields Φ_a as fields

$$\Phi_a(z) = \sum a_{(n)}z^{-n-1} \in \text{End } D\left[\left[z^{\pm}\right]\right]$$

in the sense of vertex algebras. We define $V \subset D$ to be the linear span of the set

$$E := \{a^1_{(-k_1)}a^2_{(-k_2)} \dots a^n_{(-k_n)}\Omega : a^j \in B_1, k_j \in \mathbb{Z}, k_j > 0\} \cup \{\Omega\}$$

and obtain the fields $\Phi_a, a \in B_1$, as fields in V by restriction

$$\Phi_a(z) = \sum a_{(n)}z^{-n-1} \in \text{End } V\left[\left[z^{\pm}\right]\right].$$

Now, using the properties of the energy–momentum tensor $T(z) = \sum L_n z^{-n-2}$ we obtain the endomorphism $L_{-1} : V \to V$ with the properties $[L_{-1}, \Phi_a] - \partial \Phi_a$ (the condition of primary fields (9.6) for $n = -1$) and $L_{-1}\Omega = 0$. Moreover, the fields Φ are mutually local according to the locality Axiom 1 on p. 155.

We have thus verified the requirements 1–3 of the Existence Theorem where L_{-1} has the role of the infinitesimal translation operator. If the set $E \subset V$ is a basis of V we obtain a vertex algebra V with $\Phi_a(z) = Y(a, z)$ according to the Existence Theorem reflecting the properties of the original correlation functions. If D is not linear independent we can use the above-mentioned generalization of the Existence Theorem (cf. [FKRW95*]) to obtain the same result.

We conclude this section by explaining in which sense vertex algebras are natural generalizations of associative and commutative algebras with unit.

Remark 10.25. The concept of a vertex algebra can be viewed to be a generalization of the notion of an associative and commutative algebra A over $\mathbb{C}$ with a unit 1. For such an algebra the map

$$Y : A \to \operatorname{End} A, \; Y(a).b := ab \text{ for all } a, b \in A,$$

is $\mathbb{C}$-linear with $Y(a)1 = a$ and $Y(a)Y(b) = Y(b)Y(a)$. Hence, $Y(a, z) = Y(a)$ defines a vertex algebra A with $T = 0$ and $\Omega = 1$.

Conversely, for a vertex algebra V without dependence on z, that is $Y(a, z) = Y(a)$, we obtain the structure of an associative and commutative algebra A with 1 in the following way. The multiplication is given by

$$ab := Y(a).b, \text{ for } a, b \in A := V.$$

Hence, Ω is the unit 1 of multiplication by the Vacuum Axiom. By locality $Y(a)Y(b) = Y(b)Y(a)$, and this implies $ab = Y(a)b = Y(a)Y(b)\Omega = Y(b)Y(a)\Omega = ba$. Therefore, A is commutative. In the same way we obtain $a(cb) = c(ab)$:

$$a(cb) = Y(a)Y(c)Y(b)\Omega = Y(c)Y(a)Y(b)\Omega = c(ab),$$

and this equality suffices to deduce associativity using commutativity: $a(bc) = a(cb) = c(ab) = (ab)c$.

Another close relation to associative algebras is given by the concept of a holomorphic vertex algebra.

Definition 10.26. A vertex algebra is *holomorphic* if every $Y(a, z)$ is a formal power series $Y(a, z) \in \operatorname{End} V [[z]]$ without singular terms.

The next result is easy to check.

Proposition 10.27. *A holomorphic vertex algebra is commutative in the sense that for all $a, b \in V$ the operators $Y(a, z)$ and $Y(b, z)$ commute with each other. Conversely, this kind of commutativity implies that the vertex algebra is holomorphic.*

For a holomorphic vertex algebra the constant term $a_{(-1)} \in \text{End } V$ in the expansion

$$Y(a,z) = \sum_{n<0} a_{(n)} z^{-n-1} = \sum_{k\geq 0} a_{(-(k+1))} z^k = a_{(-1)} + \sum_{k>0} a_{(-(k+1))} z^k$$

determines a multiplication by $ab := a_{(-1)}b$. Now, for $a,b \in V$ one has $[Y(a,z), Y(b,z)] = 0$ and this equality implies $a_{(-1)}b_{(-1)} = b_{(-1)}a_{(-1)}$. In the same way as above after Remark 10.25 the multiplication turns out to be associative and commutative with Ω as unit.

The infinitesimal translation operator T acts as a derivation. By Axiom V1 $[T, a_{(-1)})] = a_{(-2)}$. Because of $(Ta)_{(-1)} = a_{(-2)}$ which can be shown directly but also follows from a more general formula proven in Proposition 10.34 we obtain

$$T(ab) = Ta_{(-1)}b = a_{(-1)}Tb + (Ta)_{(-1)}b = a(Tb) + (Ta)b.$$

Proposition 10.28. *The holomorphic vertex algebras are in one-to-one correspondence to the associative and commutative unital algebras with a derivation.*

Proof. Given such an algebra V with derivation $T : V \to V$ we only have to construct a holomorphic vertex algebra in such a way that the corresponding algebra is V. We take the vacuum Ω to be the unit 1 and define the operators $Y(a,z)$ by

$$Y(a,z) := e^{zT}a = \sum_{n\geq 0} \frac{T^n a}{n!} z^n.$$

The axioms of a vertex algebra are easy to check. Moreover,

$$Y(a,z) = a + \sum_{n>0} \frac{T^n a}{n!} z^n,$$

hence $a_{(-1)} = a$ which implies that by $ab = a_{(-1)}b$ we get back the original algebra multiplication. $\square$

Note that T may be viewed as the generator of infinitesimal translations of z on the formal additive group. Thus, holomorphic vertex algebras are associative and commutative unital algebras with an action of the formal additive group. As a consequence, general vertex algebras can be regarded to be "meromorphic" generalizations of associative and commutative unital algebras with an action of the formal additive group. This point of view can be found in the work of Borcherds [Bor00*] and has been used in another generalization of the notion of a vertex algebra on the basis of Hopf algebras [Len07*].

10.5 Conformal Vertex Algebras

We begin this section by completing the example of the generating field

$$L(z) = \sum L_n z^{-n-2}$$

associated to the Virasoro algebra for which we already derived the operator product expansion (10.2) in Sect. 10.2:

$$L(z)L(w) \sim \frac{c}{2} \frac{1}{(z-w)^4} + \frac{2L(w)}{(z-w)^2} + \frac{\partial_w L(w)}{(z-w)}. \tag{10.5}$$

(We have changed the notation from $T(z)$ to $L(z)$ in order not to mix up the notation with the notation for the infinitesimal translation operator T.)

Now, we associate to Vir another example of a vertex algebra.

Virasoro Vertex Algebra. In analogy to the construction of the Heisenberg vertex algebra in Sect. 10.4 we use a suitable representation V_c of Vir where $c \in \mathbb{C}$ is the central charge. This is another induced representation, cf. Definition 10.49. V_c is defined to be the vector space with basis

$$\{v_{n_1 \ldots n_k} : n_1 \geq \ldots n_k \geq 2, n_j \in \mathbb{N}, k \in \mathbb{N}\} \cup \{\Omega\}$$

(similar to the Verma module $M(c,0)$ in Definition 6.4 and its construction in Lemma 6.5) together with the following action of Vir on V_c ($n, n_j \in \mathbb{Z}, n_1 \geq \ldots n_k \geq 2, k \in \mathbb{N}$):

$$Z := c\, \mathrm{id}_{V_c},$$

$$L_n \Omega := 0 \,, \; n \geq -1 \,, \; n \in \mathbb{Z},$$

$$L_0 v_{n_1 \ldots n_k} := (\sum_{j=1}^{k} n_j) v_{n_1 \ldots n_k},$$

$$L_{-n} \Omega := v_n, n \geq 2, \text{ and } L_{-n} v_{n_1 \ldots n_k} := v_{n n_1 \ldots n_k} \,, \; n \geq n_1.$$

The remaining actions $L_n v, v \in V_c$, are determined by the Virasoro relations, for example $L_{-1} v_n = (n-1) v_{n+1}$ or $L_n v_n = \frac{1}{12} cn(n^2-1)\Omega$ if $n > 1$, in particular $L_2 v_2 = \frac{1}{2} c\Omega$, since

$$L_{-1} v_n = L_{-1} L_{-n} \Omega = L_{-n} L_{-1} \Omega + (-1+n) L_{-1-n} \Omega = (n-1) v_{n+1},$$

and $L_n v_n = L_n L_{-n} \Omega = L_{-n} L_n \Omega + 2n L_0 \Omega + \frac{c}{12} n(n^2-1)\Omega$ with $L_n \Omega = L_0 \Omega = 0$. The definition $L(z) = \sum L_n z^{-n-2}$ directly yields that $L(z)$ is a field, since for every $v \in V_c$ there is N such that $L_n v = 0$ for all $n \geq N$.

Observe that V_c as a vector space can be identified with the space $\mathbb{C}[T_2, T_3, \ldots]$ of polynomials in the infinitely many indeterminates $T_j, j \geq 2$.

To apply Theorem 10.24 with $L(z)$ as generating field we evaluate, first of all, the "asymptotic state" $L(z)\Omega|_{z=0} =: a \in S$. Because of $L_n \Omega = 0$ for $n > -2$ and $L_{-n} \Omega = v_n$ for $n \geq 2$ we obtain

$$a = L(z)\Omega|_{z=0} = \sum_{m \leq -2} L_m \Omega z^{-m-2}|_{z=0} = L_{-2}\Omega = v_2.$$

We set $I = \{a\} = \{v_2\}$ and $\Phi_a(z) := L(z)$ in order to agree with the notation in Theorem 10.24.

Proposition 10.29. *The field* $\Phi_a(z) = L(z), a = v_2$, *generates the structure of a vertex algebra on* V_c *with* L_{-1} *as the infinitesimal translation operator.* V_c *is called the Virasoro vertex algebra with central charge* c.

Proof. Property 3 of Theorem 10.24 is satisfied, since the field $\Phi_a = L$ is local with itself according to (10.5), and property 4 holds because of the definition of V_c. As the infinitesimal translation operator T we take $T := L_{-1}$, so that property 2 is satisfied as well. Finally, $[L_{-1}, L(z)] = \partial L(z)$ (which is $[T, \Phi(z)] = \partial \Phi(z)$) can be checked directly: $[L_{-1}, L(z)] = \sum [L_{-1}, L_n] z^{-n-2} = \sum (-1 - n) L_{n-1} z^{-n-2} = \sum (-n - 2) L_n z^{-n-3} = \partial L(z)$.

As a consequence,

$$Y(v_2, z) = L(z),$$
$$Y(v_{n_1 \ldots n_k}, z) = :D^{n_1-2} T(z) \ldots D^{n_k-2} T(z):$$

define the structure of a vertex algebra which will be called the Virasoro vertex algebra with central charge c. The central charge can be recovered by $L_2 a = \frac{1}{2} c \Omega$. $\square$

V_c has the grading $V_c = \bigoplus V_N$ with V_N generated by the basis elements $\{v_{n_1 \ldots n_k} : \sum n_j = N\}$ ($\sum n_j = N = \deg v_{n_1 \ldots n_k}$), $V_0 = \mathbb{C} \Omega$. The finite-dimensional vector subspace V_N can also be described as the eigenspace of L_0 with eigenvalue N: $V_N = \{v \in V_c : L_0 v = N v\}$. The translation operator $T = L_{-1}$ is homogeneous of degree 1 and the generating field has weight 2 since each $L_{n-1} = T_{(n)}$ has degree $2 - n - 1$. Hence, V_c is a graded vertex algebra and L_0 is the degree.

This example of a vertex algebra motivates the following definition:

Definition 10.30. *(Conformal Vertex Algebra)* A field $L(z) = \sum L_n z^{-n-2}$ with the operator expansion as in (10.5) will be called a *Virasoro field with central charge* c.

A *conformal vector* with *central charge* c is a vector $v \in V$ such that $Y(v, z) = \sum v_{(n)} z^{-n-1} = \sum L_n^v z^{-n-2}$ is a Virasoro field with central charge c satisfying, in addition, the following two properties:

1. $T = L_{-1}^v$
2. L_0^v is diagonalizable.

Finally, a *conformal vertex algebra* (of *rank* c) is a vertex algebra V with a distinguished conformal vector $v \in V$ (with central charge c). In that case, the field $Y(v, z)$ is also called the *energy–momentum tensor* or energy–momentum field of the vertex algebra V.

Examples. 1. The Virasoro vertex algebras V_c are clearly conformal vertex algebras of rank c with conformal vector $v = v_2 = L_{-2} \Omega$. $L(z) = Y(v_2, z)$ is the energy–momentum tensor.

2. The vertex algebra associated to an axiomatic conformal field theory in the sense of the last chapter (cf. p. 190 under the assumptions made there) has $L_{-2}\Omega$ as a conformal vector and T is the energy–momentum tensor.

3. The Heisenberg vertex algebra S has a one-parameter family of conformal vectors

$$v_\lambda := \frac{1}{2}T_1^2 + \lambda T_2 , \ \lambda \in \mathbb{C}.$$

To see this, we have to check that the field $Y(v_\lambda,z) = \sum L_n^\lambda z^{-n-2}$ is a Virasoro field, that $T = L_{-1}^\lambda$, and that L_0^λ is diagonalizable.

That the L_n^λ satisfy the Virasoro relations and therefore determine a Virasoro field can be checked by a direct calculation which is quite involved. We postpone the proof because we prefer to obtain the Virasoro field condition as an application of the associativity of the operator product expansion, which will be derived in the next section (cf. Theorem 10.40).

The other two conditions are rather easy to verify. By the definition of the vertex operator we have $Y(T_1^2,z) = {:}\Phi(z)\Phi(z){:}$ and $Y(T_2,z) = \partial\Phi(z)$, hence

$$Y(T_1^2,z) = \sum_{k\neq 0}\sum_{n+m=k} a_n a_m z^{k}{}^{-2} + 2\sum_{n>0} a_{-n}a_n z^{-2},$$

where $\Phi(z) = \sum a_n z^{-n-1}$ with the generators a_n of the Heisenberg algebra H acting on the Fock space S, and

$$Y(T_2,z) = \sum(-k-1)a_k z^{-k-2},$$

and therefore,

$$Y(v_\lambda,z) = \frac{1}{2}\sum_{k\neq 0}\left(\sum_{n+m=k} a_n a_m - \lambda(k+1)a_k\right)z^{-k-2} + \sum_{n>0} a_{-n}a_n z^{-2}. \qquad (10.6)$$

(Recall that we defined a_0 to satisfy $a_0 = 0$ in this representation of H.) Consequently,

$$L_0 = L_0^\lambda = \sum_{n>0} a_{-n}a_n$$

and

$$L_{-1} = L_{-1}^\lambda = \sum_{n>0} a_{-n-1}a_n,$$

and both these operators turn out to be independent of λ. Now, on the monomials $T_{n_1}\ldots T_{n_k}$ we have $L_0(T_{n_1}\ldots T_{n_k}) = \sum_{j=1}^{k} n_j = \deg(T_{n_1}\ldots T_{n_k})$ and L_0 is diagonalizable with $L_0 v = \deg(v)v = nv$ for $v \in V_n$. Finally, we have already seen in (10.4) that the infinitesimal translation operator is $\sum_{n>0} a_{-n-1}a_n = L_{-1}$.

4. A fourth example of a conformal vertex algebra is given by the Sugawara vector as a conformal vector of the vertex algebra associated to a Lie algebra g.

(This example appears also in the context of associating a vertex algebra to a conformal field theory with g-symmetry in the sense of Chap. 9, but there we have not introduced the related example of a conformal field theory corresponding to a Kac–Moody algebra.)

At first, we have to describe the corresponding vertex algebra.

Affine Vertex Algebra. As a fourth example of applying the Existence Theorem 10.24 to describe vertex algebras we now come to the case of a finite-dimensional simple Lie algebra $\mathfrak{g}$ and its associated vertex algebra $V_k(\mathfrak{g}), k \in \mathbb{C}$, which will be called affine vertex algebra.

In the list of examples of central extensions in Sect. 4.1 we have introduced the affinization

$$\hat{\mathfrak{g}} = \mathfrak{g}[T, T^{-1}] \oplus \mathbb{C}Z$$

of a general Lie algebra $\mathfrak{g}$ equipped with an invariant bilinear form $(\, , \,)$ as the central extension of the loop algebra $L\mathfrak{g} = \mathfrak{g}[T^{\pm}]$ with respect to the cocycle

$$\Theta(a_m, b_n) = m(a, b)\delta_{m+n}Z,$$

where we use the abbreviation $a_m = T^m a = T^m \otimes a, b_n = T^n b$ for $a, b \in \mathfrak{g}$ and $n \in \mathbb{Z}$. The commutation relations for $a, b \in \mathfrak{g}$ and $m, n \in \mathbb{Z}$ are therefore

$$[a_m, b_n] = [a, b]_{m+n} + m(a, b)\delta_{m+n}Z, \quad [a_m, Z] = 0.$$

In the case of a finite-dimensional simple Lie algebra $\mathfrak{g}$ any invariant bilinear symmetric form $(\, , \,)$ is unique up to a scalar (it is in fact a multiple of the Killing form κ) and the resulting affinization of $\mathfrak{g}$ is called the affine Kac–Moody algebra of $\mathfrak{g}$ where the invariant form is normalized in the following way: The Killing form on $\mathfrak{g}$ is $\kappa(a, b) = \text{tr}(\text{ad } a \, \text{ad } b)$ for $a, b \in \mathfrak{g}$, where $\text{ad} : \mathfrak{g} \to \text{End}\,\mathfrak{g}$, $\text{ad } a(x) = [a, x]$ for $x \in \mathfrak{g}$ is the adjoint representation. The normalization in question now is

$$(a, b) := \frac{1}{2h^\vee} \kappa(a, b),$$

where $h^\vee$ is the dual Coxeter number of $\mathfrak{g}$ (see p. 221).

As before, we need to work in a fixed representation of the Kac–Moody algebra $\hat{\mathfrak{g}}$. Let $\{J^\rho : \rho \in \{1, \ldots, r\}\}$ be an ordered basis of $\mathfrak{g}$. Then $\{J_n^\rho : 1 \leq \rho \leq r = \dim\mathfrak{g}, n \in \mathbb{Z}\} \cup \{Z\}$ is a basis for $\hat{\mathfrak{g}}$.

We define the representation space $V_k(\mathfrak{g}), k \in \mathbb{C}$, to be the complex vector space with the basis

$$\{v_{n_1 \ldots n_m}^{\rho_1 \ldots \rho_m} : n_1 \geq \ldots n_m \geq 1, \rho_1 \leq \ldots \leq \rho_m\} \cup \{\Omega\},$$

and define the action of $\hat{\mathfrak{g}}$ on $V = V_k(\mathfrak{g})$ by fixing the action as follows ($n > 0$):

$$Z = k\text{id}_V, \quad J_n^\rho \Omega = 0,$$

$$J_{-n}^\rho \Omega = v_n^\rho, \quad J_{-n}^\rho v_{n_1 \ldots n_m}^{\rho_1 \ldots \rho_m} = v_{n n_1 \ldots n_m}^{\rho \rho_1 \ldots \rho_m},$$

if $n \geq n_1$ and $\rho \leq \rho_1$. The remaining actions of the J_n^ρ on the basis of $V_k(\mathfrak{g})$ are determined by the commutation relations

$$[J_m^\rho, J_n^\sigma] = [J^\rho, J^\sigma]_{m+n} + m(J^\rho, J^\sigma)k\delta_{m+n}.$$

The resulting representation is called the *vacuum representation of rank k*. It is again an induced representation, cf. Sect. 10.7.

The generating fields are

$$J^\rho(z) = \sum J_n^\rho z^{-n-1} \in \operatorname{End} V_k(\mathfrak{g})\left[\left[z^\pm\right]\right], 1 \leq \rho \leq r.$$

In view of the commutation relations one has

$$J_n^\rho v_{n_1 \ldots n_m}^{\rho_1 \ldots \rho_m} = 0$$

if $n > n_1$. Therefore, these formal distributions are in fact fields. Because of $J_n^\rho \Omega = 0$ for every $n \in \mathbb{Z}, n \geq 0$, we obtain

$$J^\rho(z)\Omega = \sum_{n<0} J_n^\rho \Omega z^{-n-1} = \sum_{m \geq 0} v_{m+1} z^m,$$

and thus $J^\rho(z)\Omega|_{z=0} = v_1^\rho$. Hence, to match the notation of the Existence Theorem 10.24 we should set $I = \{v_1^\rho : 1 \leq \rho \leq r\}$ and

$$\Phi_a(z) := J^\rho(z) \text{ if } a = v_1^\rho.$$

Proposition 10.31. *The fields* $\Phi_a(z), a \in I$, *resp.* $J^\rho(z), 1 \leq \rho \leq r$, *generate a vertex algebra structure on* $V_k(\mathfrak{g})$. $V_k(\mathfrak{g})$ *is the* affine vertex algebra of rank k.

Proof. In order to check locality we calculate $[J^\rho(z), J^\sigma(w)]$:

$$[J^\rho(z), J^\sigma(w)] = \sum_{m,n} [J_m^\rho, J_n^\sigma] z^{-m-1} w^{-n-1}$$

$$= \sum_{m,n} [J^\rho, J^\sigma]_{m+n} z^{-m-1} w^{-n-1} + \sum_m m(J^\rho, J\sigma)k z^{-m-1} w^{m-1}$$

$$= \sum_l [J^\rho, J^\sigma]_l w^{-l-1} \sum_m z^{-m-1} w^m + (J^\rho, J^\sigma)k \sum_m m z^{-m-1} w^{m-1}$$

$$= [J^\rho, J^\sigma](w)\delta(z-w) + (J^\rho, J^\sigma)k\partial\delta(z-w).$$

This equality implies by Theorem 10.5 that the operator product expansion is

$$J^\rho(z)J^\sigma(w) \sim \frac{[J^\rho, J^\sigma](w)}{z-w} + \frac{(J^\rho, J^\sigma)k}{(z-w)^2}, \tag{10.7}$$

and that the fields $J^\rho(z), J^\sigma(z)$ are pairwise local with respect to each other. We thus have established property 3 of the Existence Theorem 10.24, and by the construction of the space $V_k(\mathfrak{g})$ and the definition of the action of the J_n^ρ property 4 is satisfied as well.

It remains to determine the infinitesimal translation operator T which will again be defined recursively by

$$T\Omega = 0, [T, J_n^\rho] = -nJ_{n-1}^\rho.$$

$T \in \text{End } V_k(\mathfrak{g})$ is well-defined and satisfies evidently $[T, J^\rho(z)] = \partial J^\rho(z)$. Therefore, the Existence Theorem applies yielding a vertex algebra structure given by

$$Y(v_{n_1...n_m}^{\rho_1...\rho_m}, z) = :D^{n_1-1}J^{\rho_1}(z)...D^{n_m-1}J^{\rho_m}(z):.$$

$\square$

In order to determine a conformal vector of the affine vertex algebra $V_k(\mathfrak{g})$ by the Sugawara construction we denote the elements of the dual basis with respect to $\{J^1,...J^r\}$ by $J_\rho \in \mathfrak{g}$ satisfying $(J_\sigma, J^\rho) = \delta_\sigma^\rho$. Then it can be shown that the vector

$$S := \frac{1}{2}\sum_{\rho=1}^r J_{\rho,-1}J_{-1}^\rho \Omega \in V_k(\mathfrak{g})$$

is independent of the choice of the basis. We call

$$v := \frac{1}{k+h^\vee}S$$

the *Sugawara vector*.

Proposition 10.32. *Assume $k \neq -h^\vee$. Then the Sugawara vector v is a conformal vector of $V_k(\mathfrak{g})$ with central charge*

$$c = c(k) = \frac{k \dim \mathfrak{g}}{k+h^\vee}.$$

Proof. (sketch) Using the associativity of the OPE (see Theorem 10.36 in the next section) one can deduce for $Y(v,z) = L(z) = \sum L_n z^{-n-2}$ $(L_n = L_n^\vee)$ the OPE

$$L(z)J^\rho(w) \sim \frac{J^\rho(w)}{(z-w)^2} + \frac{\partial J^\rho(w)}{z-w},$$

and hence the following commutation relations

$$[L_m, J_n^\rho] = -nJ_{m+n}^\rho, m, n \in \mathbb{Z}, 1 \leq \rho \leq r.$$

These relations imply $L_{-1} = T$ and the diagonalizability of L_0 immediately. Moreover, $L_n v = 0$ for $n > 2$. Therefore, according to the above-mentioned criterion in Theorem 10.40 v is a conformal vector of central charge c where c is determined by $L_2 v = \frac{1}{2} c \Omega$. Finally,

$$L_2 v = \frac{1}{2(k+h^\vee)} L_2 \sum J_{\rho,-1} J^\rho_{-1} \Omega$$

$$= \frac{1}{2(k+h^\vee)} \sum J_{\rho,1} J^\rho_{-1} \Omega$$

$$= \frac{k \dim \mathfrak{g}}{2(k+h^\vee)} \Omega.$$

We conclude $c = \frac{k \dim \mathfrak{g}}{k+h^\vee}$. Details are in [Kac98*] and [BF01*].   □

Altogether, the coefficients L_n of the Virasoro field

$$Y(v,z) = \frac{1}{2(k+h^\vee)} \sum_{\rho=1}^{r} :J_\rho(z) J^\rho(z):$$

yield an action of the Virasoro algebra with central charge $c(k)$ on the space $V_k(\mathfrak{g})$.

Many more vertex algebras are known and many of them are not constructed by using a Lie algebra representation. It is not in the scope of this book to survey other interesting classes of vertex algebras. Instead we refer to the course of Kac [Kac98*] where the last third of the book is devoted to describe such vertex algebras as lattice vertex algebras, coset constructions, W-algebras, various $\mathbb{Z}/2\mathbb{Z}$-graded (or super) vertex algebras to include also the anticommutator in the considerations, and many more examples.

Examples are presented in the book of Frenkel and Ben-Zvi [BF01*], too, where the vertex algebras are related to algebraic curves. The first step in doing this is to formulate a theory of vertex algebras being invariant against coordinate changes $z \mapsto w(z)$. This leads eventually to vertex algebra bundles and moduli spaces as well as chiral algebras. In contrast to this local approach to algebraic curves in [Lin04*] an attempt has been made to study "global" vertex algebras on Riemann surfaces which turns out to be connected to Krichever–Novikov algebras.

Let us mention also the approach of Huang [Hua97*] who relates the algebraic approach to vertex algebras as presented here to the more geometrically and topologically inspired description of conformal field theory of Segal [Seg88a], [Seg91].

10.6 Associativity of the Operator Product Expansion

We begin with the uniqueness result of Goddard.

Theorem 10.33 (Uniqueness). *Let V be a vertex algebra and let $f \in \text{End } V[[z^\pm]]$ be a field which is local with respect to all fields $Y(a,z)$, $a \in V$. Moreover, assume that*

$$f(z)\Omega = e^{zT}b$$

for a suitable $b \in V$. Then $f(z) = Y(b,z)$.

Proof. By locality we have $(z-w)^N[f(z),Y(a,w)] = 0$, in particular,

$$(z-w)^N f(z)Y(a,w)\Omega = (z-w)^N Y(a,w)f(z)\Omega.$$

We insert the assumption $f(z)\Omega = e^{zT}b$, and the equalities $Y(a,w)\Omega = e^{wT}a$ and $Y(b,z)\Omega = e^{zT}b$ (according to Proposition 10.22), and we obtain

$$(z-w)^N f(z)e^{wT}a = (z-w)^N Y(a,w)e^{zT}b = (z-w)^N Y(a,w)Y(b,z)\Omega.$$

Since $Y(a,z)$ and $Y(b,z)$ are local to each other we have (for sufficiently large N)

$$(z-w)^N f(z)e^{wT}a = (z-w)^N Y(b,z)Y(a,w)\Omega = (z-w)^N Y(b,z)e^{wT}a.$$

Letting $w = 0$ we conclude $z^N f(z)a = z^N Y(b,z)a$ for all $a \in V$ which implies $f(z)a = Y(b,z)a$ and hence $f(z) = Y(b,z)$. $\qquad\square$

The Uniqueness Theorem yields immediately the following result:

Proposition 10.34. *The identity*

$$Y(Ta,z) = \partial Y(a,z)$$

holds in a vertex algebra.

Proof. For $f(z) = \partial Y(a,z)$ we have

$$f(z)\Omega = \sum_{n\geq 0}(n+1)a_{(-n-2)}\Omega z^n$$

and therefore $f(z)\Omega|_{z=0} = a_{(-2)}\Omega = Ta$. Using translation covariance we have $\partial(f(z)\Omega) = \partial TY(a,z)\Omega = T(f(z)\Omega)$ and we conclude $f(z)\Omega = e^{zT}Ta$ by Lemma 10.23. By Theorem 10.33 it follows that $f(z) = Y(Ta,z)$. $\qquad\square$

In a similar way as the Uniqueness Theorem 10.33 one can prove the following:

Proposition 10.35 (Quasisymmetry). *The equality*

$$Y(a,z)b = e^{zT}Y(b,-z)a$$

holds in $V((z))$.

Proof. Since $Y(a,z),Y(b,z)$ are local to each other by the Locality Axiom there exists $N \in \mathbb{N}$ with

$$(z-w)^N Y(a,z)Y(b,z)\Omega = (z-w)^N Y(b,z)Y(a,z)\Omega.$$

By $Y(a,z)\Omega)e^{zT}a$ (Proposition 10.22) and analogously for b this implies

$$(z-w)^N Y(a,z)e^{wT}b = (z-w)^N Y(b,z)e^{zT}a.$$

By Proposition 10.22 we also have $e^{zT}Y(b,w)e^{-Tz} = Y(b,z+w)$, hence, $e^{zT}Y$ $(b,w-z) = Y(b,z)e^{zT}$. Consequently,

$$(z-w)^N Y(a,z)e^{wT}b = (z-w)^N e^{zT}Y(b,w-z)a,$$

where $(w-z)^{-1}$ has to replaced by the expansion $(w-z)^{-1} = \sum_{n\geq 0} z^n w^{-n-1}$. Let N be large enough such that on the right-hand side of the above formula there appear no negative powers of $(w-z)$. Then it becomes an equality in $V((z))[[w]]$, and we can put $w = 0$ again and divide by z^N to obtain the desired identity of quasisymmetry.□

We now come to the associativity of the operator product expansion (OPE for short). To motivate the result we apply Proposition 10.22 repeatedly to obtain

$$Y(a,z)Y(b,w)\Omega = Y(a,z)e^{wT}b = e^{wT}Y(a,z-w)b, \text{ and}$$
$$e^{wT}Y(a,z-w)b = Y(Y(a,z-w)b,w)\Omega,$$

where the last expression $Y(Y(a,z-w)b,w)\Omega$ is defined by

$$Y(Y(a,z-w)b,w) := \sum_{n\in\mathbb{Z}} Y(a_{(n)}b,w)(z-w)^{-n-1}.$$

One is tempted to apply the Uniqueness Theorem 10.33 to the equality

$$Y(a,z)Y(b,w)\Omega = Y(Y(a,z-w)b,w)\Omega$$

to deduce

$$Y(a,z)Y(b,w) = Y(Y(a,z-w)b,w)$$

which is the desired "associativity" of the OPE. However, the theorem cannot be applied directly: we first have to make precise where the equality should hold. Observe that for $b \in V$ there exists n_0 such that $a_{(n)}b = 0$ for $n \geq n_0$. Consequently, $Y(Y(a,z-w)b,w) = \sum Y(a_{(n)}b,w)(z-w)^{-n-1}$ is a series in $\text{End } V[[w^{\pm}]]((z-w))$. Replacing

$$(z-w)^{-k} \mapsto \delta_-^k = \left(\sum_{n\geq 0} z^{-n-1}w^n\right)^k, k > 0,$$

we obtain an embedding

$$\text{End } V[[w^{\pm}]]((z-w)) \hookrightarrow \text{End } V[[w^{\pm},z^{\pm}]].$$

The following equalities have to be understood as identities in $\text{End } V[[w^{\pm},z^{\pm}]]$ using this embedding.

Theorem 10.36 (Associativity of the OPE). *For any vertex algebra V the following associativity property is satisfied:*

$$Y(a,z)Y(b,w) = Y(Y(a,z-w)b,w) = \sum_{n\in\mathbb{Z}} Y(a_{(n)}b,w)(z-w)^{-n-1}$$

for all $a,b \in V$. More specifically,

$$Y(a,z)Y(b,w) = \sum_{n\geq 0} Y(a_{(n)}b,w)(z-w)^{-n-1} + :Y(a,z)Y(b,w):,$$

and, equivalently,

$$[Y(a,z),Y(b,w)] = \sum_{n\geq 0} D_w^n \delta(z-w)Y(a_{(n)}b,w).$$

Proof. We use the attempt described earlier and start with

$$Y(a,z)Y(b,w)\Omega = e^{wT}Y(a,z-w)b = Y(Y(a,z-w)b,w)\Omega,$$

where the last equality can be shown in a similar way as the corresponding equality in the proof of Proposition 10.22. For arbitrary $c \in V$ we obtain the equality

$$Y(c,t)Y(a,z)Y(b,w)\Omega = Y(c,t)Y(Y(a,z-w)b,w)\Omega$$

in End $[[z^\pm,w^\pm]]$. For sufficiently large $M,N \in \mathbb{Z}$ we have by locality

$$(t-z)^M(t-w)^N Y(a,z)Y(b,w)Y(c,t)\Omega$$
$$= (t-z)^M(t-w)^N Y(c,t)Y(a,z)Y(b,w)\Omega$$

and

$$(t-z)^M(t-w)^N Y(c,t)Y(Y(a,z-w)b,w)\Omega$$
$$= (t-z)^M(t-w)^N Y(Y(a,z-w)b,w)Y(c,t)\Omega.$$

Consequently,

$$(t-z)^M(t-w)^N Y(a,z)Y(b,w)Y(c,t)\Omega$$
$$= (t-z)^M(t-w)^N Y(Y(a,z-w)b,w)Y(c,t)\Omega,$$

and by the Vacuum Axiom $Y(c,t)\Omega|_{t=0} = c$ we obtain

$$z^M w^N Y(a,z)Y(b,w)c = z^M w^N Y(Y(a,z-w)b,w)c,$$

which implies
$$Y(a,z)Y(b,w) = Y(Y(a,z-w)b,w).$$

The other two equalities follow by using the fundamental Theorem 10.5.   □

Corollary 10.37. *Each of the expansion in Theorem 10.36 is equivalent to each of the following commutation relations due to Borcherds*

$$[a_{(m)}, b_{(n)}] = \sum_{k \geq 0} \binom{m}{k} (a_{(k)})_{(m+n-k)}$$

or, equivalently,

$$[a_{(m)}, Y(b,z)] = \sum_{k \geq 0} \binom{m}{k} Y(a_{(k)}b,z) z^{m-k}.$$

We conclude that the subspace of all coefficients $a_{(n)} \in \mathrm{End}\ V, a \in V, n \in \mathbb{Z}$, is a Lie algebra Lie V with respect to the commutator.

Another direct consequence of the associativity of the OPE is the following: Note that a vertex subalgebra of a vertex algebra V is a vector subspace $U \subset V$ containing Ω such that $a_{(n)}U \subset U$ for all $a \in U$ and $n \in \mathbb{Z}$. Of course, a vertex subalgebra is itself a vertex algebra by restricting $a_{(n)}$ to U:

$$a_{(n)}^U = a_{(n)}|_U : U \to U$$

with vertex operators $Y^U(a,z) = \sum a_{(n)}^U z^{-n-1}$.

Corollary 10.38. *Let V be a vertex algebra.*

1. $a_{(0)}b = 0 \Longleftrightarrow [a_{(0)}, Y(b,z)] = 0.$
2. $\forall k \geq 0 : a_{(k)}b = 0 \Longleftrightarrow [Y(a,z), Y(b,w)] = 0.$
3. $a_{(0)}$ *is a derivation* $V \to V$ *for each* $a \in V$, *and thus* $\ker a_{(0)}$ *is a vertex subalgebra of V.*
4. *The centralizer of the field* $Y(a,z)$–*the subspace*

$$C(a) = \{b \subset V : [Y(a,z), Y(b,w)] = 0\} \subset V$$

   –*is a vertex subalgebra of V.*
5. *The fixed point set of an automorphism of V with respect to the vertex algebra structure is vertex subalgebra.*

Proof. The first two properties follow from the second equality in Corollary 10.37. Property 3 follows from the first equality in the above Corollary 10.37 for $m = 0$. 4 is implied by 2, and 5 is obvious. □

Remark 10.39. Through Corollary 10.38 the associativity of the OPE provides the possibility of obtaining new vertex algebras as subalgebras of a given vertex algebra V which are related to some important constructions of vertex algebra in physics and in mathematics.

1. The centralizer of a vector subspace $U \subset V$

$$C_V(U) = \{b \in V | \forall a \in U : [Y(a,z), Y(b,w)] = 0\}$$

is a vertex subalgebra of V by property 4 of Corollary 10.38 called the *coset model*.

2. For any subset $A \subset V$ the intersection

$$\bigcap \{\ker a_{(0)} : a \in A\}$$

is a vertex subalgebra by property 3 of Corollary 10.38 called a *W-algebra*.

3. For a subset $I \subset V$ the linear span of all the vectors

$$a^1_{(n_1)} a^2_{(n_2)} \ldots a^k_{(n_k)} \Omega, a^j \in I, n_j \in \mathbb{Z}, k \in \mathbb{N},$$

is a vertex subalgebra of V generated by the fields $Y(a,z), a \in I$.

4. Given a group G of automorphisms of a vertex algebra, the fixed point set V^G is a vertex subalgebra of V by property 5 of Corollary 10.38 called an *orbifold model* in case G is a finite group.

We finally come to the application of the associativity of the OPE to check the Virasoro field condition for the Heisenberg vertex algebra and the affine vertex algebras.

Theorem 10.40. *For a vector* $v \in V$ *denote* $L(z) := Y(v,z) = \sum_{n \in \mathbb{Z}} L_n z^{-n-2}$, *that is* $L_n = L_n^v = v_{(n+1)}$. *Suppose,* $L(z)$ *and* $c \in \mathbb{C}$ *satisfy*

$$L_{-1} = T , \ L_2 v = \frac{c}{2} \Omega , \ L_n v = 0 \text{ for } n > 2 , \ L_0 v = 2v.$$

Then $L(z)$ *is a Virasoro field with central charge* c. *If, in addition,* L_0 *is diagonalizable on* V, *then* v *is a conformal vector with central charge* c.

Proof. By the OPE (Theorem 10.36)

$$Y(v,z)Y(v,w) \sim \sum_{n \geq 0} \frac{Y(v_{(n)}v, w)}{(z-w)^{n+1}} = \sum_{n \geq -1} \frac{Y(L_n v, w)}{(z-w)^{n+2}}.$$

By the assumptions on $L_n v$ we obtain

$$L(z)L(w) \sim \frac{1}{2} c \frac{Y(\Omega, w)}{(z-w)^4} + \frac{Y(L_1 v, w)}{(z-w)^3} + \frac{Y(2v, w)}{(z-w)^2} + \frac{Y(Tv, w)}{(z-w)}.$$

It remains to show that the term $Y(L_1 v, z)$ vanishes, because in that case by inserting $Y(Tv, z) = \partial Y(v, w)$ (according to Corollary 10.34) and using $Y(\Omega, w) = \mathrm{id}_V$, one obtains the desired expansion

$$L(z)L(w) \sim \frac{1}{2} c \frac{1}{(z-w)^4} + \frac{2L(w)}{(z-w)^2} + \frac{\partial L(w)}{(z-w)}.$$

In order to show $a(z) := Y(L_1 v, z) = 0$ one interchanges z and w and obtains

$$L(w)L(z) \sim \frac{1}{2} c \frac{1}{(z-w)^4} - \frac{a(z)}{(z-w)^3} + \frac{2L(z)}{(z-w)^2} - \frac{\partial L(z)}{(z-w)},$$

hence, by Taylor expansion

$$L(w)L(z) \sim \frac{1}{2}c\frac{1}{(z-w)^4} - \frac{a(w)+Da(w)(z-w)+D^2a(w)(z-w)^2}{(z-w)^3}$$
$$+2\frac{L(w)+DL(w)(z-w)}{(z-w)^2} - \frac{\partial L(w)}{(z-w)}.$$

By locality, the two expansions of $L(z)L(w)$ and $L(w)L(z)$ have to be equal and this implies

$$\frac{a(w)}{(z-w)^3} = 0$$

and thus $a(z) = 0$. □

We are now in the position to apply the associativity of the OPE in order to show that the vectors v_λ resp. v_k are conformal vectors in our examples of the Heisenberg vertex algebra S resp. of the affine vertex algebra $V_k(\mathfrak{g})$.

We focus on the Heisenberg case since the corresponding equalities for the affine vertex algebra have been established already on page 198. We already know that $L_0 = \deg$ and $L_1 = T$. It remains to show that $L(z) = \sum L_n z^{-n-2}, L_n = L_n^v$, is a Virasoro field which means by Theorem 10.40 that only $L_2 v_\lambda = \frac{1}{2}c\Omega$ and $L_n v_\lambda = 0$ for $n \geq 3$ have to be checked. By using the expansion (10.6) of $Y(v_\lambda, z)$ we obtain

$$L_n = \frac{1}{2}\sum_{m\in\mathbb{Z}} a_{n-m}a_m - \lambda(n+1)a_n.$$

Now, $a_2(v_\lambda) = 2\lambda\Omega$ and $a_{n-m}a_m(v_\lambda) = 0$ for $m > 2$ or $m < n-2$ (because then $n - m > 2$). In the case of $n > 2$ we have $a_n(v_\lambda) = 0$ and only for $n = 3, n = 4$ there exist m with $n - 2 \leq m \leq 2$. It follows that $L_n v_\lambda = 0$ for $n \geq 5$. Because of $a_2 a_1 v_\lambda = 0$ and $a_2 a_2 v_\lambda = 0$ we also have $L_3 v_\lambda = 0 = L_4 v_\lambda$. For $n = 2$ we get $L_2 v_\lambda = \frac{1}{2}a_1 a_1(v_\lambda) + a_2 a_0(v_\lambda) - 6\lambda^2\Omega = (\frac{1}{2} - 6\lambda^2)\Omega$, and the central charge is $c = 1 - 12\lambda^2$. □

Remark 10.41. The Fock space representations of the Virasoro algebra which we have studied in the context of the quantization of the bosonic string on p. 116 are in perfect analogy with the observation that the $\frac{1}{2}T_1^2 + \lambda T_2$ are conformal vectors. We can show that

$$L_{-2}\Omega = \frac{1}{2} + 2\mu T_2$$

for

$$L_{-2} = \frac{1}{2}a_{-1}^2 + \sum_{k>0} a_{-k-1}a_{k-1},$$

where μ is the eigenvalue of a_0 to Ω. This yields another way of constructing a vertex algebra from the Heisenberg algebra using the calculations made there.

Indeed, $a_{-1}^2\Omega = T_1^2$ and $\sum_{k>0} a_{-k-1}a_{k-1}\Omega = a_{-2}a_0\Omega = 2\mu T_2$, hence

$$L_{-2}\Omega = \frac{1}{2}T_1^2 + 2\mu T_2. \qquad\qquad \square$$

Primary Fields. The conformal vector v of a conformal vertex algebra V provides, in particular, the diagonalizable endomorphism $L_0 : V \to V$. For each eigenvector $a \in V$ of L_0 with $L_0 a = ha$ the OPE (cf. Theorem 10.36) yields

$$Y(v,z)Y(a,w) \sim \sum_{n \geq -1} \frac{Y(L_n a, w)}{(z-w)^{n+2}},$$

and therefore begins with the following terms

$$Y(v,z)Y(a,w) \sim \frac{\partial Y(a,w)}{(z-w)} + \frac{hY(a,w)}{(z-w)^2} + \cdots.$$

Here, we use $L_{-1} = T$ and $Y(Ta,w) = \partial Y(a,w)$ (according to Corollary 10.34) and $L_0 a = ha$.

Definition 10.42 (Primary Field). A field $Y(a,z)$ of a conformal vertex algebra V with conformal vector v is called *primary* of *(conformal) weight* h if there are no other terms in the above OPE, that is

$$Y(v,z)Y(a,w) \sim \frac{\partial Y(a,w)}{(z-w)} + \frac{hY(a,w)}{(z-w)^2}.$$

Equivalently, $Y(L_n a, z) = 0$ for all $n > 0$.

The following is in accordance with Definition 9.7.

Corollary 10.43. *The field $Y(a,z)$ is primary of weight h if and only if one of the following equivalent conditions holds:*

1. *$L_0 a = ha$ and $L_n a = 0$ for all $n > 0$.*

2. *$[L_n, Y(a,z)] = z^{n+1} \partial Y(a,z) + h(n+1)z^n Y(a,z)$ for all $n \in \mathbb{Z}$.*

3. *$[L_n, a_{(m)}] = ((h-1)n - m)a_{(m+n)}$ for all $n, m \in \mathbb{Z}$.*

Proof. We have already stated the equivalence with 1. To show the second property for a primary field $Y(a,z)$ we compare

$$[Y(v,z)Ya,w)] = \sum_{n \in \mathbb{Z}} [L_n, Y(a,w)]z^{-n-2}$$

with

$$[Y(v,z)Ya,w)] = \partial Y(a,w)\delta(z-w) + hY(a,w)\partial\delta(z-w) =$$

$$= \sum_{m\in\mathbb{Z}}(-m-1)a_{(m)}w^{-m-2}\sum_{n\in\mathbb{Z}}z^{-n-1}w^n$$

$$+h\sum_{m\in\mathbb{Z}}a_{(m)}w^{-m-1}\sum_{n\in\mathbb{Z}}nz^{-n-1}w^{n-1}$$

$$= \sum_{m\in\mathbb{Z}}\sum_{n\in\mathbb{Z}}(-m-1+h(n+1))a_{(m)}w^{n-m-1}z^{-n-2},$$

and obtain for all $n \in \mathbb{Z}$

$$[L_n,Y(a,w)] = (-m-1+h(n+1))a_{(m)}w^{n-m-1}$$

$$= w^{n+1}\sum_{m\in\mathbb{Z}}(-m-1)a_{(m)}w^{-m-2} + w^n h(n+1)\sum_{m\in\mathbb{Z}}a_{(m)}w^{-m-1}$$

$$= w^{n+1}\partial Y(a,w) + z^n h(n+1)Y(a,z).$$

Hence, a primary field $Y(a,z)$ satisfies 2, and the converse is true since the implications above can be reversed.

To deduce 3 from 2 we use

$$[L_n,Y(a,z)] = \sum_{m\in\mathbb{Z}}[L_n,a_{(m)}]z^{-m-1}$$

$$= z^{n+1}\sum_{m\in\mathbb{Z}}a_{(m)}z^{-m-2} + z^n h(n+1)\sum_{m\in\mathbb{Z}}a_{(m)}z^{-m-1}$$

$$= \sum_{m\in\mathbb{Z}}(-m-n-1+h(n+1))a_{(m+n)}z^{-m-1}$$

to obtain $[L_n,a_{(m)}] = ((h-1)(n-1)-m)a_{(m+n)}$ by comparing coefficients. Hence, 2 implies 3 and vice versa. $\qquad\square$

Correlation Functions. Let us end this short introduction to vertex algebra theory by presenting the fundamental properties of correlation functions of a vertex algebra which have not been discussed so far although they play an important role in the axiomatic theory of quantum field theory and of conformal field theory as explained in Sections 8 and 9.

Let V^* denote the dual of V that is the space of linear functions $\mu : V \to \mathbb{C}$. Given $a_1,\dots,a_n \in V$ and $v \in V$ we consider

$$\langle\mu,Y(a_1,z_1)\dots Y(a_n,z_n)v\rangle := \mu(Y(a_1,z_1)\dots Y(a_n,z_n)v)$$

as a formal power series in $\mathbb{C}[[z_1^\pm,\dots,z_m^\pm]]$. These series are called n-point functions or *correlation functions*. Since $v = Y(v,z)|_{z=0}\Omega$ it is enough to study the case of $v = \Omega$ only.

Theorem 10.44. *Let* (V,Y,T,Ω) *be a vertex algebra and let* $\mu \in V^*$ *be a linear functional on* V. *For any* $a_1,\dots,a_n \in V$ *there exists a series*

$$f^{\mu}_{a_1 \dots a_n}(z_1, \dots, z_n) \in \mathbb{C}[[z_1 \dots z_n]][(z_i - z_j)^{-1}, i \neq j]$$

such that the following properties are satisfied:

1. For any permutation π of $\{1, \dots, n\}$ the correlation function

$$\langle \mu, Y(\pi(a_1), z_{\pi(1)}) \dots Y(\pi(a_n), z_{\pi(n)})\Omega \rangle$$

is the expansion in $\mathbb{C}((z_{\pi(1)})) \dots ((z_{\pi(n)}))$ of $f^{\mu}_{a_1 \dots a_n}(z_1, \dots, z_n)$.
2. For $i < j$ we have

$$f^{\mu}_{a_1 \dots a_n}(z_1, \dots z_n) = f_{(Y(a_i, z_i - z_j)a_j)a_1 \dots \hat{a}_i \dots \hat{a}_j \dots a_n}(z_1 \dots \hat{z}_i \dots z_j \dots z_n),$$

where $(z_i - z_j)^{-1}$ has to be replaced by its expansion $\sum\limits_{k \geq 0} z_i^{-k-1} z_j^k$ into positive powers of $\frac{z_j}{z_i}$.
3. For $1 \leq j \leq n$ we have

$$\partial_{z_j} f^{\mu}_{a_1 \dots a_n}(z_1, \dots z_n) = f^{\mu}_{a_1 \dots Ta_j \dots a_n}(z_1, \dots z_n).$$

Proof. Since $Y(a, z)$ is a field by the defining properties of a vertex algebra we have $\langle \mu, Y(a, z)v \rangle \in \mathbb{C}((z))$ for all $a, v \in V$, and by induction

$$\langle \mu, Y(\pi(a_1), z_{\pi(1)}) \dots Y(\pi(a_n), z_{\pi(n)})\Omega \rangle \in \mathbb{C}((z_{\pi(1)})) \dots ((z_{\pi(n)})).$$

By the Locality Axiom V2 there exist integers $N_{ij} > 0$ such that

$$(z_i - z_j)^{N_{ij}}[Y(a_i, z_i), Y(a_j, z_j)] = 0.$$

Hence, the series

$$\prod_{i<j}(z_i - z_j)^{N_{ij}} \langle \mu, Y(\pi(a_1), z_{\pi(1)}) \dots Y(\pi(a_n), z_{\pi(n)})\Omega \rangle$$

is independent of the permutation π. Moreover, it contains only non-negative powers of all the variables $z_i, 1 \leq i \leq n$, because of $Y(a, z)\Omega \in V[[z]]$ (Vacuum Axiom V3). Consequently,

$$\prod_{i<j}(z_i - z_j)^{N_{ij}} \langle \mu, Y(\pi(a_1), z_{\pi(1)}) \dots Y(\pi(a_n), z_{\pi(n)})\Omega \rangle$$

coincides with

$$\prod_{i<j}(z_i - z_j)^{N_{ij}} \langle \mu, Y(a_1, z_1) \dots Y(a_n, z_n)\Omega \rangle \in \mathbb{C}[[z_1, \dots, z_n]]$$

as a series in $\mathbb{C}[[z_1, \dots, z_n]]$. Dividing this series by $\prod_{i<j}(z_i - z_j)^{N_{ij}}$ yields the series $f^{\mu}_{a_1 \dots a_n} \in \mathbb{C}[[z_1 \dots z_n]][(z_i - z_j)^{-1}, i \neq j]$ with property 1.

The second property follows directly from 1 and the associativity of the OPE (Theorem 10.36). For example, in the case of $n = 2$ it has the form

$$f^{\mu}_{a_1 a_2}(z_1, z_2) = f_{(Y(a_1, z_1 - z_2)a_2)}(z_2)$$

and this equality is the same as

$$\langle \mu, Y(a_1, z_1)Y(a_2, z_2)\Omega \rangle = \langle \mu, Y(Y(a_1, z_1 - z_2)a_2, z_2)\Omega \rangle.$$

The third property is a consequence of the equality $Y(Ta, z) = \partial Y(a, z)$ proven in Corollary 10.34. $\qquad\square$

10.7 Induced Representations

In the course of these notes we have used Fock spaces and representation spaces for Lie algebras which all look very similar to each other and mostly have been given as vector spaces of polynomials. The unifying principle behind this observation is that all these representation spaces can be understood as certain induced representations which are mostly induced by a one-dimensional representation of a Lie subalgebra of the Lie algebra in question. This has to do with the fact that our representation spaces are cyclic in the sense that they can be generated by a suitable vector.

In order to describe induced representations we use the concept of a universal enveloping algebra. For any associative algebra A let $L(A)$ denote the Lie algebra with A as the underlying vector space and with the commutator as the Lie bracket.

Definition 10.45. A *universal enveloping algebra* of a Lie algebra $\mathfrak{g}$ is a pair (U, i) of an associative algebra U with unit 1 and a Lie algebra homomorphism $i : \mathfrak{g} \to L(U)$, such that the following universal property is fulfilled. For any associative algebra A with unit 1 and any Lie algebra homomorphism $j : \mathfrak{g} \to L(A)$ there exists a unique algebra homomorphism $h : U \to A$ with $h(1) = 1$ such that $h \circ i = j$.

Observe that a representation of the Lie algebra $\mathfrak{g}$, that is a Lie algebra homomorphism $\mathfrak{g} \to L(\text{End } W)$ (where End W is considered as an associative algebra) has a natural extension to $U(\mathfrak{g})$ as a homomorphism of associative algebras by the universal property. Conversely, a homomorphism $U(\mathfrak{g}) \to \text{End } W$ of associative algebras can be restricted to $\mathfrak{g}$ in order to obtain a Lie algebra homomorphism, that is a representation. We have shown:

Lemma 10.46. *The representations* $\mathfrak{g} \to \text{End } W$ *are in one-to-one correspondence with the representations* $U(\mathfrak{g}) \to \text{End } W$.

Lemma 10.47. *To each Lie algebra there corresponds a universal enveloping algebra unique up to isomorphism.*

Proof. The uniqueness of such a pair (U, i) is easy to show. In order to establish the existence let

$$T(W) = \bigoplus_{n=0}^{\infty} W^{\otimes n}$$

be the tensor algebra of a vector space W, where $W^{\otimes n}$ is n-fold tensor product of W with itself. The tensor algebra has the universal property that every linear map $W \to A$ into an associative algebra A with unit has a unique extension $T(W) \to A$ as an algebra homomorphism sending 1 to 1. Let $J \subset T(\mathfrak{g})$ be the two-sided ideal generated by the elements of the form $a \otimes b - b \otimes a - [a,b]$, $a,b \in \mathfrak{g}$. Let $U(\mathfrak{g}) := T(\mathfrak{g})/J$ be the quotient algebra with projection $p : T(\mathfrak{g}) \to U(\mathfrak{g})$. The map i is then defined by the restriction of p to $\mathfrak{g}$ with respect to its natural embedding $\mathfrak{g} \subset U(\mathfrak{g})$.

To show that $(U(\mathfrak{g}),i)$ fulfills the universal property, let A be an associative algebra with unit 1 and let $j : \mathfrak{g} \to L(A)$ be a Lie algebra homomorphism. Then, by the universal property of the tensor algebra $T(\mathfrak{g})$, there exists a unique algebra homomorphism $H : T(\mathfrak{g}) \to A$ extending the linear map j and satisfying $H(1) = 1$. Each generating element $a \otimes b - b \otimes a - [a,b]$ of J is annihilated by H since $H(a \otimes b - b \otimes a) = H(a)H(b) - H(b)H(a) = j(a)j(b) - j(b)j(a) = j([a,b]) = H([a,b])$. Hence, the ideal J is contained in the kernel of H. Consequently, H has a factorization h through p, that is there is an algebra homomorphism $h : U(\mathfrak{g}) \to A$ respecting the units with $H = h \circ p$ and thus $j = H|_{\mathfrak{g}} = h \circ p|_{\mathfrak{g}} = h \circ i$. $\qquad\square$

Neither the definition nor the above proof yields the injectivity of i. However, using the construction of $U(\mathfrak{g})$ this follows from the Poincaré–Birkhoff–Witt theorem which can be found in many books, e.g., [HN91]. We state one essential consequence of this theorem which is of special interest regarding the various descriptions of representation spaces.

Proposition 10.48 (Poincaré–Birkhoff–Witt). *Let $(a_i)_{i \in I}$ be an ordered basis of the Lie algebra $\mathfrak{g}$. Then the elements $p(a_{i_1} \otimes \ldots \otimes a_{i_m}), m \in \mathbb{N}, i_1 \leq \ldots \leq i_m$, together with 1 form a basis of $U(\mathfrak{g})$.*

As a consequence we obtain an isomorphism of vector spaces from the symmetric algebra

$$S(\mathfrak{g}) := \bigoplus_{n=0}^{\infty} \mathfrak{g}^{\odot n} \longrightarrow U(\mathfrak{g})$$

to $U(\mathfrak{g})$, where $W^{\odot n}$ is the n-fold symmetric product of a vector space, that is the subspace of symmetric tensors in $W^{\otimes n}$. $S(W)$ can also be understood as the quotient $T(W)/S$ with respect to the two-sided ideal $S \subset T(W)$ generated by all elements of the form $v \otimes w - w \otimes v$, $v,w \in W$. So far $S(\mathfrak{g})$ is the enveloping algebra of an abelian Lie algebra $\mathfrak{g}$.

Note that the symmetric algebra $S(W)$ can be identified with the algebra of polynomials $\mathbb{C}[T_i : i \in I]$ whenever $(a_i)_{i \in I}$ is an ordered basis of the vector space W.

Consequently, as a vector space the universal enveloping algebra $U(\mathfrak{g})$ of $\mathfrak{g}$ is isomorphic to the vector space $\mathbb{C}[T_i : i \in I]$ of polynomials:

$$1 \mapsto 1, \ T_{i_1} \ldots T_{i_m} \mapsto p(a_{i_1} \otimes \ldots \otimes a_{i_m}), \ m \in \mathbb{N}, \ i_1 \leq \ldots \leq i_m,$$

provides an isomorphism.

Now, let $\mathfrak{b}$ be a Lie subalgebra of the Lie algebra $\mathfrak{g}$ and let $\pi : \mathfrak{b} \to \mathrm{End}\, W$ a Lie algebra homomorphism, that is a representation of $\mathfrak{b}$ in the vector space W.

Definition 10.49. The *induced representation* (induced by π) is given by the *induced $\mathfrak{g}$-module*

$$\mathrm{Ind}_{\mathfrak{b}}^{\mathfrak{g}} = U(\mathfrak{g}) \otimes_{U(\mathfrak{b})} W,$$

that is

$$\mathrm{Ind}_{\mathfrak{b}}^{\mathfrak{g}} = (U(\mathfrak{g}) \otimes W)/U(\mathfrak{g})\{b \otimes w - 1 \otimes \pi(b)w : (b,w) \in \mathfrak{b} \times W\},$$

where $\mathfrak{g}$ acts by left multiplication in the first factor.

It is straightforward to check that this prescription defines a representation. In fact, the action of $a \in U(\mathfrak{g})$ on $U(\mathfrak{g}) \otimes W$, $x \otimes w \mapsto ax \otimes w$, descends to a linear action $\rho(a) \in \mathrm{End}\ (\mathrm{Ind}_{\mathfrak{b}}^{\mathfrak{g}})$ since $J_\pi := U(\mathfrak{g})\{b \otimes w - 1 \otimes \pi(b)w : (b,w) \in \mathfrak{b} \times W\}$ is a left ideal, in particular $a(J_\pi) \subset J_\pi$. In addition, $\rho(a)([x \otimes w]) = [ax \otimes w]$ defines a homomorphism $a \mapsto \rho(a)$ of associative algebras, again since J_π a left ideal in $U(\mathfrak{g}) \otimes W$. The restriction of ρ to $\mathfrak{g}$ is therefore a Lie algebra homomorphism.

An elementary example is the Fock space representation of the Heisenberg algebra described on p. 114. The Heisenberg algebra H is generated by $a_n, n \in \mathbb{Z}$, and the central element Z. The inducing representation π is defined on the abelian Lie subalgebra B $\subset$ H generated by the $a_n, n \geq 0$ and Z, with $W = \mathbb{C}$, and this representation $\pi : \mathrm{P} \to \mathrm{End}\ \mathbb{C} \cong \mathbb{C}$ is determined by

$$\rho(Z) = \mathrm{id}_\mathbb{C} = 1, \rho(a_0) = \mu \mathrm{id}_\mathbb{C} = \mu,\ \rho(a_n) = 0 \text{ for } n > 0.$$

Let $\Omega := 1 \otimes 1$. Then $a_n \in J_p i$ for $n > 0$, since $a_n \Omega = a_n \otimes 1 = 1 \otimes \pi(a_n)1 = 0$, $a_0 \Omega = a_0 \otimes 1 = 1 \otimes \mu = \mu \Omega$, and $Z(\Omega) = 1 \otimes \pi(Z) = \Omega$. Hence, $a_n \in J_\pi, n > 0$, and a_0, Z depend on Ω modulo J_π.

Consequently, $\mathrm{Ind}_{\mathfrak{b}}^{\mathfrak{g}}(\mathbb{C})$ is generated by the classes

$$[a_{i_1} \otimes \ldots \otimes a_{i_m} \Omega], m \in \mathbb{N}, i_1 \leq \ldots \leq i_m < 0,$$

and Ω according to Proposition 10.48. These elements remain linearly independent, since the a_{-n}, a_{-m} commute with each other for $m, n \geq 0$, so that $\mathrm{Ind}_{\mathfrak{b}}^{\mathfrak{g}}(\mathbb{C})$ is isomorphic to the vector space $\mathbb{C}[T_n : n > 1]$ with the action $\rho(a_{-n})\Omega = T_n$ for $n > 0$, and, more generally,

$$\rho(a_{-n})P = T_n P,$$

for any polynomial $P \in \mathbb{C}[T_n : n > 1]$. Similarly, because of the other commutation relations, for $n > 0$ we obtain $\rho(a_n)T_m = 0$ if $n \neq m$ and $\rho(a_n)T_n = n\Omega$, and, more generally, $\rho(a_n)P = n\partial_{T_n}P$. This, of course, is exactly the representation on p. 114.

The example is typical, in the cases considered in these notes, we have $W = \mathbb{C}$ and an ordered basis $(a_i)_{i \in I}$ with a division $I = I_+ \cup I_-$ such that $a_i, i \in I_+$ is a basis of J_π and $\mathrm{Ind}_{\mathfrak{b}}^{\mathfrak{g}}(\mathbb{C})$ is isomorphic to the space of polynomials $\mathbb{C}[T_n : n \in I_-]$. The action of the $a_i, i \in I$, is then essentially determined by $a_i \Omega = T_i$ if $i \in I_-$ and the commutation relations of all the a_i.

In this way we obtain similarly the description of a Verma module with respect to given numbers $c, h \in C$ on p. 94, the representation of the string algebra on p. 119, the representation V_c of the Virasoro algebra Vir used for the Virasoro vertex algebras on p. 193, the representation of the Kac–Moody algebras on p. 196, and in a certain sense even the free boson representation on p. 136 where, however, the Hilbert space structure has to be respected as well. Analogously, the fermionic Fock space on p. 52 can be described as an induced representation. To do this, we have to extend the consideration to the case of Lie superalgebras in order to include the anticommutation relations.

References

[BD04*]   A. Beilinson and V. Drinfeld. *Chiral Algebras*. AMS Colloquium Publications **51**, AMS, Providence, RI, 2004.

[BF01*]   D. Ben-Zvi and E. Frenkel. *Vertex Algebras and Algebraic Curves*. AMS, Providence, RI, 2001.

[Bor86*]   R.E. Borcherds. Vertex algebras, Kac-Moody algebra and the monster. *Proc. Natl. Acad. Sci. USA* **83** (1986), 3068–3071.

[Bor00*]   R.E. Borcherds, Quantum vertex algebras. In: *Taniguchi Conference on Mathematics Nara '98*. Advanced Studies in Pure Mathematics **31**, 51–74. Math. Soc. Japan, 2000.

[FKRW95*] E. Frenkel, V. Kac, A. Radul, and W. Wang. $\mathcal{W}_{1+\infty}$ and $W(gl_N)$ with central charge N. *Commun. Math. Phys.* **170** (1995), 337–357.

[FLM88*]   I. Frenkel, J. Lepowsky, and A. Meurman. *Vertex Operator Algebras and the Monster*. Academic Press, New York, 1988.

[HN91]   J. Hilgert and K.-H. Neeb. *Lie Gruppen und Lie Algebren*. Vieweg, Braunschweig, 1991.

[Hua97*]   Y-Z. Huang. *Two-Dimensional Conformal Geometry and Vertex Operator Algebras*. Progress in Mathematics 148, Birkhuser, Basel, 1997.

[Kac98*]   V. Kac. *Vertex Algebras for Beginners*. University Lecture Series 10, AMS, Providencs, RI, 2nd ed., 1998.

[Len07*]   S. Lentner. *Vertex Algebras Constructed from Hopf Algebra Structures*. Diplomarbeit, LMU München, 2007.

[Lin04*]   K. Linde. Global vertex operators on Riemann surfaces. Dissertation. LMU München, 2004.

[Seg88a]   G. Segal. The definition of conformal field theory. Unpublished Manuscript, 1988. Reprinted in *Topology, Geometry and Quantum Field Theory*, U. Tillmann (Ed.), 432–574, Cambridge University Press, Cambridge, 2004.

[Seg91]   G. Segal. Geometric aspects of quantum field theory. *Proc. Intern. Congress Kyoto 1990, Math. Soc.* Japan, 1387–1396, 1991.

Chapter 11
Mathematical Aspects of the Verlinde Formula

The Verlinde formula describes the dimensions of spaces of conformal blocks (cf. Sect. 9.3) of certain rational conformal field theories (cf. [Ver88]). With respect to a suitable mathematical interpretation, the Verlinde formula gives the dimensions of spaces of generalized theta functions (cf. Sect. 11.1). These dimensions and their polynomial behavior (cf. Theorem 11.6) are of special interest in mathematics. Prior to the appearance of the Verlinde formula, these dimensions were known for very specific cases only, e.g., for the classical theta functions (cf. Theorem 11.5).

The Verlinde formula has been presented by E. Verlinde in [Ver88] as a result of physics. Such a result is, of course, not a mathematical result, it will be considered as a conjecture in mathematics. However, the physical insights leading to the statement of the formula and its justification can be of great help in proving it. Several mathematicians have worked on the problem of proving the Verlinde formula, starting with [TUY89] and coming to a certain end with [Fal94]. These proofs are all quite difficult to understand. For a recent review on general theta functions we refer to the article [Fal08*] of Faltings.

In this last chapter of the present notes we want to explain the Verlinde formula in the context of stable holomorphic bundles on a Riemann surface, that is as a result in function theory or in algebraic geometry. Furthermore, we will sketch a strategy for a proof of the Verlinde formula which uses a kind of fusion for compact Riemann surfaces with marked points. This strategy is inspired by the physical concept of the fusion of fields in conformal field theory as explained in the preceding chapter. We do not explain the interesting transformation from conformal field theory to algebraic geometry. Instead we refer to [TUY89], [Uen95], [BF01*], [Tyu03*].

11.1 The Moduli Space of Representations and Theta Functions

In the following, S is always an oriented and connected compact surface of genus $g = g(S) \in \mathbb{N}_0$ without boundary. The *moduli space of representations* for the group G is

$$\mathcal{M}^G := \mathrm{Hom}(\pi_1(S), G)/G.$$

Schottenloher, M.: *Mathematical Aspects of the Verlinde Formula*. Lect. Notes Phys. **759**, 213–233 (2008)
DOI 10.1007/978-3-540-68628-6_12   © Springer-Verlag Berlin Heidelberg 2008

The equivalence relation indicated by "$/G$" is the conjugation

$$g \sim g' \Longleftrightarrow \exists h \in G : g = hgh^{-1}.$$

Theorem 11.1. $\mathcal{M}^G$ *has a number of quite different interpretations. In the case of* $G = \mathrm{SU}(r)$ *these interpretations can be formulated in form of the following one-to-one correspondences (denoted by "$\cong$"):*

1. $\mathcal{M}^{\mathrm{SU}(r)} = \mathrm{Hom}(\pi_1(S), \mathrm{SU}(r))/\mathrm{SU}(r).$

   Topological *interpretation: the set* $\mathcal{M}^{\mathrm{SU}(r)}$ *is a topological invariant, which carries an amount of information which interpolates between the fundamental group* $\pi_1(S)$ *and its abelian part*

$$H_1(S) = \pi_1(S)/[\pi_1(S), \pi_1(S)],$$

   the first homology group of S.

2. $\mathcal{M}^{\mathrm{SU}(r)} \cong$ *set of equivalence classes of flat* $\mathrm{SU}(r)$*-bundles.*

   Geometric *interpretation: there are two related (and eventually equivalent) interpretations of "flat"* $\mathrm{SU}(r)$*-bundles; "flat" in the sense of a flat vector bundle with constant transition functions and "flat" in the sense of a vector bundle with a flat connection (corresponding to* $\mathrm{SU}(r)$ *in both cases). Two such bundles are called* equivalent *if they are isomorphic as flat bundles.*

3. $\mathcal{M}^{\mathrm{SU}(r)} \cong \check{H}^1(S, \mathrm{SU}(r)) \cong H^1(\pi_1(S), \mathrm{SU}(r)).$

   Cohomological *interpretation:* $\check{H}^1(S, \mathrm{SU}(r))$ *denotes the first Čech cohomology set with values in* $\mathrm{SU}(r)$ *(this is not a group in the non-abelian case) and* $H^1(\pi_1(S), \mathrm{SU}(r))$ *denotes the group cohomology of* $\pi_1(S)$ *with values in* $\mathrm{SU}(r)$.

4. $\mathcal{M}^{\mathrm{SU}(r)} \cong \mathscr{A}_0/\mathscr{G}.$

   Interpretation as a phase space: $\mathscr{A}$ *is the space of differentiable connections on the trivial bundle* $S \times \mathrm{SU}(r) \to S$, $\mathscr{A}_0 \subset \mathscr{A}$ *is the subspace of flat connections and* $\mathscr{G}$ *is the corresponding gauge group of bundle automorphisms, that is*

$$\mathscr{G} \cong \mathscr{C}^\infty(S, \mathrm{SU}(r)).$$

$\mathscr{A}_0/\mathscr{G}$ *appears as the phase space of a three-dimensional Chern–Simons theory with an internal symmetry group* $\mathrm{SU}(r)$ *with respect to a suitable gauge (cf.* [Wit89]).

5. $\mathcal{M}^{\mathrm{SU}(r)} \cong$ *moduli space of semi-stable holomorphic vector bundles* E *on* S *of rank* r *with* $\det E = \mathcal{O}_S$.

   Complex analytical *interpretation: here, one has to introduce a complex structure* J *on the surface* S *such that* S *equipped with* J *is a Riemann surface* S_J. *The vector bundles in the above moduli space are holomorphic with respect to this complex structure and the sheaf* $\mathcal{O}_S$ *is the structure sheaf on* S_J. *To emphasize the dependence on the complex structure* J *on* S, *we denote this moduli space by* $\mathcal{M}_J^{\mathrm{SU}(r)}$.

To prove the above bijections "$\cong$" in the cases 2., 3., and 4. is an elementary exercise for understanding the respective concepts. Case 5. is a classical theorem of Narasimhan and Seshadri [NS65] and is much more involved.

In each of these cases, "$\cong$" is just a bijection of sets. However, the different interpretations yield a number of different structures on $\mathcal{M}^{SU(r)}$. In 1., for instance, $\mathcal{M}^{SU(r)}$ obtains the structure of a subvariety of $SU(r)^{2g}/SU(r)$ (because of the fact that $\pi_1(S)$ is a group of $2g$ generators and one relation, cf. (11.4) below), in 4. the set $\mathcal{M}^{SU(r)}$ obtains the structure of a symplectic manifold and in 5., according to [NS65], the structure of a Kähler manifold outside the singular points of $\mathcal{M}^{SU(r)}$.

Among others, there are three important generalizations of Theorem 11.1:

- to other Lie groups G instead of $SU(r)$,
- to higher-dimensional compact manifolds M instead of S and, in particular, to Kähler manifolds in connection with 5.
- to $S \setminus \{P_1,\ldots,P_m\}$ instead of S with points $P_1,\ldots,P_m \in S$ (cf. Sect. 11.3) and a suitable fixing of the vector bundle structure near the points $P_1,\ldots,P_m \in S$.

To begin with, we do not discuss these more general aspects, but rather concentrate on $\mathcal{M}^{SU(r)}$. The above-mentioned structures induce the following properties on $\mathcal{M}^{SU(r)}$:

- $\mathcal{M}^{SU(r)}$ has a natural symplectic structure, which is induced by the following 2-form ω on the affine space

$$\mathscr{A} \cong \mathscr{A}^1(S,\mathfrak{su}(r))$$

of connections:

$$\omega(\alpha,\beta) = c \int_S \mathrm{tr}(\alpha \wedge \beta) \qquad (11.1)$$

for $\alpha,\beta \in \mathscr{A}^1(S,\mathfrak{su}(r))$ with a suitable constant $c \in \mathbb{R} \setminus \{0\}$.

Here,

$$\mathrm{tr} : \mathfrak{su}(r) \to \mathbb{R}$$

is the trace of the complex $r \times r$-matrices with respect to the natural representation. In what sense this defines a symplectic structure on $\mathscr{A}$ and on $\mathscr{A}_0/\mathscr{G}$ will be explained in more detail in the following.

In fact, for a connection $A \in \mathscr{A}$ the tangent space $T_A\mathscr{A}$ of the affine space $\mathscr{A}$ can be identified with the vector space $\mathscr{A}^1(S,\mathfrak{su}(r))$ of $\mathfrak{su}(r)$-valued differentiable 1-forms. Hence, a 2-form on $\mathscr{A}$ is given by a family $(\omega_A)_{A \in \mathscr{A}}$ of bilinear mappings ω_A on $\mathscr{A}^1(S,\mathfrak{su}(r)) \times \mathscr{A}^1(S,\mathfrak{su}(r))$ depending differentiably on $A \in \mathscr{A}$. Now, the map

$$\omega : \mathscr{A}^1(S,\mathfrak{su}(r)) \times \mathscr{A}^1(S,\mathfrak{su}(r)) \to \mathbb{C}$$

defined by (11.1) is independent of $A \in \mathscr{A}$ with respect to the natural trivialization of the cotangent bundle

$$T^*\mathscr{A} = \mathscr{A} \times \mathscr{A}^1(S,\mathfrak{su}(r))^*.$$

Consequently, ω with (11.1) is a closed 2-form. It is nondegenerate since $\omega(\alpha, \beta) = 0$ for all α implies $\beta = 0$. Hence, it is a symplectic form on $\mathscr{A}$ defining the symplectic structure. Moreover, it can be shown that the pushforward of $\omega|_{\mathscr{A}_0}$ with respect to the projection $\mathscr{A}_0 \to \mathscr{A}_0/\mathscr{G}$ gives a symplectic form $\omega_{\mathscr{M}}$ on the regular part of $\mathscr{A}_0/\mathscr{G}$. Indeed, $\mathscr{A}_0/\mathscr{G}$ is obtained by a general Marsden–Weinstein reduction of $(\mathscr{A}, \omega)$ with respect to the action of the gauge group $\mathscr{G}$ where the curvature map turns out to be a moment map.

This symplectic form $\omega_{\mathscr{M}}$ is also induced by Chern–Simons theory (cf. [Wit89]). $\mathscr{A}_0/\mathscr{G}$ with this symplectic structure is the phase space of the classical fields.

- Moreover, on $\mathscr{M}^{\mathrm{SU}(r)}$ there exists a natural line bundle $\mathscr{L}$ (the determinant bundle) – which is uniquely determined up to isomorphism – together with a connection ∇ on $\mathscr{L}$ whose curvature is $2\pi i \omega$. With a fixed complex structure J on S, for instance, the line bundle $\mathscr{L}$ has the following description:

$$\Theta := \left\{ [E] \in \mathscr{M}_J^{\mathrm{SU}(r)} : \dim_{\mathbb{C}} H^0(S, E) \geq 1 \right\}$$

is a Cartier divisor (the "theta divisor") on $\mathscr{M}_J^{\mathrm{SU}(r)}$, for which the sheaf

$$\mathscr{L} = \mathscr{L}_\Theta = \mathscr{O}(\Theta) = \text{sheaf of meromorphic functions } f \text{ on } \mathscr{M}_J^{\mathrm{SU}(r)}$$
$$\text{with } (f) + \Theta \geq 0$$

is a locally free sheaf of rank 1. Hence, $\mathscr{L}$ is a complex line bundle, which automatically is holomorphic with respect to the complex structure on the moduli space induced by J. ($H^0(S, E)$ is the vector space of holomorphic sections on the compact Riemann surface $S = S_J$ with values in the holomorphic vector bundle E and $[E]$ denotes the equivalence class represented by E.)

Definition 11.2. The space of holomorphic sections in $\mathscr{L}^k$, that is

$$H^0\left(\mathscr{M}_J^{\mathrm{SU}(r)}, \ \mathscr{L}^k \right),$$

is the space of *generalized theta functions* of level $k \in \mathbb{N}$.

Here, $\mathscr{L}^k$ is the k-fold tensor product of $\mathscr{L}$: $\mathscr{L}^k = \mathscr{L} \otimes \ldots \otimes \mathscr{L}$ (k-fold). Since $\mathscr{M}_J^{\mathrm{SU}(r)}$ is compact, $H^0(\mathscr{M}_J^{\mathrm{SU}(r)}, \mathscr{L}^k)$ is a finite-dimensional vector space over $\mathbb{C}$. In the context of geometric quantization, the space

$$H^0\left(\mathscr{M}_J^{\mathrm{SU}(r)}, \ \mathscr{L} \right)$$

can be interpreted as the quantized state space for the phase space $(\mathscr{M}^{\mathrm{SU}(r)}, \omega)$, prequantum bundle $\mathscr{L}$ and holomorphic polarization J. A similar result holds for $H^0(\mathscr{M}_J^{\mathrm{SU}(r)}, \mathscr{L}^k)$. To explain this we include a short digression on geometric quantization (cf. [Woo80] for a comprehensive introduction):

Geometric Quantization. Geometric quantization of a classical mechanical system proceeds as follows. The classical mechanical system is supposed to be represented by a symplectic manifold (M, ω). For quantizing (M, ω) one needs two additional geometric data, a prequantum bundle and a polarization. A *prequantum bundle* is a complex line bundle $L \to M$ on M together with a connection ∇ whose curvature is $2\pi i\omega$. A *polarization* F on M is a linear subbundle F of (that is a distribution on) the complexified tangent bundle $TM^{\mathbb{C}}$ fulfilling some compatibility conditions. An example is the bundle F spanned by all "y-directions" in $M = \mathbb{R}^2$ with coordinates (x, y) or on $M = \mathbb{C}^n$ the complex subspace of $TM^{\mathbb{C}}$ spanned by the directions $\frac{\partial}{\partial z_j}, j = 1, \ldots, n$. This last example is the holomorphic polarization which has a natural generalization to arbitrary complex manifolds M. Now the (uncorrected, see (11.3)) state space of geometric quantization is

$$Z := \{s \in \Gamma(M, L) : s \text{ is covariantly constant on } F\}.$$

Here, $\Gamma(M, L)$ denotes the $\mathscr{C}^\infty$-sections on M of the line bundle L and the covariance condition means that $\nabla_X s = 0$ for all local vector fields $X : U \to F \subset TM^{\mathbb{C}}$ with values in F. In case of the holomorphic polarization the state space Z is simply the space $H^0(M, L)$ of holomorphic sections in L.

Back to our moduli space $\mathscr{M}_J^{\mathrm{SU}(r)}$ with symplectic form $\omega_{\mathscr{M}}$, the holomorphic line bundle $\mathscr{L} \to \mathscr{M}_J^{\mathrm{SU}(r)}$, and holomorphic polarization one gets the following: for every $k \in \mathbb{N}$, $\mathscr{L}^k$ is a prequantum bundle of $(\mathscr{M}_J^{\mathrm{SU}(r)}, k\omega_{\mathscr{M}})$. Consequently, $H^0(\mathscr{M}_J^{\mathrm{SU}(r)}, \mathscr{L}^k)$ is the (uncorrected) state space of geometric quantization.

In order to have a proper quantum theory constructed by geometric quantization it is necessary to develop the theory in such a way that the state space Z obtains an inner product. By an appropriate choice of the prequantum bundle and the polarization one has to try to represent those observables one is interested in as self-adjoint operators on the completion of Z (see [Woo80]). We are not interested in these matters and only want to point out that the space of generalized theta functions has an interpretation as the state space of a geometric quantization scheme: The space

$$H^0\left(\mathscr{M}_J^{\mathrm{SU}(r)}, \mathscr{L}^k\right)$$

is the (uncorrected) quantized state space of the phase space

$$\left(\mathscr{M}_J^{\mathrm{SU}(r)}, k\omega\right),$$

for the prequantum bundle $\mathscr{L}^k$ and for the holomorphic polarization on $\mathscr{M}_J^{\mathrm{SU}(r)}$.

Before continuing the investigation of the spaces of generalized theta functions we want to mention an interesting connection of geometric quantization with representation theory of compact Lie groups which we will use later for the description of parabolic bundles. In fact, to a large extent, the ideas of geometric quantization developed by Kirillov, Kostant, and Souriau have their origin in representation theory.

Let G be a compact, semi-simple Lie group with Lie algebra $\mathfrak{g}$ and fix an invariant nondegenerate bilinear form $<,>$ on $\mathfrak{g}$ by which we identify $\mathfrak{g}$ and the dual $\mathfrak{g}^*$ of $\mathfrak{g}$. For simplicity we assume G to be a matrix group. Then G acts on $\mathfrak{g}$ by the *adjoint action*

$$\mathrm{Ad}_g : \mathfrak{g} \to \mathfrak{g}, \; X \to gXg^{-1},$$

$g \in G$, and on $\mathfrak{g}^*$ by the *coadjoint action*

$$\mathrm{Ad}_g^* : \mathfrak{g}^* \to \mathfrak{g}^*, \xi \to \xi \circ \mathrm{Ad}_g,$$

$g \in G$. The orbits $\mathscr{O} = G\xi = \{\mathrm{Ad}_g^*(\xi) : g \in G\}$ of the coadjoint action are called *coadjoint orbits*. They carry a natural symplectic structure given as follows. For $A \in \mathfrak{g}$ let $X_A : \mathscr{O} \to T\mathscr{O}$ be the Jacobi field, $X_A(\xi) = \frac{d}{dt}(\mathrm{Ad}_{e^{tA}}^* \xi)\big|_{t=0}$. Then by

$$\omega_\xi(X_A, X_B) := \xi([A,B])$$

for $\xi \in \mathscr{O}$, $A, B \in \mathfrak{g}$, we define a 2-form which is nondegenerate and closed, hence a symplectic form.

The coadjoint orbits have another description using the isotropy group $G_\xi = \{g \in G : \mathrm{Ad}_g^*\xi = \xi\}$, namely

$$\mathscr{O} \cong G/G_\xi \cong G^{\mathbb{C}}/B,$$

where $G^{\mathbb{C}}$ is the complexification of G and $B \subset G^{\mathbb{C}}$ is a suitable Borel subgroup. In this manner $\mathscr{O} \cong G^{\mathbb{C}}/B$ is endowed with a complex structure induced from the complex homogeneous (flag) manifold $G^{\mathbb{C}}/B$. ω turns out to be a Kähler form with respect to this complex structure, such that $(\mathscr{O}, \omega)$ is eventually a Kähler manifold. Assume now that we find a holomorphic prequantum bundle on $\mathscr{O}$. Then G acts in a natural way on the state space $H^0(\mathscr{O}, \mathscr{L})$. Based on the Borel–Weil–Bott theorem we have the following result.

Theorem 11.3 (Kirillov [Kir76]). *Geometric quantization of each coadjoint orbit of maximal dimension endowed with a prequantum bundle yields an irreducible unitary representation of G. Every irreducible unitary representation of G appears exactly once amongst these (if one takes account of equivalence classes of prequantum bundles $\mathscr{L} \to \mathscr{O}$ only).*

To come back to our moduli spaces and spaces of holomorphic sections in line bundles we note that a close connection of the spaces of generalized theta functions with conformal field theory is established by the fact that $H^0(\mathscr{M}_J^{\mathrm{SU}(r)}, \mathscr{L}^k)$ is isomorphic to the space of conformal blocks of a suitable conformal field theory with gauge symmetry (cf. Sect. 9.3). This is proven in [KNR94] for the more general case of a compact simple Lie group G.

At the end of this section we want to discuss the example $G = \mathrm{U}(1)$ which does not completely fit into the scheme of the groups $\mathrm{SU}(r)$ or groups with a simple complexification. However, it has the advantage of being relatively elementary, and it explains why the elements of $H^0(\mathscr{M}_J^{\mathrm{SU}(1)}, \mathscr{L}^k)$ are called generalized theta functions:

Example 11.4. (e.g. in [Bot91*]) Let G be the abelian group $U(1)$ and let J be a complex structure on the surface S. Then $\mathcal{M}_J^{U(1)}$ is isomorphic (as a set) to

1. the moduli space of holomorphic line bundles on the Riemann surface $S = S_J$ of degree 0.
2. the set of equivalence classes of holomorphic vector bundle structures on the trivial C^∞ vector bundle $S_J \times \mathbb{C} \to S_J$.
3. $\mathrm{Hom}(\pi_1(S), U(1)) \cong \check{H}^1(S, U(1)) \cong H^1(S_J, \mathcal{O})/H^1(S, \mathbb{Z})$, which is a complex g-dimensional torus where $\mathcal{O}$ is the sheaf of germs of holomorphic functions in S_J.
4. $\mathbb{C}^g/\Gamma \cong$ Jacobi variety of S_J.

Let $\mathcal{L} \to \mathcal{M}_J^{U(1)}$ be the theta bundle, given by the theta divisor on the Jacobi variety. Then

- $H^0(\mathcal{M}_J^{U(1)}, \mathcal{L}) \cong \mathbb{C}$ is the space of classical theta functions and
- $H^0(\mathcal{M}_J^{U(1)}, \mathcal{L}^k)$ is the space of classical theta functions of level k.

Theorem 11.5. $\dim_\mathbb{C} H^0(\mathcal{M}_J^{U(1)}, \mathcal{L}^k) = k^g$ *(independently of the complex structure).*

The Verlinde formula is a generalization of this dimension formula to other Lie groups G instead of $U(1)$. Here we will only treat the case of the Lie groups $G = SU(r)$.

11.2 The Verlinde Formula

Theorem 11.6 (Verlinde Formula). *Let*

$$z_k^{SU(r)}(g) := \dim_\mathbb{C} H^0\left(\mathcal{M}_J^{SU(r)}, \mathcal{L}^k\right).$$

Then

$$z_k^{SU(2)}(g) = \left(\frac{k+2}{2}\right)^{g-1} \sum_{j=1}^{k+1} \left(\sin^2 \frac{j\pi}{k+2}\right)^{1-g} \quad and \quad (11.2)$$

$$z_k^{SU(r)}(g) = \left(\frac{r}{k+r}\right)^g \sum_{\substack{S \subset \{1,\ldots,k+r\} \\ |S|=r}} \prod_{\substack{s \in S, t \notin S \\ 1 \le t \le k+r}} \left| 2\sin \pi \frac{s-t}{r+k} \right|^{g-1}$$

for $r \ge 2$.

The theorem (cf. [Ver88], [TUY89], [Fal94], [Sze95], [Bea96], [Bea95], [BT93], [MS89], [NR93], [Ram94], [Sor95]) has a generalization to compact Lie groups for which the complexification is a simple Lie group $G^\mathbb{C}$ of one of the types A, B, C, D, or G ([BT93], [Fal94]).

Among other aspects the Verlinde formula is remarkable because

- the expression on the right of the equation actually defines a natural number,
- it is polynomial in k, and
- the dimension does not depend on the complex structure J.

Even the transformation of the second formula into the first for $r = 2$ requires some calculation. Concerning the independence of J: physical insights related to rational conformal field theory imply that the space of conformal blocks does not depend on the complex structure J on S. This makes the independence of the dimension formula of the structure J plausible. However, a mathematical proof is still necessary.

From a physical point of view, the Verlinde formula is a consequence of the fusion rules for the operator product expansion of the primary fields (cf. Sect. 9.3). We will discuss the fusion mathematically in the next section. Using the fusion rules formulated in that section, the Verlinde formula will be reduced to a combinatorical problem, which is treated in Sect. 11.4.

There is a shift $k \to k + r$ in the Verlinde formula which also occurs in other formulas on quantum theory and representation theory. This shift has to do with the quantization of the systems in question and it is often related to a central charge or an anomaly (cf. [BT93]). In the following we will express the shift within geometric quantization or rather metaplectic quantization. This is based on the fact that $H^0(\mathcal{M}_J^{SU(r)}, \mathcal{L}^k)$ can be obtained as the state space of geometric quantization. Indeed, the shift has an explanation as to arise from an incomplete quantization procedure. Instead of the ordinary geometric quantization one should rather take the metaplectic correction.

Metaplectic Quantization. In many known cases of geometric quantization, the actual calculations give rise to results which do not agree with the usual quantum mechanical models. For instance, the dimensions of eigenspaces turn out to be wrong or shifted. This holds, in particular, for the Kepler problem (hydrogen atom) and the harmonic oscillator. Because of this defect of the geometric quantization occurring already in elementary examples one should consider the metaplectic correction which in fact yields the right answer in many elementary classical systems, in particular, in the two examples mentioned above. To explain the procedure of metaplectic correction we restrict to the case of a Kähler manifold (M, ω) with Kähler form ω as a symplectic manifold. In this situation a *metaplectic structure* on M is given by a spin structure on M which in turn is given by a square root $K^{\frac{1}{2}}$ of the canonical bundle K on M. (K is the holomorphic line bundle $\det T^*M$ of holomorphic n-forms, when n is the complex dimension of M.) The metaplectic correction means – in the situation of the holomorphic polarization – taking the spaces

$$Z^m = H^0\left(M, L \otimes K^{\frac{1}{2}}\right) \tag{11.3}$$

as the state spaces replacing $Z = H^0(M, L)$.

In the context of our space of generalized theta functions the metaplectic correction is

$$Z^m = H^0 \left(\mathcal{M}_J^{\text{SU}(r)}, \mathcal{L}^k \otimes \mathcal{K}^{\frac{1}{2}} \right),$$

where $\mathcal{K}$ is the canonical bundle of $\mathcal{M}_J^{\text{SU}(r)}$.

Now, the canonical bundle of $\mathcal{M}_J^{\text{SU}(r)}$ turns out to be isomorphic to the dual of $\mathcal{L}^{2r}$, hence a natural metaplectic structure in this case is $\mathcal{K}^{\frac{1}{2}} = \mathcal{L}^{-r}$ (:= dual of $\mathcal{L}^r$). As a result of the metaplectic correction the shift disappears:

$$Z^m = H^0 \left(\mathcal{M}_J^{\text{SU}(r)}, \mathcal{L}^k \otimes \mathcal{L}^{-r} \right) = H^0 \left(\mathcal{M}_J^{\text{SU}(r)}, \mathcal{L}^{k-r} \right).$$

The dimension of the corrected state space Z^m is

$$d_k^{m,\text{SU}(r)}(g) = \dim H^0 \left(\mathcal{M}_J^{\text{SU}(r)}, \mathcal{L}^k \otimes \mathcal{L}^{-r} \right)$$

and we see

$$d_k^{m,\text{SU}(r)}(g) = d_{k-r}^{\text{SU}(r)}(g).$$

This explanation of the shift is not so accidental as it looks at first sight. A similar shift appears for a general compact simple Lie group G. To explain the shift in this more general context one has to observe first that r is the dual Coxeter number of SU(r) and that the shift for general G is $k \to k + h^\vee$ where $h^\vee$ is the dual Coxeter number of G (see [Fuc92], [Kac90] for the dual Coxeter number which is the Dynkin index of the adjoint representation of G). Now, the metaplectic correction again explains the shift because the canonical bundle on the corresponding moduli space $\mathcal{M}_J^G$ is isomorphic to $\mathcal{L}^{-2h}$.

Another reason to introduce the metaplectic correction appears in the generalization to higher-dimensional Kähler manifolds X instead of S_J. In order to obtain a general result on the deformation independence of the complex structure generalizing the above independence result it seems that only the metaplectic correction gives an answer at all. This has been shown in [Sche92], [ScSc95].

A different but related explanation of the shift by the dual Coxeter number of a nature closer to mathematics uses the Riemann–Roch formula for the evaluation of the $d_k^G(g)$ where h appears in the Todd genus of $\mathcal{M}_J^G$ because of $\mathcal{L}^{-2h} = \mathcal{K}$.

11.3 Fusion Rules for Surfaces with Marked Points

In this section G is a simple compact Lie group which we assume to be SU(2) quite often for simplification.

As above, let $S_J =: \Sigma$ be a surface S of genus g with a complex structure J. We fix a *level* $k \in \mathbb{N}$.

Let $P = (P_1, \ldots, P_m) \in S^m$ be (pairwise different) points of the surface, which will be called the *marked points*. We choose a labeling $R = (R_1, \ldots, R_m)$ of the marked points, that is, we associate to each point P_j an (equivalence class of an) irreducible representation R_j of the group G as a *label*.

From Theorem 11.3 of Kirillov we know that these representations R_j correspond uniquely to quantizable coadjoint orbits $\mathcal{O}_j$ of maximal dimension in $\mathfrak{g}^*$. Using the invariant bilinear form on $\mathfrak{g}$ the $\mathcal{O}_j$s correspond to adjoint orbits in $\mathfrak{g}$ and these, in turn, correspond to conjugacy classes $C_j \subset G$ by exponentiation. The analogue of the moduli space $\mathcal{M}^G$ will be defined as

$$\mathcal{M}^G(P,R) := \{\rho \in \mathrm{Hom}(\pi_1(S \setminus P), G) : \rho(c_j) \in C_j\}/G.$$

Here, c_j denotes the representative in $\pi_1(S \setminus P)$ of a small positively oriented circle around P_j.

Note that the fundamental group $\pi_1(S \setminus P)$ of $S \setminus P$ is isomorphic to the group generated by

$$a_1, \ldots, a_g, b_1, \ldots, b_g, c_1, \ldots, c_m$$

with the relation

$$\prod_{j=1}^{g} a_j b_j a_j^{-1} b_j^{-1} \prod_{i=1}^{m} c_i = 1. \qquad (11.4)$$

In the case of $G = \mathrm{SU}(2)$ the R_j correspond to conjugacy classes C_j generated by

$$\begin{pmatrix} e^{2\pi i \theta_j} & 0 \\ 0 & e^{-2\pi i \theta_j} \end{pmatrix} =: g_j. \qquad (11.5)$$

Let us suppose the θ_j to be rational numbers. This condition is no restriction of generality (see [MS80]). Hence, we obtain natural numbers N_j with $g_j{}^{N_j} = 1$ which describe the conjugacy classes C_j. We now define the *orbifold fundamental group* $\pi_1^{orb}(S) = \pi_1(S, P, R)$ as the group generated by

$$a_1, \ldots, a_g, b_1, \ldots, b_g, c_1, \ldots, c_m$$

with the relations

$$\prod_{j=1}^{g} a_j b_j a_j^{-1} b_j^{-1} \prod_{i=1}^{m} c_i = 1 \quad \text{and} \quad c_i^{N_i} = 1 \qquad (11.6)$$

for $i = 1, \ldots, m$, where N_j depends on θ_j. Then $\mathcal{M}^{\mathrm{SU}(2)}(P,R)$ can be written as

$$\mathrm{Hom}(\pi_1^{orb}(S), \mathrm{SU}(2))/\mathrm{SU}(2).$$

Theorem 11.1 has the following generalization to the case of surfaces with marked points.

Theorem 11.7. *Let S be marked by P with labeling R. The following three moduli spaces are in one-to-one correspondence:*

1. $\mathscr{M}^{\text{SU}(2)}(P,R) = \text{Hom}(\pi_1^{orb}(S), \text{SU}(2))/\text{SU}(2)$.
2. *The set of gauge equivalence classes (that is gauge orbits) of singular* $\text{SU}(2)$-*connections, flat on* $S \setminus P$ *with holonomy around* P_j *fixed by the conjugacy class* C_j *induced by* R_j, $j = 1, \ldots, m$.
3. *The moduli space* $\mathscr{M}_J^{\text{SU}(2)}(P,R)$ *of semi-stable parabolic vector bundles of rank* 2 *with paradegree* 0 *and paradeterminant* $\mathscr{O}_S$ *for* (P,R).

We have to explain the theorem. To begin with, the moduli space of singular connections in 2. can again be considered as a phase space of a classical system. The classical phase space $\mathscr{A}_0/\mathscr{G}$ (cf. 4. in Theorem 11.1) is now replaced with the quotient

$$\mathscr{M} := \mathscr{A}_{\mathscr{O}}/\mathscr{G}.$$

Here, $\mathscr{A}_{\mathscr{O}}$ is the space of *singular unitary connections* A on the trivial vector bundle of rank 2 over the surface S subject to the following conditions: over $S \setminus P$ the curvature of A vanishes and at the marked points P_i the curvature is (up to conjugation) locally given by

$$m(A) = \sum T_i \delta(P_i - x)$$

(with the Dirac δ-functional $\delta(P_i - x)$ in P_i) where $T_i \in \mathfrak{su}(2)$ belongs to the adjoint orbit determined by $\mathscr{O}_j$. Hence, $\mathscr{A}_{\mathscr{O}}$ can be understood as the inverse image $m^{-1}(\mathscr{O})$ of a product $\mathscr{O}$ of suitable coadjoint orbits of the dual $(\text{Lie}\mathscr{G})^*$ of the Lie algebra of the gauge group $\mathscr{G}$. Regarding m as a moment map, $\mathscr{M} = \mathscr{A}_{\mathscr{O}}/\mathscr{G}$ turns out to be a generalized Marsden–Weinstein reduction.

A related interpretation of $\mathscr{M}$ in this context is as follows: the differentiable $\text{SU}(2)$-connections A on the trivial rank 2 vector bundle over $S \setminus P$ define a parallel transport along each closed curve γ in $S \setminus P$. Hence, each A determines a group element $W(A, \gamma)$ in $\text{SU}(2)$ up to conjugacy. If A is flat in $S \setminus P$ one obtains a homomorphism $W(A) : \pi_1(S \setminus P) \to \text{SU}(2)$ up to conjugacy (see (11.4) for $\pi_1(S \setminus P)$) since for a flat connection the parallel transport from one point to another is locally independent of the curve connecting the points. Now, the labels R_j at the marked points P_j fix the conjugacy classes C_j assigned by $W(A)$ to the simple circles (represented by c_j in the description (11.4) of the fundamental group $\pi_1(S \setminus P)$) around the marked points: $W(A)(c_j)$ has to be contained in C_j. Hence, the elements of $\mathscr{M}$ define conjugacy classes of representations in $\mathscr{M}^{\text{SU}(2)}(P,R)$ yielding a bijection.

This explains the first bijection of the theorem. The second bijection has been shown by Mehta and Seshadri [MS80] as a generalization of the theorem of Narasimhan and Seshadri [NS65] (cf. Theorem 11.1). To understand it, we need the following concepts:

Definition 11.8. A *parabolic structure* on a holomorphic vector bundle E of rank r over a marked Riemann surface $\Sigma = S_J$ with points $P_1, \ldots, P_m \in \Sigma$ is given by the following data:

- a *flag* of proper subspaces in every fiber E_i of E over P_i:

$$E_i = F_i^{(0)} \supset \cdots \supset F_i^{(r_i)} \supset \{0\}$$

with $k_i^{(s)} := \dim F_i^{(s)} / F_i^{(s+1)}$ as multiplicities, and

- a sequence of weights $\alpha_i^{(s)}$ corresponding to every flag with

$$0 \le \alpha_i^{(0)} \le \ldots \le \alpha_i^{(r_i)} \le 1.$$

The *paradegree* of such a parabolic bundle E is

$$\text{paradeg } E := \deg(E) + \sum_i d_i \qquad \text{with } d_i := \sum_s \alpha_i^{(s)} k_i^{(s)}.$$

A parabolic bundle E is semi-stable if for all parabolic subbundles F of E one has:

$$(\text{rg}(F))^{-1}\text{paradeg } F \le (\text{rg}(E))^{-1}\text{paradeg } E.$$

E is *stable* if "$\le$" can be replaced with "$<$".

The *paradeterminant* for this parabolic structure (resp. for these weights at the marked points) is the usual determinant $\det E = \bigwedge^r E$ tensored with the holomorphic line bundle given by $\mathcal{O}_\Sigma(-\sum d_i x_i)$ for the divisor $-\sum d_i P_i$ if d_i is an integer. Otherwise the paradeterminant is undefined.

The second bijection in Theorem 11.7 has the following significance: one collects those equivalence classes of parabolic vector bundles over $\Sigma = S_J$, whose weights $\alpha_i^{(s)}$ are rational and for which all $d_j := \sum_s \alpha_j^{(s)} k_j^{(s)}$ are integers. Then the $\alpha_j^{(s)}$ fix suitable conjugacy classes in $\text{SU}(r)$ and hence a labeling through irreducible representations R_j. Conversely, given the labels R_j attached to the points, only those parabolic bundles are considered where the weights fit the labels. Now the space

$$\mathcal{M}_J^{\text{SU}(r)}(P,R)$$

consists of the equivalence classes of such parabolic vector bundles, which, in addition, are semi-stable with paradegree 0 and trivial paradeterminant. For instance, for $r = 2$ the representation ρ belonging to $[E] \in \mathcal{M}_J^{\text{SU}(2)}(P,R)$ is given on the c_j by

$$\rho(c_j) = \begin{cases} \exp 2\pi i \, \text{diag}\left(\alpha_j^{(0)}, \alpha_j^{(0)}\right) & \text{for } k_j^{(0)} = 2 \\ \exp 2\pi i \, \text{diag}\left(\alpha_j^{(0)}, \alpha_j^{(1)}\right) & \text{for } k_j^{(0)} = 1 = k_j^{(1)}. \end{cases}$$

The moduli space $\mathcal{M}_J^{\text{SU}(2)}(P,R)$ is according to [MS80] in a one-to-one correspondence to

$$\text{Hom}(\pi_1^{\text{orb}}(S), \text{SU}(2))/\text{SU}(2).$$

Furthermore,

$$\mathcal{M}_J^{SU(2)}(P,R)$$

has the structure (depending on J) of a projective variety over $\mathbb{C}$. In this variety, the stable parabolic vector bundles correspond to the regular points. An analogous theorem holds for parabolic vector bundles of rank r (cf. [MS80]).

In the case of $P = \emptyset$ the moduli space

$$\mathcal{M}_{J,g}^{SU(2)}(P,R) := \mathcal{M}_J^{SU(2)}(P,R)$$

coincides with the previously introduced moduli space $\mathcal{M}_J^{SU(2)}$ (cf. Sect. 11.1). Recall that $\mathcal{M}_J^{SU(2)}$ has a natural line bundle $\mathcal{L}$ which is used to introduce the generalized theta functions or conformal blocks. This has a generalization to the case $P \neq \emptyset$: $\mathcal{M}_{J,g}^{SU(2)}(P,R)$ possesses a natural line bundle $\mathcal{L}$ – the determinant bundle or the theta bundle – together with a connection whose curvature is $2\pi i \omega_{\mathcal{M}}$. Here, $\omega_{\mathcal{M}}$ is the Kähler form on the regular locus of $\mathcal{M}_{J,g}^{SU(2)}(P,R)$. Now, the finite-dimensional space of holomorphic sections

$$H^0\left(\mathcal{M}_{J,g}^{SU(2)}(P,R), \mathcal{L}^k\right)$$

is the *space of generalized theta functions* of level k with respect to (P,R).

For our special case of the group $G = SU(2)$ let us denote by the number $n \in \mathbb{N}$ the (up to isomorphism) uniquely determined irreducible representation $n : SU(2) \to GL(V_n)$ with $\dim_{\mathbb{C}} V_n = n+1$. With respect to the level $k \in \mathbb{N}$ only those labels $R = (n_1,\ldots,n_m)$ are considered in the following which satisfy $n_j \leq k$ for $j = 1,\ldots,m$.

Theorem 11.9. (Fusion Rules)

0. $z_k(g;n_1,\ldots,n_m) := \dim_{\mathbb{C}} H^0(\mathcal{M}_{J,g}^{SU(2)}(P,R), \mathcal{L}^k)$ *does not depend on J and on the position of the points* $P_1,\ldots,P_m \in S$. *Here, $R = (n_1,\ldots,n_m)$. Let $\mathcal{M}_{g,m}$ be the moduli space of marked Riemann surfaces of genus g with m points and let $\overline{\mathcal{M}}_{g,m}$ be the Deligne–Mumford compactification of $\mathcal{M}_{g,m}$. Then, the bundle $\pi : Z_{g,k}(R) \to \mathcal{M}_{g,m}$ with fiber*

$$\pi^{-1}(J,P) = H^0\left(\mathcal{M}_{J,g}^{SU(2)}(P,R), \mathcal{L}^k\right)$$

has a continuation $\overline{Z}_{g,k}(R) \to \overline{\mathcal{M}}_{g,m}$ *to* $\overline{\mathcal{M}}_{g,m}$ *as a locally free sheaf of rank* $z_k(g;n_1,\ldots,n_m)$.

1. $z_k(g;n_1,\ldots,n_m) = \sum_{n=0}^k z_k(g-1;n_1,\ldots,n_m,n,n)$.
2. *For $1 \leq s \leq m$ one has*

$$z_k(g'+g'';n_1,\ldots,n_m)$$
$$= \sum_{n=0}^k z_k(g';n_1,\ldots,n_s,n)z_k(g'';n,n_{s+1},\ldots,n_m).$$

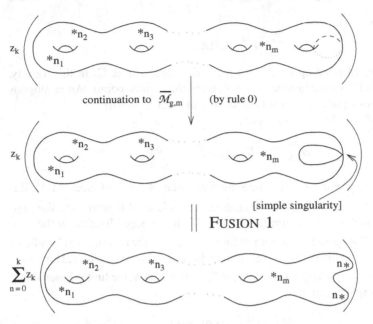

Fig. 11.1 Fusion rule 1

The formulation of the fusion rules for SU(2) in Theorem 11.9 is special since every representation ρ of the group SU(2) is equivalent to its conjugate representation ρ^* (Figs. 11.1 and 11.2). For more general Lie groups G instead of SU(2), one of the two representations (n,n) in the fusion rules has to be replaced with its conjugate.

A proof of the fusion rules 1 and 2 in approximately this form can be found in [NR93] together with [Ram94].

Even in the case of $P = \emptyset$ it is quite difficult to show that the dimensions of $H^0\left(\mathcal{M}_J^{\mathrm{SU}(r)}, \mathcal{L}^k\right)$ do not depend on the complex structure J. This can be deduced from a stronger property which states that the spaces

$$H^0\left(\mathcal{M}_J^{\mathrm{SU}(r)}, \mathcal{L}^k\right)$$

as well as

$$H^0\left(\mathcal{M}_{J,g}^{\mathrm{SU}(r)}(P,R), \mathcal{L}^k\right)$$

are essentially independent of the complex structure. This is in agreement with physical requirements since these spaces are considered to be the result of a quantization which only depends on the topology of S or $S \setminus P$. For this reason the resulting quantum field theory is called a *topological quantum field theory* (cf. [Wit89]). In particular, the state spaces – more precisely their projectivations – should not depend on any metric or complex structure.

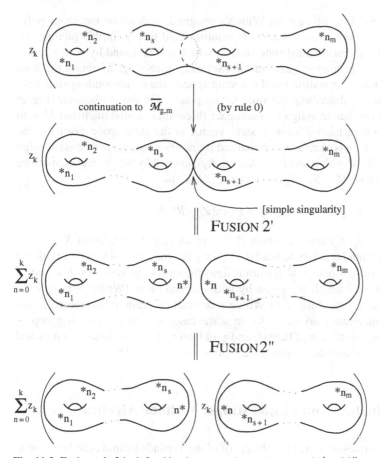

Fig. 11.2 Fusion rule 2 is defined by the successive application of 2' and 2"

That the above state spaces do not depend on the complex structure has been proven in [APW91] and [Hit90] in the case of $P = \emptyset$. Hitchin's methods carry over to the case of $P \neq \emptyset$ using some results of non-abelian Hodge theory [Sche92], [ScSc95]. The strategy of the proof is to consider the bundle $Z_{g,k}(R) \to \mathcal{M}_{g,m}$ over the moduli space $\mathcal{M}_{g,m}$ of Riemann surfaces of genus g and m marked points with fiber $H^0 (\mathcal{M}_{g,J}^{SU(r)}(P,R), \mathcal{L}^k)$ over $(J,P) \in \mathcal{M}_{g,m}$. On this bundle $Z_{g,k}(R)$ one constructs a natural projectively flat connection. Incidentally, the existence of such a natural projectively flat connection is again motivated by considerations from conformal field theory. Then the fibers of the bundle can be identified in a natural way by parallel transport with respect to this connection up to a constant, that is they are projectively identified. It is remarkable that in the course of the construction in the general case of $P \neq \emptyset$ it seems to be necessary to use the metaplectic correction instead of the uncorrected geometric quantization (see p. 221 and [ScSc95]).

The case $P \neq \emptyset$ is significant for Witten's program, to describe the Jones polynomials of knot theory in the context of quantum field theory. In this picture, the $Z_{k,g}(R)$ are quantum mechanical state spaces, which can be found by path integration [Wit89] or by geometric quantization [Sche92], [ScSc95]. To obtain the knot invariants, one needs, in addition to these state spaces, the corresponding state vectors ("propagators") describing the time development. On the mathematical level this means that one has to assign to a compact three-dimensional manifold M with boundary containing labeled knots a state vector in the state space given by the boundary of M which is a surface with marked points. For instance, one has to assign to such a manifold M with knots $K = (K_1, \ldots, K_s)$, labeled by $SU(r)$-representations and with boundary $\partial M = S_g \cup S'_{g'}$, a vector $Z_k(M, K)$ in

$$Z_{k,g}(R)^* \otimes Z_{k,g'}(R') \cong \mathrm{Hom}(Z_{k,g}(R), Z_{k,g'}(R')).$$

The points in S_g, $S'_{g'}$ and the labels R, R' are induced by the knots $K_1, \ldots, K_s$, which may run from boundary to boundary. Only the state spaces together with the state vectors yield a topological quantum field theory. A rigorous construction of these state vectors – which are given by path integration in [Wit89] – is still not known. In the meantime, instead of Witten's original program, other constructions of topological quantum field theories – in some cases by using quantum groups – have been proposed (cf., e.g., [Tur94]) and yield interesting invariants of knots and three manifolds. Related developments are presented in [BK01*].

11.4 Combinatorics on Fusion Rings: Verlinde Algebra

Using the fusion rules of Sect. 11.3, the proof of the Verlinde formula can be reduced to the determination of

$$z_k(0; n), z_k(0; n, m), z_k(0; n, m, l)$$

for $n, m, l \in \{0, \ldots, k\}$. This combinatorical reduction has an algebraification, which also has a meaning for more general groups than $SU(r)$ (cf. [Bea96], [Bea95], [Sze95]).

Definition 11.10 (Fusion Algebra). Let F be a finite-dimensional complex vector space with an element $1 \in F$. For every $g \in \mathbb{Z}, g \geq 0$ and $v_1, \ldots, v_m \in F$ let

$$Z(g)_{v_1, \ldots, v_m} \in \mathbb{C}$$

be given. $(F, 1, Z)$ is a *fusion ring* if the following fusion rules hold:

(F1) $Z(g)_{1, \ldots, 1} = 1$.
(F2) $Z(g)_{v_1, \ldots, v_m} = Z(g)_{1, v_1, \ldots, v_m}$ does not depend on the order of the $v_1, \ldots, v_m$.
(F3) $v \to Z(0)_{v_1, \ldots, v_j, v, v_{j+1}, \ldots, v_m}$ is $\mathbb{C}$-linear.
(F4) $(v, w) \to Z(0)_{v,w}$ is not degenerated.

We use the notation

$$\int v := Z(0)_v, \quad \langle v, w \rangle := Z(0)_{v,w} \quad and \quad \eta(v,w,u) := Z(0)_{v,w,u}.$$

Let $(b_j), (b^j)$ be a pair of bases with $\delta_j^i = \langle b_j, b^i \rangle$. Then, additionally, the following rules hold

(F5) $Z(g)_{v_1,\dots,v_m} = \sum Z(g-1)_{b_j, b^j, v_1, \dots, v_m}$, $g \geq 1$ (Fusion 1).
(F6) $Z(g+g')_{v_1,\dots,v_m,v'_1,\dots,v'_m} = \sum Z(g)_{v_1,\dots,v_m,b_j} Z(g')_{b^j, v'_1,\dots,v'_m}$, (Fusion 2).

One easily proves

Lemma 11.11. *The product $v \cdot w := \sum \eta(v,w,b_j) b^j$ for $v,w \in F$ induces on F the structure of a commutative and associative complex algebra with 1.*

Lemma 11.12. *The bilinear form $\langle , \rangle$ satisfies the trace condition $\langle v \cdot w, x \rangle = \langle v, w \cdot x \rangle$. Therefore, F is a Frobenius algebra.*

Proof. $\langle v \cdot w, x \rangle = \sum \eta(v,w,b_i) \langle b^i, x \rangle$ by definition and linearity. Thus $\langle v \cdot w, x \rangle = \eta(v,w,x)$, since $x = b_i \langle b^i, x \rangle$. In the same way, we obtain $\langle v, w \cdot x \rangle = \langle w \cdot x, v \rangle = \eta(w,x,v) = \eta(v,w,x)$ by (F2). □

Both results need the axioms for $g = 0$ only. With similar arguments one can prove the following version of the Verlinde formula using the fusion rules for general g.

Lemma 11.13. *With $\alpha := \sum b_j b^j = \sum \eta(b_i, b^i, b_k) b^k \in F$ the abstract Verlinde formula holds:*

$$Z(g)_{v_1,\dots,v_m} = \int \alpha^g v_1 \cdot \dots \cdot v_m.$$

Proof. By induction on m we show

$$Z(g)_{v_1,\dots,v_m} = Z(g)_{v_1 \cdot \dots \cdot v_m}.$$

The case $m = 1$ is trivial. For $m \geq 2$ we have

$$Z(g)_{v_1,\dots,v_m} = \sum Z(0)_{v_1,v_2,b_j} Z(g)_{b^j,v_3,\dots,v_m} \quad \text{by(F6)}$$

$$= \sum \eta(v_1,v_2,b_j) Z(g)_{b^j,v_3,\dots,v_m}$$

$$= Z(g)_{\sum \eta(v_1,v_2,b_j) b^j, v_3,\dots,v_m} \quad \text{by(F3)}$$

$$= Z(g)_{v_1 \cdot v_2, v_3,\dots,v_m} \quad \text{by the definition of the product}$$

$$= Z(g)_{v_1 \cdot v_2 \cdot v_3 \cdot \dots \cdot v_m} \quad \text{by the induction hypothesis.}$$

This implies

$$Z(g)_v = \sum Z(g-1)_{b_j, b^j, v} = Z(g-1)_{\sum b_j b^j v} = Z(g-1)_{\alpha v}$$

and

$$Z(g)_v = Z(g-1)_{\alpha v} = Z(g-2)_{\alpha^2 v} = Z(0)_{\alpha^g v}.$$

Hence for $v = v_1 \cdot \ldots \cdot v_m$ the claimed statement follows.   □

For the derivation of the Verlinde formula (Theorem 11.6) from the fusion rules using Lemma 11.13 we refer to [Sze95], where general simple Lie groups instead of $SU(2)$ are treated.

To indicate the role of the above formula as an abstract Verlinde formula let us represent F as the algebra of functions on the spectrum $\Sigma = \mathrm{Spec}\, F$, that is the finite set of algebra homomorphisms $h : F \to \mathbb{C}$ satisfying, in particular, $h(1) = 1$. With the aid of the Gelfand map $v \mapsto \hat{v}, \hat{v}(h) = h(v)$, we identify F and the function algebra $\mathrm{Map}(\Sigma)$. The structure map $Z(0) : F \to \mathbb{C}$ induces on $F = \mathrm{Map}(\Sigma)$ a complex measure μ which is given by a map $\mu : \Sigma \to \mathbb{C}$. We have $Z(0)_v = \int v d\mu = \sum_{h \in \Sigma} v(h)\mu(h)$ and conclude $1 = Z(0)_1 = \int d\mu = \sum \mu(h)$ and $\mu(h) \neq 0$ for all $h \in \Sigma$.

In order to determine the element $\alpha \in F$ from Lemma 11.13 one uses the characteristic functions e_h of the points $h \in \Sigma$ as a basis: $e_h(k) = \delta_{h,k}$. The dual basis e^h is given by $e^h = \mu(h)^{-1} e_h$ because of

$$\langle e_h, e^h \rangle = Z(0)_{e_h, e^h} = \int e_h e^h d\mu = \mu(h).$$

Therefore, $\alpha = \sum \mu(h)^{-1} e_h$ and $\alpha^g = \sum \mu(h)^{-g} e_h$. Inserting this term into the abstract Verlinde formula in 11.13 gives

$$\int \alpha^g d\mu = \sum \mu(h)^{-g} \mu(h) = \sum \mu(h)^{1-g}.$$

Hence, for $Z(g) = Z(g)_1$ we obtain the following formula which is much closer in its appearance to the Verlinde formula (11.2).

Lemma 11.14.
$$Z(g) = \sum_{h \in \Sigma} (\mu(h))^{g-1}.$$

The fusion rules have their origin in the operator product expansion (cf. p. 168). In the case of the conformal field theory associated to a simple Lie group G (like $SU(2)$ as considered above) the fusion rules are also related to basic properties of the group and its representations. In fact, the fusion rules have a manifestation in the tensor product of representations of G and the fusion algebras considered above turn out to be isomorphic to certain quotients of the representation ring $R(G)$. These quotients are called *Verlinde algebras* (cf. [Wit93*]).

We describe the Verlinde algebra $V_k(G)$ explicitly in the case of the group $G = SU(2)$. The representation ring $R(G)$, that is the ring of (isomorphism classes of) finite-dimensional representations of G with the tensor product as multiplication, is in the case of $G = SU(G)$ generated by the standard two-dimensional representation V_1. All other irreducible representations are known to be isomorphic to some V_m where V_m is the symmetric product:

$$V_m := V_1^{\odot m} = V_1 \odot \ldots \odot V_1.$$

V_m is the $(m+1)$-dimensional irreducible representation of SU(2), unique up to isomorphism, in particular, V_0 is the trivial one-dimensional representation. Let b_n denote the isomorphism class of V_n in R(SU(2)) (denoted by n in the last section). We regard R(SU(2)) as a vector space over $\mathbb{C}$ and observe that (b_j) is a basis of R(SU(2)). In particular, R(G) is an algebra over $\mathbb{C}$.

The multiplication "×" on R(G) induced by the tensor product is given by the Clebsch–Gordan formula

$$V_m \otimes V_n \cong V_{m+n} \oplus V_{m+n-2} \oplus \ldots \oplus V_{|m-n|}.$$

Hence, on R(G) we have

$$b_{m+p} \times b_m = \sum_{j=0}^{m} b_{2m+p-2j}.$$

The truncated multiplication of level $k \in \mathbb{N}$ is

$$b_{m+p} \cdot b_m = b_{m+p} \times b_m, \text{ if } 2m+p \le k,$$

and

$$b_{m+p} \cdot b_m = \sum_{j \ge 2m+p-k}^{m} b_{2m+p-2j} = b_{2k-2m-p} + \ldots + b_p,$$

if $2m+p > k$ and $m+p \le k$. The definition implies that no terms b_n with $n > k$ can appear in the summation on the right-hand side. The resulting algebra, the *Verlinde algebra* V_k(SU(2)) *of level* k, is the quotient R(G)/(b_{k+1}) with respect to the ideal (b_{k+1}) generated by $b_{k+1} \in$ R(G). It is a Frobenius algebra and a fusion algebra in the sense of Definition 11.10. It describes the fusion in the level k case for SU(2).

The Verlinde algebra has a direct description with respect to the basis $b_0, \ldots, b_k$ in the form

$$b_i \cdot b_j = \sum_{m=0}^{k} N_{ij}^m b_m$$

with coefficients $N_{ij}^m \in \{0,1\}$.

Now, the homomorphisms of V_k(SU(2)) can be determined using the fact that all complex homomorphisms on R(SU(2)) have the form

$$h_z(b_n) = \frac{\sin(n+1)z}{\sin z},$$

where $z \in \mathbb{C}$ is a complex number. Such a homomorphism h_z vanishes on (b_{k+1}) if $\sin(k+2)z = 0$. We conclude that the homomorphisms of V_k(SU(2)) are precisely the $k+1$ maps $h_p : V_k$(SU(2)) $\to \mathbb{C}$ satisfying

$$h_p(b_j) = \frac{\sin(j+1)z_p}{\sin z_p}, z_p = \frac{p\pi}{k+2}, p = 1, \ldots, k+1.$$

Using

$$Z(0)_{b_j} = \int b_j = \sum_{n=1}^{k+1} b_j(h_n)\mu(h_n),$$

an elementary calculation yields

$$\mu(h_n) = \frac{2}{k+2}\sin^2\frac{n\pi}{k+2}, n = 1,\ldots,k+1,$$

from which the Verlinde formula (11.2) follows by Lemma 11.14.

Recently, a completely different description of the Verlinde algebra using equivariant twisted K-theory has been developed by Freed, Hopkins, and Teleman [FHT03*] (see also [Mic05*], [HJJS08*]).

References

[APW91]   S. Axelrod, S. Della Pietra, and E. Witten. Geometric quantization of Chern-Simons gauge theory. *J. Diff. Geom.* **33** (1991), 787–902.

[BK01*]   B. Bakalov and A. Kirillov, Jr. *Lectures on Tensor Categories and Modular Functors*, AMS University Lecture Series **21**, AMS, Providence, RI, 2001.

[Bea95]   A. Beauville. Vector bundles on curves and generalized theta functions: Recent results and open problems. In: *Current Topics in Complex Algebraic Geometry*. Math. Sci. Res. Inst. Publ. **28**, 17–33, Cambridge University Press, Cambridge, 1995.

[Bea96]   A. Beauville. Conformal blocks, fusion rules and the Verlinde formula. In: *Proceedings of the Hirzebruch 65 Conference on Algebraic Geometry (Ramat Gan, 1993)*, 75–96, Bar-Ilan University, Ramat Gan, 1996.

[BF01*]   D. Ben-Zvi and E. Frenkel. *Vertex Algebras and Algebraic Curves*. AMS, Providence, RI, 2001.

[BT93]   M. Blau and G. Thompson. Derivation of the Verlinde formula from Chern-Simons theory. *Nucl. Phys.* **B 408** (1993), 345–390.

[Bot91]   R. Bott. Stable bundles revisited. *Surveys in Differential Geometry (Supplement to J. Diff. Geom.)* **1** (1991), 1–18.

[Fal94]   G. Faltings. A proof of the Verlinde formula. *J. Alg. Geom.* **3** (1994), 347–374.

[Fal08*] G. Faltings. Thetafunktionen auf Modulräumen von Vektorbündeln. *Jahresbericht der DMV* **110** (2008), 3–18.

[FHT03*] D. Freed, M. Hopkins, and C. Teleman. Loop groups and twisted K-theory III. arXiv:math/0312155v3 (2003).

[Fuc92]   J. Fuchs. *Affine Lie Algebras and Quantum Groups*. Cambridge University Press, Cambridge, 1992.

[Hit90]   N. Hitchin. Flat connections and geometric quantization. *Comm. Math. Phys.* **131** (1990), 347–380.

[HJJS08*] Husemöller, D., Joachim, M., Jurco, B., Schottenloher, M.: *Basic Bundle Theory and K-Cohomological Invariants*. Lect. Notes Phys. **726**. Springer, Heidelberg (2008)

[Kac90]   V. Kac. *Infinite Dimensional Lie Algebras*. Cambridge University Press, Cambridge, 3rd ed., 1990.

[Kir76]   A. A. Kirillov. *Theory of Representations*. Springer Verlag, Berlin, 1976.

[KNR94]   S. Kumar, M. S. Narasimhan, and A. Ramanathan. Infinite Grassmannians and moduli spaces of G-bundles. *Math. Ann.* **300** (1994), 41–75.

[MS80]   V. Mehta and C. Seshadri. Moduli of vector bundles on curves with parabolic structures. *Ann. Math.* **248** (1980), 205–239.

[Mic05*] J. Mickelsson. Twisted K Theory Invariants. *Letters in Mathematical Physics* **71** (2005), 109–121.

[MS89] G. Moore and N. Seiberg. Classical and conformal field theory. *Comm. Math. Phys.* **123** (1989), 177–254.

[NR93] M.S. Narasimhan and T. Ramadas. Factorization of generalized theta functions I. *Invent. Math.* **114** (1993), 565–623.

[NS65] M.S. Narasimhan and C. Seshadri. Stable and unitary vector bundles on a compact Riemann surface. *Ann. Math.* **65** (1965), 540–567.

[Ram94] T. Ramadas. Factorization of generalized theta functions II: The Verlinde formula. Preprint, 1994.

[Sche92] P. Scheinost. Metaplectic quantization of the moduli spaces of at and parabolic bundles. Dissertation, LMU München, 1992.

[ScSc95] P. Scheinost and M. Schottenloher. Metaplectic quantization of the moduli spaces of at and parabolic bundles. *J. Reine Angew. Math.* **466** (1995), 145–219.

[Sor95] C. Sorger. La formule de Verlinde. Preprint, 1995. (to appear in Sem. Bourbaki, année 1994–95, no 793)

[Sze95] A. Szenes. The combinatorics of the Verlinde formula. In: *Vector Bundles in Algebraic Geometry*, Hitchin et al. (Eds.), 241–254. Cambridge University Press, Cambridge, 1995.

[TUY89] A. Tsuchiya, K. Ueno, and Y. Yamada. Conformal field theory on the universal family of stable curves with gauge symmetry. In: Conformal field theory and solvable lattice models. *Adv. Stud. Pure Math.* **16** (1989), 297–372.

[Tur94] V.G. Turaev. *Quantum Invariants of Knots and 3-Manifolds*. DeGruyter, Berlin, 1994.

[Tyu03*] A. Tyurin. *Quantization, Classical and Quantum Field Theory and Theta Functions*, CRM Monograph Series **21** AMS, Providence, RI, 2003.

[Uen95] K. Ueno. On conformal field theory. In: *Vector Bundles in Algebraic Geometry*, N.J. Hitchin et al. (Eds.), 283–345. Cambridge University Press, Cambridge, 1995.

[Ver88] E. Verlinde. Fusion rules and modular transformations in two-dimensional conformal field theory. *Nucl. Phys.* **B 300** (1988), 360–376.

[Wit89] E. Witten. Quantum field theory and the Jones polynomial. *Commun. Math. Phys.* **121** (1989), 351–399.

[Wit93*] E. Witten. The Verlinde algebra and the cohomology of the Grassmannian. hep-th/9312104 In: *Geometry, Topology and Physics*, Conf. Proc.Lecture Notes in Geom. Top, 357–422. Intern. Press, Cambridge MA (1995).

[Woo80] N. Woodhouse. *Geometric Quantization*. Clarendon Press, Oxford, 1980.

Appendix A
Some Further Developments

Due to the character of these notes with the objective to present and explain the basic principles of conformal field theory on a mathematical basis in a rather detailed manner there has been nearly no room to mention further developments.

In this appendix we concentrate on boundary conformal field theory (BCFT) and on stochastic Loewner evolution (SLE) as two developments which lead to new structures not being part of conformal field theory (CFT) as described in these notes but strongly connected with CFT.

We only give a brief description and some references.

Boundary Conformal Field Theory. Boundary conformal field theory is essentially conformal field theory on domains with a boundary. As an example, let us consider strings moving in a background Minkowski space M as in Chap. 7. For a closed string, that is a closed loop moving in M, one gets a closed surface. After quantization one obtains the corresponding CFT on this surface as developed in Chap. 7. In case of an open string, that is a connected part of a closed loop (which is the image of an interval under an injective embedding) with two endpoints, the string weeps out an open surface or better a surface with boundary. The boundary is given by the movement of the two endpoints of the string. We obtain the corresponding CFT in the interior of the surface, the bulk CFT, together with compatibility conditions on the boundary of the surface.

BCFT has important applications in string theory, in particular, in the physics of open strings and D-branes (cf. [FFFS00b*], for instance), and in condensed matter physics in boundary critical behavior.

BCFT is in some respect simpler than CFT. For instance, in the case of the upper half plane H with the real axis as its boundary one possible boundary condition is that the energy–momentum tensor T satisfies $T(z) = \overline{T}(\bar{z})$. This implies that the correlation functions of $\overline{T}$ are the same as those of T, analytically continued to the lower halfplane. This simplifies among other things the conformal Ward identities. Moreover, there is only one Virasoro algebra.

For general reviews on BCFT we refer to [Zub02*] and [Car04*]. See also [Car89*] and [FFFS00a*].

Stochastic Loewner Evolution. There is a deep connection between BCFT and conformally invariant measures on spaces of curves in a simply connected domain

Schottenloher, M.: *Some Further Developments*. Lect. Notes Phys. **759**, 235–237 (2008)
DOI 10.1007/978-3-540-68628-6_13   © Springer-Verlag Berlin Heidelberg 2008

H in $\mathbb{C}$ which start at the boundary of the domain. This has been indicated in both the survey articles of Cardy [Car04*] on BCFT and [Car05*] on SLE and in a certain sense already in [LPSA94]. Such measures arise naturally in the continuum limit of certain statistical mechanics models.

For instance, in the case of the upper half plane H a measure of this type can be constructed using a family of conformal mappings g_t, $t \geq 0$. In such a construction one uses the stochastic Loewner evolution (SLE) first described by [Schr00*]. More precisely, for a constant $\kappa \in \mathbb{R}$, $\kappa \geq 0$, the so-called SLE_κ curve $\gamma : [0, \infty[\to \mathbb{C}$ in the upper half plane H is generated as follows: $\gamma : [0, \infty[\to \mathbb{C}$ is continuous with $\gamma(0) = 0$ and $\gamma(]0, \infty[) \subset H$. γ is furthermore determined by the unique conformal diffeomorphism

$$g_t : H \setminus \gamma(]0, t]) \to H, \ t \geq 0,$$

satisfying the Loewner equation

$$\frac{\partial g_t(z)}{\partial t} = \frac{2}{g_t(z) - \sqrt{\kappa} b_t}, \ g_0(z) = z,$$

normalized by the condition $g_t(z) = z + o(1)$ for $z \to \infty$. Here, b_t, $t \geq 0$, is an ordinary brownian motion starting at $b_0 = 0$. Hence, $\gamma(t) = \gamma_t$ is precisely the point satisfying $g_t^\sim(\gamma_t) = \sqrt{\kappa} b_t$ for the continuous extension $g_t^\sim$ of g_t to $H \setminus \gamma(]0, t[)$ that is into the boundary point $\gamma(t)$ of $H \setminus \gamma(]0, t[)$.

A comprehensive introduction to SLE is given in Lawler's book [Law05*]. A first exact application to the critical behavior of statistical mechanics models can be found in [Smi01*].

The relation of SLE to CFT is not easy to detect. It has been uncovered in the articles [BB03*] and [FW03*].

Modularity. Modularity properties have been studied in the articles on vertex algebras and CFT from the very beginning, in particular with respect to the examples of large finite simple groups (see [Bor86*] and [FLM88*], for instance). A comprehensive survey can be found in [Gan06*].

References

[BB03*]   M. Bauer and D. Bernard. Conformal field theories of stochastic Loewner evolutions. *Comm. Math. Phys.* **239** (2003) 493–521.

[Bor86*]   R. E. Borcherds. Vertex algebras, Kac-Moody algebra and the monster. *Proc. Natl. Acad. Sci. USA* **83** (1986), 3068–3071.

[Car89*]   J.L. Cardy. Boundary conditions, fusion rules and the Verlinde formula. *Nucl. Phys.* **B324** (1989), 581–596.

[Car04*]   J.L. Cardy. *Boundary Conformal Field Theory*. [arXiv:hepth/ 0411189v2] (2004) (To appear in *Encyclopedia of Mathematical Physics*, Elsevier).

[Car05*]   J.L. Cardy. SLE for theoretical physicists. *Ann. Phys.* **318** (2005), 81–118.

[FFFS00a*] G. Felder, J. Fröhlich, J. Fuchs, and C. Schweigert. Conformal boundary conditions and three-dimensional topological field theory. *Phys. Rev. Lett.* **84** (2000), 1659–1662.

[FFFS00b*] G. Felder, J. Fröhlich, J. Fuchs and C. Schweigert. The geometry of WZW branes. *J. Geom. Phys.* **34** (2000), 162–190.

[FLM88*] I. Frenkel, J. Lepowsky, and A. Meurman. *Vertex Operator Algebras and the Monster.* Academic Press, New York, 1988.

[FW03*] R. Friedrich and W. Werner. Conformal restriction, highest weight representations and SLE. *Comm. Math. Phys.* **243** (2003), 105–122.

[Gan06*] T. Gannon. *Moonshine Beyond the Monster. The Bridge Connecting Algebra, Modular Forms and Physics.* Cambridge University Press, Cambridge, 2006.

[LPSA94] R. Langlands, P. Pouliot, and Y. Saint-Aubin. Conformal invariance in two-dimensional percolation. *Bull. Am. Math. Soc.* **30** (1994), 1–61.

[Law05*] G.F. Lawler. *Conformally Invariant Processes in the Plane. Mathematical Surveys and Monographs* **114**. AMS, Providence, RI, 2005.

[Schr00*] O. Schramm. Scaling limits of loop-erased random walks and uniform spanning trees. *Israel J. Math.* **118** (2000), 221–288.

[Smi01*] S. Smirnov. Critical percolation in the plane: Conformal invariance, Cardy's formula, scaling limits. *C. R. Acad. Sci. Paris* **333** (2001), 239–244.

[Zub02*] J.B. Zuber. CFT, BCFT, ADE and all that. In: *Quantum Symmetries in Theoretical Physics and Mathematics*, Coquereaux et alii (Eds.), Contemporary Mathematics **294**, 233–271, AMS, Providence, RI, 2002.

References

The "*" indicates that the respective reference has been added to the list of references in the second edition of these notes.

[APW91] S. Axelrod, S. Della Pietra, and E. Witten. Geometric quantization of Chern-Simons gauge theory. *J. Diff. Geom.* **33** (1991), 787–902.

[BK01*] B. Bakalov and A. Kirillov, Jr. *Lectures on tensor categories and modular functors.* AMS University Lecture Series **21**, AMS, Providence, RI, 2001.

[Bar54] V. Bargmann. On unitary ray representations of continuous groups. *Ann. Math* **59** (1954), 1–46.

[Bar64] V. Bargmann. Note on Wigner's theorem on symmetry operations. *J. Math. Phys.* **5** (1964), 862–868.

[BB03*] M. Bauer and D. Bernard. Conformal field theories of stochastic Loewner evolutions. *Comm. Math. Phys.* **239** (2003) 493–521.

[BD04*] A. Beilinson and V. Drinfeld *Chiral algebras.* AMS Colloquium Publications **51** AMS, Providence, RI, 2004.

[BEG67*] J. Bros, H. Epstein, and V. Glaser. On the Connection Between Analyticity and Lorentz Covariance of Wightman Functions. *Comm. Math. Phys,* **6** (1967), 77–100

[BF01*] D. Ben-Zvi and E. Frenkel *Vertex Algebras and Algebraic Curves.* AMS, Providence, RI, 2001.

[Bea95] A. Beauville. Vector bundles on curves and generalized theta functions: Recent results and open problems. In: *Current topics in complex algebraic geometry* Math. Sci. Res. Inst. Publ. **28**, 17–33, Cambridge Univ. Press, 1995.

[Bea96] A. Beauville. Conformal blocks, fusion rules and the Verlinde formula. In: *Proceedings of the Hirzebruch 65 Conference on Algebraic Geometry (Ramat Gan, 1993),* 75–96, Bar-Ilan Univ., Ramat Gan, 1996.

[Bor86*] R. E. Borcherds. Vertex algebras, Kac-Moody algebra and the monster. *Proc. Natl. Acad. Sci. USA.* **83** (1986), 3068–3071.

[Bor00*] R. E. Borcherds. Quantum vertex algebras. In: *Taniguchi Conference on Mathematics Nara '98* Adv. Stud. Pure Math. **31**, 51–74. Math. Soc. Japan, 2000.

[Bot91*] R. Bott. On E. Verlinde's formula in the context of stable bundles. In: *Topological Methods in Quantum Field Theories*, W. Nahm, S. Randjbar-Daemi, E. Sezgin, E. Witten (Eds.), 84–95. World Scientific, 1991.

[Bot91] R. Bott. Stable bundles revisited. *Surveys in Differential Geometry (Supplement to J. Diff. Geom.)* **1** (1991), 1–18.

[BLT75*] N.N. Bogolubov, A.A. Logunov, and I.T. Todorov. *Introduction to Axiomatic Quantum Field Theory.* Benjamin, Reading, Mass., 1975.

[BPZ84] A. A. Belavin, A. M. Polyakov, and A. B. Zamolodchikov. Infinite conformal symmetry in two-dimensional quantum field theory. *Nucl. Phys.* **B 241** (1984), 333–380.

[BR77] A. O. Barut and R. Raczka. *Theory of group representations and applications.* PWN – Polish Scientific Publishers, 1977.

[BT93]   M. Blau and G. Thompson. Derivation of the Verlinde formula from Chern-Simons
            theory. *Nucl. Phys.* **B 408** (1993), 345–390.
[Car89*]   J.L. Cardy. Boundary conditions, fusion rules and the Verlinde formula. *Nucl. Phys.*
            B324 (1989), 581–596.
[Car04*]   J.L. Cardy. *Boundary conformal field theory.* [arXiv:hep-th/0411189v2] (2004).
            (To appear in *Encyclopedia of Mathematical Physics*, Elsevier)
[Car05*]   J.L. Cardy. SLE for theoretical physicists. *Ann. Physics* **318** (2005), 81–118.
[CdG94]   F. Constantinescu and H. F. de Groote. *Geometrische und Algebraische Meth-
            oden der Physik: Supermannigfaltigkeiten und Virasoro-Algebren.*   Teubner,
            Stuttgart, 1994.
[Del99*]   P. Deligne et al. *Quantum Fields and Strings: A Course for Mathematicians I, II.*
            AMS, Providence, RI, 1999.
[DFN84]   B. A. Dubrovin, A. T. Fomenko and S. P. Novikov. *Modern geometry - methods and
            applications I.* Springer-Verlag, 1984.
[Dic89]   R. Dick. Conformal Gauge Fixing in Minkowski Space. *Letters in Mathematical
            Physics* **18** (1989), 67–76.
[Die69]   J. Dieudonné. *Foundations of Modern Analysis, Volume 10-I.* Academic Press, New
            York-London, 1969.
[Diec91*]   T. tom Dieck. *Topologie.* de Gruyter, Berlin, 1991.
[Dir36*]   P. A. M. Dirac. Wave equations in conformal space. *Ann. of Math.* **37** (1936),
            429–442.
[DMS96*]   P. Di Francesco, P. Mathieu and D. Sénéchal. *Conformal Field Theory.* Springer-
            Verlag, 1996.
[Fal94]   G. Faltings. A proof of the Verlinde formula. *J. Alg. Geom.* **3** (1994), 347–374.
[Fal08*]   G. Faltings. Thetafunktionen auf Modulräumen von Vektorbündeln. *Jahresbericht
            der DMV* **110** (2008), 3–18.
[FFFS00a*] G. Felder, J. Fröhlich, J. Fuchs and C. Schweigert. Conformal boundary condi-
            tions and three-dimensional topological field theory. *Phys. Rev. Lett.* **84** (2000),
            1659–1662.
[FFFS00b*] G. Felder, J. Fröhlich, J. Fuchs and C. Schweigert. The geometry of WZW branes.
            J. Geom. Phys. **34** (2000), 162–190.
[FFK89]   G. Felder, J. Fröhlich, and J. Keller. On the structure of unitary conformal field
            theory, I. Existence of conformal blocks. *Comm. Math. Phys.* **124** (1989), 417–463.
[FLM88*]   I. Frenkel, J. Lepowsky, and A. Meurman. *Vertex operator algebras and the monster.*
            Academic Press, 1988.
[FHT03*]   D. Freed, M. Hopkins, and C. Teleman. Loop groups and twisted K-theory III.
            arXiv:math/0312155v3 (2003).
[FKRW95*]   E. Frenkel, V. Kac, A. Radul, and W. Wang. $\mathscr{W}_{1+\infty}$ and $\mathscr{W}_{(glN)}$ with central charge
            N. *Commun. Math. Phys.* **170** (1995), 337–357.
[FQS84]   D. Friedan, Z. Qiu, and S. Shenker. Conformal invariance, unitarity and two-
            dimensional critical exponents. In: *Vertex operators in Mathematics and Physics,*
            Eds. Lepowsky et al., 419–449. Springer Verlag, 1984.
[FQS86]   D. Friedan, Z. Qiu, and S. Shenker. Details of the non-unitary proof for highest
            weight representations of the Virasoro algebra. *Comm. Math. Phys.* **107** (1986),
            535–542.
[FS87]   D. Friedan and S. Shenker. The analytic geometry of two-dimensional conformal
            field theory. *Nucl. Phys.* **B 281** (1987), 509–545.
[FW03*]   Friedrich and W. Werner. Conformal restriction, highest-weight representations and
            SLE. *Comm. Math. Phys.* **243** (2003), 105–122.
[Fuc92]   J. Fuchs. *Affine Lie Algebras and Quantum Groups.* Cambridge University
            Press, 1992.
[Gan06*]   T. Gannon. *Moonshine beyond the Monster. The bridge connecting algebra, modular
            forms and physics.* Cambridge University Press, Cambridge, 2006.
[Gaw89]   K. Gawedzki. Conformal field theory. *Sém. Bourbaki 1988–89*, Astérisque 177-178
            (no 704), 95–126, 1989.

[GF68]   I. M. Gelfand and D. B. Fuks. Cohomology of the Lie algebra of vector fields of a circle. *Funct. Anal. Appl.* **2** (1968), 342–343.

[Gin89]   P. Ginsparg. Introduction to conformal field theory. *Fields, Strings and Critical Phenomena*, Les Houches 1988, Elsevier, Amsterdam 1989.

[GKO86]   P. Goddard, A. Kent, and D. Olive. Unitary representations of the Virasoro and Super-Virasoro algebras. *Comm. Math. Phys.* **103** (1986), 105–119.

[GO89]   P. Goddard and D. Olive. Kac-Moody and Virasoro algebras in relation to quantum mechanics. *Int. J. Mod. Physics* **A1** (1989), 303–414.

[GR05*]   L. Guieu and C. Roger. *L'algèbre et le groupe de Virasoro: aspects géometriques et algébriques, généralisations.* Preprint, 2005.

[GSW87]   M. B. Green, J. H. Schwarz, and E. Witten. *Superstring Theory, Vol. 1.* Cambridge University Press, Cambridge, 1987.

[GW85]   R. Goodman and N. R. Wallach. Projective unitary positive-energy representations of Diff($\mathbb{S}$). *Funct. Analysis* **63** (1985), 299–321.

[Haa93*]   R. Haag. *Local Quantum Physics.* Springer-Verlag, 2nd ed., 1993 .

[Her71]   M.-R. Herman. Simplicité du groupe des difféomorphismes de classe C^∞, isotope à l'identité, du tore de dimension n. *C.R. Acad. Sci. Paris* **273** (1971), 232–234.

[Hit90]   N. Hitchin. Flat connections and geometric quantization. *Comm. Math. Phys.* **131** (1990), 347–380.

[HJJS08*] D. Husemöller, M. Joachim, B. Jurco, and M. Schottenloher. *Basic Bundle Theory and K-Cohomological Invariants. Lect. Notes in Physics* **726**, Springer, 2008.

[HN91]   J. Hilgert and K.-H. Neeb. *Lie Gruppen und Lie Algebren.* Vieweg, 1991.

[HS66]   N. S. Hawley and M. Schiffer. Half-order differentials on Riemann surfaces. *Acta Math.* **115** (1966), 175–236.

[Hua97*]   Y-Z. Huang. *Two-dimensional conformal geometry and vertex operator algebras.* Progress in Mathematics **148**, Birkhuser, Basel, 1997.

[IZ80]   C. Itzykson and J.-B. Zuber. *Quantum Field Theory.* McGraw-Hill, 1980.

[JW06*]   A. Jaffe and E. Witten. Quantum Yang-Mills theory. In: *The millennium prize problems*, 129–152. Clay Math. Inst., Cambridge, MA, 2006.

[Kac80]   V. Kac. Highest weight representations of infinite dimensional Lie algebras. In: *Proc. Intern. Congress Helsinki*, Acad. Sci. Fenn., 299–304, 1980.

[Kac90]   V. Kac. *Infinite dimensional Lie algebras.* Cambridge University Press, 3rd ed., 1990.

[Kac98*]   V. Kac. *Vertex algebras for beginners.* University Lecture Series **10**, AMS, Providencs, RI, 2nd ed., 1998.

[KR87]   V. Kac and A. K. Raina. *Highest Weight Representations of Infinite Dimensional Lie Algebras.* World Scientific, Singapore, 1987.

[Kak91]   M. Kaku. *Strings, Conformal Fields and Topology.* Springer Verlag, 1991.

[Kir76]   A. A. Kirillov. *Theory of Representations.* Springer Verlag, 1976.

[KJ88]   A. A. Kirillov and D. V. Juriev. Representations of the Virasoro algebra by the orbit method. *J. Geom. Phys.* **5** (1988), 351–363.

[KNR94]   S. Kumar, M. S. Narasimhan, and Ramanathan. Infinite Grassmannians and moduli spaces of G-bundles. *Math. Ann.* **300** (1994), 41–75.

[Kui65*]   N. Kuiper. The homotopy type of the unitary group of Hilbert space. *Topology* **3**, (1965), 19–30,

[Law05*]   G.F. Lawler. *Conformally invariant processes in the plane. Mathematical Surveys and Monographs* **114**. AMS, Providence, RI, 2005.

[Lem97*]   L. Lempert. The problem of complexifying a Lie group. In: *Multidimensional complex analysis and partial differential equations*, Eds. P.D. Cordaro et al., *Contemporary Mathematics* **205** (1997), 169–176.

[Len07*]   S. Lentner. *Vertex Algebras Constructed from Hopf Algebra Structures.* Diplomarbeit, LMU München, 2007.

[Lin04*]   K. Linde. *Global Vertex Operators on Riemann Surfaces* Dissertation, LMU München, 2004.

[LM76]   M. Lüscher and G. Mack. The energy-momentum tensor of critical quantum field
            theory in $1 + 1$ dimensions. Unpublished Manuscript, 1976.

[LPSA94]   R. Langlands, P. Pouliot, and Y. Saint-Aubin. Conformal invariance in two-
            dimensional percolation. *Bull. Am. Math. Soc.* **30** (1994), 1–61.

[Mic05*]   J. Mickelsson, Twisted K Theory Invariants. *Letters in Mathematical Physics* **71**
            (2005), 109–121.

[Mil84]   J. Milnor. Remarks on infinite dimensional Lie groups. In: *Relativity, Groups and
            Topology II, Les Houches 1983*, 1007–1058. North-Holland, 1984.

[MS80]   V. Mehta and C. Seshadri. Moduli of vector bundles on curves with parabolic struc-
            tures. *Ann. Math.* **248** (1980), 205–239.

[MS89]   G. Moore and N. Seiberg. Classical and conformal field theory. *Comm. Math. Phys.*
            123 (1989), 177–254.

[MZ55]   D. Montgomory and L. Zippin. *Topological transformation Groups*. Interscience,
            New York, 1955.

[Nas91]   C. Nash. *Differential Topology and Quantum Field Theory*. Academic Press, 1991.

[Nit06*]   T. Nitschke. *Komplexifizierung unendlichdimensionaler Lie-Gruppen*. Diplomar-
            beit, LMU München, 2006.

[NR93]   M.S. Narasimhan and T. Ramadas. Factorization of generalized theta functions I.
            Invent. Math. **114** (1993), 565–623.

[NS65]   M.S. Narasimhan and C. Seshadri. Stable and unitary vector bundles on a compact
            Riemann surface. *Ann. Math.* **65** (1965), 540–567.

[OS73]   K. Osterwalder and R. Schrader. Axioms for Euclidean Green's functions I. *Comm.
            Math. Phys* **31** (1973), 83–112.

[OS75]   K. Osterwalder and R. Schrader. Axioms for Euclidean Green's functions II. *Comm.
            Math. Phys* **42** (1975), 281–305.

[Pal65*]   R.S. Palais. On the homotopy type of certain groups of operators. *Topology* **3** (1965),
            271–279.

[PS86*]   A. Pressley and G. Segal. *Loop Groups*. Oxford Univ. Press, 1986.

[Ram94]   T. Ramadas. Factorization of generalized theta functions II: The Verlinde formula.
            Topology **3** (1996), 641–654. *Preprint*, 1994.

[RS80*]   M. Reed and B. Simon. *Methods of modern Mathematical Physics, Vol. 1: Func-
            tional Analysis*. Academic Press, 1980.

[Rud73*]   W. Rudin. *Functional Analysis*. McGraw-Hill, 1973.

[Sche92]   P. Scheinost. Metaplectic quantization of the moduli spaces of flat and parabolic
            bundles. *Dissertation, LMU München*, 1992.

[ScSc95]   P. Scheinost and M. Schottenloher. Metaplectic quantization of the moduli spaces
            of flat and parabolic bundles. *J. reine angew. Math.* **466** (1995), 145–219.

[Scho95]   M. Schottenloher. *Geometrie und Symmetrie in der Physik*. Vieweg, 1995.

[Schr00*]   O. Schramm. Scaling limits of loop-erased random walks and uniform spanning
            trees. *Israel J. Math.* **118** (2000), 221–288.

[Schw57*]   L. Schwartz. *Théorie des distributions*. Hermann, Paris, 1957.

[Seg88a]   G. Segal. The Definition of Conformal Field Theory. Unpublished Manuscript,
            1988. Reprinted in *Topology, Geometry and Quantum Field Theory*, ed. U. Tillmann.
            Cambridge Univ. Press, 2004, 432–574.

[Seg88b]   G. Segal. Two dimensional conformal field theories and modular functors. In: *Proc.
            IXth Intern. Congress Math. Phys. Swansea*, 22–37, 1988.

[Seg91]   G. Segal. Geometric aspects of quantum field theory. *Proc. Intern. Congress Kyoto
            1990, Math. Soc. Japan*, 1387–1396, 1991.

[Sim68]   D. Simms. *Lie Groups and Quantum Mechanics*. *Lecture Notes in Math.* **52**,
            Springer Verlag, 1968.

[Simo74*]   B. Simon. *The $P(\phi)_2$ Euclidian (Quantum) Field Theory*. *Princeton Series in
            Physics*, Princeton University Press, 1974.

[Smi01*]   S. Smirnov. Critical percolation in the plane: conformal invariance, Cardy's formula,
            scaling limits. *C. R. Acad. Sci. Paris* **333** (2001), 239–244.

[Sor95]   C. Sorger. La formule de Verlinde. *Semin. Bourbaki* **95** (1994). *Preprint*, 1995. (to appear in Sem. Bourbaki, année **95** (1994), no. 793)

[ST65*]   D. Shale and W.F. Stinespring. Spinor representations of infinite orthogonal groups. *J. Math. Mech.* **14** (1965), 315–322.

[SW64*]   R. F. Streater and A. S. Wightman. *PCT, spin and statistics, and all that.* Princeton University Press, 1964 (Corr. third printing 2000).

[Sze95]   A. Szenes. The combinatorics of the Verlinde formula. In: *Vector Bundles in Algebraic Geometry*, Eds. Hitchin et al., 241–254. Cambridge Univ. Press, 1995.

[Tho84]   C.B. Thorn. A proof of the no-ghost theorem using the Kac determinant. In: *Vertex operators in Mathematics and Physics, Eds. Lepowsky et al.*, 411–417. Springer Verlag, 1984.

[TUY89]   A. Tsuchiya, K. Ueno, and Y. Yamada. Conformal field theory on the universal family of stable curves with gauge symmetry. in: Conformal field theory and solvable lattice models. *Adv. Stud. Pure Math.* **16** (1989), 297–372.

[Tur94]   V.G. Turaev. *Quantum invariants of knots and 3-manifolds.* DeGruyter, 1994.

[Tyu03*]   A. Tyurin. *Quantization, classical and quantum field theory and theta functions.* CRM Monograph Series **21** AMS, Providence, RI, 2003.

[Uen95]   K. Ueno. On conformal field theory. In: *Vector Bundles in Algebraic Geometry*, Eds. Hitchin et al., 283–345. Cambridge Univ. Press, 1995.

[Ver88]   E. Verlinde. Fusion rules and modular transformations in two-dimensional conformal field theory. *Nucl. Phys.* **B 300** (1988), 360–376.

[Wig31]   E. Wigner. *Gruppentheorie.* Vieweg, 1931.

[Wit89]   E. Witten. Quantum field theory and the Jones polynomial. *Commun. Math. Phys.* **121** (1989), 351–399.

[Wit93*]   E. Witten. The Verlinde algebra and the cohomology of the Grassmannian. hep-th/9312104. In: *Geometry, Topology and Physics*, Conf. Proc. Lecture Notes in Geom. Top., 357–422. Intern. Press, Cambridge MA (1995).

[Woo80]   N. Woodhouse. *Geometric Quantization.* Clarendon Press, 1980.

[Wur01*]   T. Wurzbacher. Fermionic second quantization and the geometry of the restricted Grassmannian. *Infinite dimensional Kähler manifolds* (Oberwolfach 1995), DMV Sem. **31**, 351–399. Birkhäuser, Basel, 2001.

[Wur06*]   T. Wurzbacher. An elementary proof of the homotopy equivalence between the restricted general linear group and the space of Fredholm operators. In: *Analysis, geometry and topology of elliptic operators*, 411–426, World Sci. Publ., Hackensack, NJ, 2006.

[Zub02*]   J.B. Zuber. CFT, BCFT, ADE and all that. In: *Quantum Symmetries in Theoretical Physics and Mathematics*, eds.: Coquereaux et alii. *Contemporary Mathematics* **294**, 233–271, AMS, 2002.

Index